Population and Environment

This book is based on a project on population
and the environment sponsored jointly by
Development Alternatives with Women for a New Era (DAWN),
the International Social Science Council,
and the Social Science Research Council.

Population and Environment

Rethinking the Debate

EDITED BY

Lourdes Arizpe, M. Priscilla Stone, and David C. Major

Routledge
Taylor & Francis Group
LONDON AND NEW YORK

First published 1994 by Westview Press

Published 2019 by Routledge
52 Vanderbilt Avenue, New York, NY 10017
2 Park Square, Milton Park, Abingdon, Oxon OX14 4RN

Routledge is an imprint of the Taylor & Francis Group, an informa business

Library of Congress Cataloging-in-Publication Data
Population and Environment : rethinking the debate / edited by
　　Lourdes Arizpe, M. Priscilla Stone, and David C. Major.
　　　　p. cm.
　　Includes bibliographical references and index.
　　1. Population—Economic aspects. 2. Population—Social aspects.
3. Population—Environmental aspects. I. Arizpe S., Lourdes.
II. Stone, Margaret Priscilla. III. Major, David C.,
1938-.
HB849.41.P646　　　　1994
304.2'8—dc20
94-2805
CIP

ISBN 13: 978-0-367-28385-8 (hbk)
ISBN 13: 978-0-367-29931-6 (pbk)

Contents

PART TWO
Population and Environment:
Reviews and Case Studies

PART THREE
Population and Environment:
Conclusions

Preface

This volume originated in an initiative of the John D. and Catherine T. MacArthur Foundation, which in 1990 brought together representatives of the International Social Science Council (ISSC), the Social Science Research Council (SSRC) and Development Alternatives with Women for a New Era (DAWN) in a joint effort to rethink the population-environment debate. The core of this effort became the challenge of identifying and examining the micro-level linkages between population and environment and relating these to macro-level considerations. This book is part of that continuing effort on the part of these organizations as well as many other scholars and activists.

This book is also aimed at bringing the experiences and perspectives of poor Third World women to the front and center of the population and environment debate by encouraging a broadening of the discussion and a reorientation of priorities. Through these efforts it is hoped that a more insightful view of population and environment and their interrelationships will be obtained and a more useful and equitable approach to development will emerge. This project and similar undertakings have assumed special importance and urgency given the issues facing the United Nations Conference on Environment and Development in Rio de Janeiro in 1992, the follow-up activities to that conference and the 1994 International Conference on Population and Development in Cairo.

The work reported here was done collaboratively by representatives of ISSC, SSRC and DAWN. These organizations represent quite different constituencies and diverse yet complementary perspectives, resulting in an unusual mix of academic and grassroots perspectives and insights. The group was chaired by Lourdes Arizpe of the Universidad Nacional Autónoma de México, who represented ISSC, a UNESCO funded organization headquartered in Paris with a focus on international research on the environment. Alberto Palloni of the University of Wisconsin represented the SSRC, an international association dedicated to the advancement of interdisciplinary scholarship in the social sciences. Rosina Wiltshire of the Caribbean Conservation Association represented DAWN, a network of activists, researchers and policy-makers principally drawn from Africa, Asia and Latin America that bases much of its understanding of development processes on the experiences of poor women throughout the Third World.

This volume draws principally on the papers and discussions from a workshop held in Cocoyoc, near Mexico City, in 1992. A selection of the chapters, which were commissioned by the project for the workshop, was edited for and forms the core of this volume, together with additional chapters by researchers invited to attend the workshop. The editors have provided an introduction, which is drawn in part from the original project proposal to the MacArthur Foundation, and a conclusion, which summarizes much of the discussion at the workshop. Discussions at the workshop were greatly assisted by a talented and lively group of commentators. They included: Peggy Antrobus (University of the West Indies), Jane Collins (University of Wisconsin, Madison), Sonia Correa (SOS Corpo, Brazil), Magali Daltabuit (CRIM-Universidad Nacional Autónoma de México, Cuernavaca), Haudea Izazola (Mexican Demographic Society), Elizabeth Jelín (Instituto de Investigaciones Sociales, Facultad de Ciencias Sociales, Universidad de Buenos Aires), Francis Lelo (Egerton University, Kenya), Cheywa Spindel (Instituto de Estudos Econômicos, Sociais e Políticos de São Paulo, Brazil) and Margarita Velázquez (Universidad Nacional Autónoma de México, Mexico).

This undertaking was generously funded by the MacArthur Foundation, with additional funding from the John Merck Fund and the Rockefeller Foundation.

The editors are grateful for the substantial contributions of Rosina Wiltshire and Alberto Palloni in crafting the foundation proposals and planning the workshop. Much of the material in Chapters 1 and 11, as well as the overall project design, reflects their contributions. The editors are also grateful for the work of Richard Rockwell, then with the SSRC and now with the Inter-University Consortium for Political and Social Research, who shepherded the project in its early stages and contributed greatly to the early written formulations of the project, some of which are reflected in Chapter 1. The editors also benefitted from comments in the early stages of the project from individual members of SSRC's committees and from the insightful reviews of Emilio Moran and Roger Kasperson. Valuable editorial and administrative assistance was provided by Alexandra Cordero, Susan Forester, Sarah Gordon, Susan Merryman and Felicia Sullivan of the SSRC.

Lourdes Arizpe
M. Priscilla Stone
David C. Major

Rethinking the Population-Environment Debate

There is general agreement that important linkages exist between population and environment. Much of the debate about this relationship, however, has been based on the macro level of analysis, using worldwide historical statistics and aggregate simulations. This approach has neglected the findings and perspectives of micro level research on specific communities and regions. These findings, if analyzed systematically and comprehensively, could greatly contribute to our understanding of the population-environment relationship and of the many intermediary factors that condition it. The goal of this volume is to assist in a rethinking of the population and environment debate by helping to place population processes in their social, political and economic settings, whether they be local, regional or global, and by linking them to environmental outcomes.

When population and environmental processes are modelled independently of the more complex social contexts in which they are embedded, a series of polemical arguments emerges in which people and the environment are necessarily antagonistic. This results in a polarized debate which ultimately poses an impossible choice for policy makers--a choice between people's needs and wants, and the conservation of the environment. Rarely in the public or policy debates are demographic processes understood as part of the same complex system as environmental ones. Research in the social sciences in population and the environment has largely followed traditional disciplinary boundaries, with separate analyses of demographic and environmental phenomena. As more studies have accumulated, the interactions between these two phenomena have been in many ways less rather than more understood. The understanding of these interactions, seen in all their appropriate richness and complexity, provides daunting challenges. The chapters and reflections in this volume are offered as a series of beginning perspectives in what must be an extended effort by social scientists to improve knowledge of and insights into the full dimensions of the interactions between population and environment, macro as well as micro. The participants in the workshop described in the preface (and the authors of the chapters in this volume) have differing views of population and environment relationships, but agree that intellectual progress has been slowed by substantial deficiencies in the ways social scientists have dealt with the true complexities of these issues.

For social scientists to better understand the human dimensions of environmental change, both at the global and the local level, they must ultimately adapt their models and analyses by collaborating with natural scientists. As a first step, however, social science approaches to understanding population and the environment must themselves be reevaluated.*

Ultimately, it is hoped that the lively debates among scholars and practitioners about a range of development issues such as sustainability will benefit from these new approaches. Sustainable development has been defined as "development that meets the needs of the present without compromising the ability of future generations to meet their own needs" (World Commission on Environment and Development, 1987: 343). For this to be achieved, however, both economic development and environmental protection must be incorporated into the same theoretical model. Population and environmental processes are certainly at the center of many of these new frameworks, but it is as yet unclear how to incorporate them so as to truly reflect the complexity of the interactions.

Poles in the Debate

Despite scientific advances in the various disciplines, the debate about the global relationship between population and the environment has progressed relatively little in the past twenty years. Much of the debate remains polarized between two extreme positions, each of which has its roots in a partial scientific understanding of the problem and is based on macro data. One position holds that an increasing population is the principal threat to the environment because of the planet's finite resources; the other that human creativity will continue to find solutions through improved technology to expand the planet's carrying

---

* Any discussion of population and the environment needs to recognize the substantial amount of work done in both the natural and social sciences on global environmental change and its human dimensions. The natural science research needs are set out in International Geosphere-Biosphere Programme (1990). The natural science of global warming is well described in Schneider (1990); see also Schneider (1994). Agenda-setting documents for the social sciences include Jacobson and Price (1990) and Stern et al. (1992); a path-breaking overview is in B.L. Turner et al. (1990). The references in these documents can be seen for further detail. Available data and a wealth of references are provided in World Resources Institute (1994).

capacity. These positions, captured most starkly in popular renditions of the Ehrlich/Simon debate (Simon 1990, Ehrlich and Ehrlich 1991), focus on uni-causal explanations and offer little hope for progress in untangling the truly complex interaction of population and environment (see Arizpe and Velázquez, Chapter 1 of this volume, for discussion of these positions).

Complicating the Relationships

Because it has largely been confined to the macro level, the debate has failed to benefit from the wealth of data generated at the micro level--data which provide rich information on the social and economic factors that mediate the relation between population and the environment.

The principal purpose of this volume is to help bring an appropriate degree of complexity to the understanding of the basic propositions linking population and the environment. Specifically, it is hoped that this work will encourage people working on population and the environment to confront the fact that "the population problem" does not just involve absolute numbers of people nor even just population densities or overall rates of increase, but also, in important ways, social, political, and institutional factors. Complex patterns of human relationships overlay, alter and distort the relation of people to the land and to the cities.

Models used to analyze population must not only take into account population size, density, rate of increase, mortality rates, age distributions, and sex ratios but also gender, migration, type of human settlements, and mobility. Such data, furthermore, must be analyzed within frameworks that take into account comparative data at the local, national and international level. This will provide a richer and more complex scientific approach than has heretofore been widely applied (see Lutz, Chapter 2 of this volume).

Similarly, the authors of this volume argue that "environment" is a term which must be critically examined in its own right. It cannot just be viewed as a set of objects in nature to be handled--or mishandled--by social actors (see Chapter 11 for further discussion on this point), but must be understood as having a global dimension as well as a great variety of local and very particular characteristics. The cultural, social and political filters through which the environment is interpreted and viewed (see, for example, Little in Chapter 7 on the concept of "desertification") are also crucial to the social science understanding of ecology and environment.

It is clear that new research models, concepts, and analyses will ultimately be required to adequately grapple with these increasingly complex ways of linking population to environment. Several new research approaches are described in the following chapters. Arizpe and Velázquez describe examples including a study by Clark and his colleagues presented at the workshop. This work examined the relation between pollution and population using national level data, finding considerable variation within the sample studied. The researchers concluded that global statements about the relation of population and the environment should be abandoned. Instead, they call for local studies of causal relations among specific combinations of population, consumption and production, noting that these local studies need to aim for a general theory that will account for the great variety of local experience. Another study by Kolsrud and Torrey (1993) illustrates an understanding that the environmental effect of adding a new person depends on consumption levels, which in turn are heavily influenced by whether that person lives in a less or more developed country. The approaches applied by Palloni (Chapter 5) also suggest a new, and more nuanced, way to model population-environment interactions.

In such analyses, the debate shows real progress. The lesson is clear; excessive reductionism inevitably leads to models that relate population to environment in ways that short circuit causal connections and omit important intervening processes. The scientific debate, while thus beginning to move towards a more complex understanding of the population-environment issue, has yet to explore fully the empirical contribution that micro level case studies can make in deciphering the intricate social, economic and political processes linking population and environment.

Rethinking the Debate in New Directions

The aim of this volume is thus to assist in rethinking the scholarly and public debate from the perspectives of the social sciences along the lines of new and more complex propositions. The debate might still in future years ask the question: "How much environmental damage does the addition of a new person add?" But it will add: "What social, political, or economic factors shape how that person uses and manages natural resources?" In this volume, the question of how such a more nuanced debate might be fostered is examined; this is a matter of rejecting false dichotomies, avoiding extrapolations from global observations to local actions, adopting new and more complex definitions, and structuring the inquiry so that it speaks directly to issues of policy. New directions in the debate are suggested in Chapter 11.

The construction of a new framework can begin from a specific stance: the rejection of the analytical divide in the population-environment debate which rigidly separates the actions of the North from those of the South, those of the rich from those of the poor. Political divides may be, but are not always, intrinsically significant as scientific boundaries. Key environmental problems--such as global warming--may have differential impacts in different parts of the world, but they will require global solutions. Rosenzweig and Parry (1994), using crop growth and world food trade models, found that although an increase in atmospheric carbon dioxide concentration will lead to only a small decrease in global crop production, developing countries are likely to bear the brunt of the problem. Clearly, solutions for this problem and others will not be formulated properly if the North and the South are seen as contending over the fate of the earth, for that dichotomy will likely produce unsolvable conflicts among interest groups. In addition, that approach reifies as a single dichotomy the many differences among countries on complex continua including national income, industrial development, and distribution of poverty.

Adopting a different view will serve as a corrective in itself: much of the present debate is about population policies to implement in the South rather than about policies to curb equally damaging environmental phenomena in the industrialized North. Much of the public attention to specific cases of degradation relates to events in the South, such as the destruction of the Amazonian rainforest, with policies being advanced to halt such environmental depredations. Curbing wasteful consumption of the earth's resources in the North is not being given equal public attention.

Policies developed on the presumption that population increase is the root cause of environmental degradation affect the peoples of the South most directly, especially poor women, because it is there that population is increasing most rapidly. Policies developed on the presumption that human ingenuity will solve all problems tend to benefit the North more quickly and directly. As a consequence, under the terms of the current policy debate, much of the cost of environmental protection would be borne by the world's most vulnerable people, and the blame for the problems would be laid at their door. This prospect does not offer much hope for the success of international negotiations to address environmental problems. A new global understanding and conceptualization are needed.

The effort to theorize about women's relationships to the environment and its protection, as well as their role in population growth and control, is critical to advancing these debates and remains a relatively new area of research. Sen's essay in this volume (Chapter 3) examines the different perspectives on population and environment held by environmental scientists and environmental activists on the one hand, and women's health researchers and feminist activists on the other. She identifies the positions taken by these two broad groupings within the larger discourse on population and the environment, and proposes possible bases for greater mutual understanding. In the course of this she provides a highly useful summary of the development of views on population, women, poverty and the environment.

In another analysis presented at the workshop (not included here) Emberson Bain examined the effects of mining and nuclear testing on the environment and on population dynamics in the islands and atolls of the Pacific. She pointed out that migration, depopulation and urbanization together with high fertility rates, poverty and losses in cultural systems remain important obstacles to achieving pamphlets, graffiti, even jokes—became more common as Chile entered the 1980s. *Radio Chilena* and *Radio Cooperativa* consistently and bravely reported news that was omitted on the state-controlled television networks.[4] Although the staff and reporters of the stations were of the middle class, the bulk of the listening audience was from the *poblaciones*. Opposition newspapers such as *Analisis, Cauce,* and *APSI* began to offer a print alternative to the progovernment *El Mercurio*. Theater productions were offered informally in every form imaginable, from informal, one-man and one-woman monologues on busy downtown streetcorners to *peñas* (songfests) in neighborhoods.

In the religious sphere, the Catholic church established the justly famous *Vicaría de la Solidaridad* in 1976. Its purpose was to offer legal, medical, and other aid to the victims of political persecution. Attempts were made to set up centers for human rights in a variety of locations, especially in the *poblaciones,* where violations were the most numerous. In 1982, open protests began to break out, particularly in the poorer sectors. The protests were universally suppressed, often with considerable brutality, but their very existence demonstrated a new spirit of opposition. In the years after the 1973 takeover, the churches—first hesitantly and then with a firmer voice—began to speak out in defense of those who were suffering through no fault of their own. Led by Raúl Cardinal Silva, the Catholic bishops publicly declared their disapproval of a number of government measures.

More important for the theme of this book, localized church groups, both Protestant and Catholic, began to support popular initiatives in various ways. Sometimes the support involved simply allowing church facilities to be used by various self-help groups. Gradually, though, churches and smaller subunits, such as chapels, became vehicles where religious leaders as well as members of the congregations could articulate pain (in public prayer, for example) and voice criticism of official policies. The term *popular church,* briefly employed in the 1971–1972 movement Christians for Socialism when Chile was led by Salvador Allende, came into use again. Now, however, the meaning was changed to suggest that the church was responsive to the needs of the poor. Fernando Castillo, in an important book *(Iglesia Liberadora y política,* published in 1986), even speaks of the "liberating" church. One should be careful, though, before describing the church as unequivocally standing up for the poor.[5] The church reflects the same divisions as society as a whole. Its official leaders, like many other leaders, have been more notable for hesitating to speak than for speaking with an unequivocal and unified voice. Philip Berryman, a well-known commentator on liberation theology and Christian-base communities, estimates that even in countries where the commitment to liberation theology appears to be strong, perhaps only 10 research on population and environment but on several of the most important ones in terms of near-term effects upon the majority population of the earth--the poor population. Thus, Bilsborrow and Geores (Chapter 6) focus on agricultural intensification, Little (Chapter 7) on land degradation, Schmink (Chapter 8) on deforestation (noted above), Bunker (Chapter 9) on extractive economies, and Roberts (Chapter 10) on urbanization.

Bilsborrow and Geores provide an overview of ways in which population change can influence land use, in particular agricultural intensification; review the literature on recent changes in land use; and lay out what currently available cross-country data indicate about the underlying relationships. In the process, they move the analysis beyond the current state of almost entirely descriptive, ad hoc knowledge. They show that although there are some positive relationships between population growth and land intensification, these relationships are complicated by other responses to population pressures such as out-migration as well as by the large number of factors besides population pressure that influence the decision to intensify. Farmers may intensify to raise their standard of living, to obtain higher incomes, or to avoid the need to bring new lands under cultivation. Similarly, Little in dealing with the problem of land degradation in dry regions questions the common emphasis on population growth as the primary cause of desertification. He instead emphasizes the role of population distribution and describes a variety of social and economic variables that contribute to land degradation in dry regions.

Bunker studies problems of population and the environment in extractive economies, cases in which human activities can have substantial negative environmental impacts. He outlines a general model of natural resource extraction and applies this to population and environment problems. He finds in general that the relationship between the severity of social and economic costs of extractive activities and absolute population is tenuous. Roberts considers the relationships of population, urbanization, and the environment. He provides an historical review of the impacts of urbanization; examines urbanization in the developing world; and then lays out the impacts on the environment of urbanization and its attendant population movements.

As all of these authors conclude, the links between environmental problems, human activities and issues of population are rarely direct. It is clear that social scientists must carefully reexamine social, economic and political processes from the point of view of their potential environmental impacts. Models that accomplish this would include mechanisms that govern the use, access and control of resources, as well as the allocation of costs and benefits of human activities exploiting those resources (for a discussion of these issues in reference to land use, see National Research Council 1993). The fate of the "commons" may be the most stark, but is certainly not the only, instance in which the competing interests of individuals can work to the detriment, or the protection, of the environment.

Institutions which mediate between people and resources--whether legal, political, or economic--are central to many of the papers presented here. Some institutions relate to state control of markets, including pricing and the import and export of goods and services. Others emerge from the effects of international markets and reflect consumption patterns. Overexploitation of the environment may have less to do with feeding a growing local population than with servicing international debt. Cultural attitudes and values that people have about the environment or about population growth and control are central aspects of many micro level case studies and appear as important, yet understudied, parameters.

Reflections on the Volume

The authors of the chapters in this volume represent a range of theoretical perspectives and employ different methodologies, yet the papers as a whole cover only a limited part of the universe of possible approaches to the population-environment issue; indeed this is inherent in the difficulty and complexity of the issues confronted. In the proposal from which the contributions in this volume stem, it was initially suggested (Social Science Research Council, 1991) that a series of empirically-grounded generalizations drawn from micro-level case studies could be produced as a first step. What was found, however (see for example Palloni in this volume), was that there were few case studies which could adequately serve this purpose. To a substantial degree existing case studies reported on insufficient numbers of variables, did not present their quantitative or even qualitative data in accessible forms, or were overly driven by theoretical imperatives. Many of the discussions at the workshop focused on the particular data needs and models for designing the comprehensive data collection efforts which these issues demand. In a presentation to the workshop, for example, Monte-Mor on behalf of the Center for Development and Regional Planning (CEDEPLAR) in Brazil gave an illustration of one multidisciplinary program's efforts to create an eco-demographic model for generating the kinds of data needed for systematic understanding.

A principal contribution of the studies in this volume, therefore, is to help chart new ways of thinking about these data needs and linkages and to point out directions for future research. By illustrating the range of micro-linkages between population and the environment and by critically examining the role of population as a principal force driving various forms of environmental degradation, the authors point out areas that will benefit from further theoretical development and empirical exploration.

The primary lesson of the volume is that population trends must be analyzed in relation to other processes, whether those processes be environmental or political, economic or social. By stretching the concept of population from simple growth or pressure to include ideas such as urbanization and extraction of resources and by expanding the concept of environment to include local as well as long term and global environmental contexts, social scientists begin a crucial reconceptualization of the population and environment debate.

References

Ehrlich, Paul R. and Anne H. Ehrlich, *The Population Explosion* (New York: Simon and Schuster, 1991).

International Geosphere-Biosphere Programme, *The Initial Core Projects*, Global Change Report No. 12 (Stockholm, 1990).

Jacobson, Harold K., and Martin F. Price, *A Framework for Research on the Human Dimensions of Global Environmental Change* (Paris, 1990).

Kolsrud, Gretchen and Barbara Boyle Torrey, "The Importance of Population Growth in Future Commercial Energy Consumption," in *Global Climate Change*, edited by James White (New York: Plenum Press, 1993), 127-141.

National Research Council. *Population and Land Use in Developing Countries: Report of a Workshop.* Carole L. Jolly and Barbara Boyle Torrey (eds.) (Washington, DC: National Academy Press, 1993).

Rosenzweig, Cynthia and Martin L. Parry. "Potential impact of climate change on world food supply." *Nature 367*:6459, 13 January 1994:133-38.

Schneider, Stephen H., *Global Warming* (New York, 1990).

Schneider, Stephen H., "Detecting Climatic Change Signals: Are There Any 'Fingerprints'?," *Science, 263*, 21 January 1994, 341-347.

Simon, Julian L., *Population Matters: People, Resources, Environment, and Immigration.* (New Brunswick, NJ: Transaction Publishers, 1990).

Social Science Research Council, "Recasting the Population-Environment Debate: A Proposal for a Research Program," New York, 1991.

Stern, Paul C., Oran R. Young, and Daniel Druckman, eds., *Global Environmental Change* (Washington, DC: National Academy Press, 1992).

Turner, B. L. II et al., *The Earth as Transformed by Human Action* (Cambridge and New York: Cambridge University Press, 1990).

World Commission on Environment and Development, *Our Common Future* (New York: Oxford University Press, 1987).

World Resources Institute (in collaboration with the United Nations Environment Programme and the United Nations Development Programme), *World Resources 1994-95* (New York: Oxford University Press, 1994).

Population and Environment: Overviews and Methodologies

Introduction to Part One

The debate on the relation between population and environment needs to include an understanding of the true complexity of the issues. The chapters in this section of the volume are intended to begin the process of bringing order and conceptual clarity to our understanding of that complexity.

Arizpe and Velázquez (Chapter 1) consider the reconceptualization of fundamental terms such as population. They, as well as Lutz (Chapter 2), evaluate population trends and further an ongoing discussion of what is known and what needs to be known about the role of population (growth, density, distribution) as a driving force in environmental degradation.

Chapters 3 and 4 pay special attention to the role of gender in discussions of population and its future. Sen (Chapter 3) situates the debate about population growth and its control in the history of gender theory and politics. Agarwal's companion piece (Chapter 4) examines the link between gender and environment, drawing on studies of rural India.

These authors argue that direct population-environment linkages hold, if they hold at all, only at aggregate levels. At local levels it seems clear that historical, social, political and cultural processes explain the patterns better than demographic processes alone.

The final chapter in this section (Chapter 5) grapples with the methodological challenges posed by this new, more complicated view of the issues. Palloni's chapter is an effort to sort out some of the conceptual issues involved in generalizing from case studies and to address the complex web of causality derived from these empirical works.

Taken together, these chapters reinforce a central argument of the volume-- that population trends must be analyzed in relation to many other, non-demographic, processes--and set the stage for the cases presented in Part Two.

1

The Social Dimensions of Population

Lourdes Arizpe and Margarita Velázquez

While the scientific understanding of environmental and demographic change, as studied separately, is increasing dramatically, our ability to link the two in any synthetic and holistic manner lags behind. The central argument of this paper will be that the scientific community cannot use current models and methodologies for understanding the dynamic relationship between population and environment, but needs a new framework. This new framework will need to extend key definitions of issues and concepts and propose new methods for researching them. Population, for example, cannot be limited to population size, density, rate of increase, age distribution and sex ratios, but must also include access to resources, livelihoods, social dimensions of gender, and structures of power. New models have to be explored in which population control is not simply a question of family planning but of social and political planning (United Nations 1990:202-216; Jacobson 1988:152-54) in which the wasteful use of resources is not simply a question of finding new substitutes but of reshaping affluent life-styles (Meadows 1988; Repetto 1987) and in which pollution control is not simply a matter of "polluter pays" but also of emission controls, which in turn are associated with political and social processes. These will need to be models in which sustainability is seen not only as a global aggregate process but

Many of the ideas expressed in this section were discussed at meetings of the SSRC/ISSC/DAWN project described in the preface. Our thanks to Richard Rockwell, Gita Sen, William Clark, Rosina Wiltshire and Alberto Palloni.

one that incorporates the policy goals of sustainable livelihoods for a majority of local peoples.

This is no small undertaking, and yet the theoretical and empirical challenges posed by global environmental change are in themselves a whole new order of magnitude. Although at a global level many recognize the challenge of harmonizing population growth and human expectations with the rate at which the planet's natural resources are being used or polluted, we lack the models with which to understand and plan for these changes. Human control of the environment is being overridden by unexpected new phenomena--the greenhouse effect, leading to climate change, and ozone depletion--or by the cumulative effects of old phenomena--desertification, loss of biological and cultural diversity, and soil erosion, among others. Humans are vulnerable now to natural and human-made hazards of a different order than ever before.

Three factors distinguish what we face today from challenges of the past. First, the scale of such phenomena is much larger and the number of people who will be affected by these changes is historically unprecedented. Second, while ecological mismanagement did occur in the past, populations could opt for outmigration. Now, however, there is nowhere left to go. Third, the natural inequities in the geographical distribution of resources have been further aggravated by the concentration of human-made capital in industrialized nations and in elite circles of less developed countries.

A challenge such as this, of a higher magnitude and complexity than humanity has had to face in the past, requires conceptualization and planning at a more inclusive and complex level. But we lack the appropriate scientific and political frameworks; issues tend to be constructed, and dealt with, around single factor explanations and ensuing simplistic actions. Believing, for example, that population is the key cause of environmental degradation is a reductionist argument that leads to narrowly conceived policies. The complexity of the issues involved actually requires a debate on the political and economic planning for a global world.

This chapter contends that the population-environment debate has become deadlocked because it has become a question of taking sides instead of delving deeply into the complexity of the issues. Also the tendency to use mechanistic, predictive models is inappropriate given the level of uncertainty. Population issues have been decontextualized from actual social environments as well as from the broader and more profound issues concerning the new, emerging economic and political structure of the world and its relationship to the resource base of the planet.

Some may argue that engaging in the analysis of such broad issues may distract from the urgent need to act on population problems. Experience shows, however, that policy solutions focused exclusively on deterring population growth in the short term are ineffective when compared to more encompassing economic and social reforms. The most urgent task, then, is to establish a hierarchy of goals--economic, ecological, social, and cultural--to better direct the already existing potential for action.

At present, the disarray in the debate on population and resource use has been attributed to the lack of reliable data and the uncertainty of predictions. But it is also associated with the failure to analyze population trends *in relationship to other processes*. This chapter contends, accordingly, that all demographic transitions have been embedded in broader socioeconomic transitions; that population growth is not a *driving* force but an *accelerating* force except under rare circumstances where all other conditions remain static; and that population growth can only be understood by analyzing it in relation to rates of growth in the consumption of natural and human-made resources. Finally, we argue, as many others have, that curbing population growth can only occur in the long term with plans for *sustainable* development at a national, regional and global scale (Ehrlich 1989; Ehrlich and Erhlich 1991; Keyfitz 1991 and 1991b; Costanza 1991; Leff 1990; Little and Horowitz 1987; Toledo 1990; Maihold and Urquidi 1990).

Population Trends at the Threshold of the New Millennium

The demographic transitions in North America and Western Europe at the end of the nineteenth century were linked to improved medical services and nutrition levels, which led to the decline in mortality due to infectious diseases (Demeny 1990; Lutz and Prinz 1991). But they were made possible by a number of interrelated changes, including the shift from an agricultural to an urban-industrial society and by associated changes in family composition, age at marriage, and education.

In contrast, mortality decline in less developed countries in the second half of the twentieth century has come about mainly as a result of improved medical and health care services in many cases without the accompanying social, economic, and political transformations. Since these socioeconomic transitions have occurred unevenly, frequently inequitably, and sometimes have even been reversed, the demographic transitions in such countries have not been completed, especially in Africa.

Some authors believe a general demographic transition is already underway. Julian Simon, for example, argues that fertility rates have decreased in countries all over the world (1990). Others reject this optimistic view or believe that such transitions are occurring too slowly (Ehrlich 1989; Ehrlich and Ehrlich 1991; Grant and Tanton 1981). Recent figures, in fact, show that worldwide the crude birth rate decreased from 33.9 (1950-70) to 27.1 (1985-90), while the total fertility rate fell from 5.9 to 3.3 during that same period (World Resources Institute 1990:256) (see Table 1.1).

TABLE 1.1   Fertility Trends in the Developing World

Year	Average No. Births/Woman
1950-55	6.1
1955-60	6.0
1960-65	6.1
1965-70	6.0
1970-75	5.4
1975-80	4.5
1980-85	4.2

Source: Bongaarts, Mauldin, and Phillips 1990.

Lutz and Prinz state that projections for the next 30 years are actually rather reliable, since they are insensitive to minor changes in mortality, migration, and fertility (1991). In Figure 1.1, they summarize projections according to different scenarios. It is estimated that world population will reach around 8 billion by the year 2010 (United Nations Population Fund 1991:3, 48; United Nations Department of International Economic and Social Affairs, 1989; Demeny 1990:41; Sánchez 1989:16; United Nations Development Programme 1990:166).

As Lutz and Prinz emphasize, the *momentum* of population growth has to be taken into account--the age structure of a fast growing population is so young that even if fertility per woman declined to a very low level, the increasing number of young women entering reproductive ages will cause the population to grow further for quite some time (1991). For the year 2050 and beyond, projections begin to vary from 8 to 14 billion and they diverge even more widely for the next century after that (Lutz and Prinz 1991).

What Do Population Numbers Mean?

The concept of population as numbers of human bodies is of very limited use in understanding the future of societies in a global context. It is what these bodies do, what they extract and give back to the environment, what use they make of land, trees, and water, and what impact their commerce and industry have on their social and ecological systems that are crucial (Demeny 1988:217; Harrison 1990; Durning 1991).

An early attempt to establish such a link was by estimating the planet's carrying capacity. Approximations varied widely, ranging from 7.5 billion

FIGURE 1.1 Total Projected World Population 1990-2100 According to Scenario

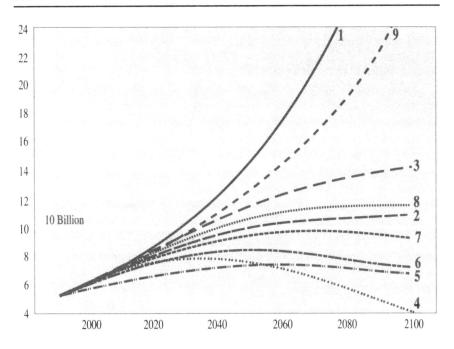

Scenarios

1: Constant Rates; constant 1985-1990 fertility and mortality rates

2: UN Medium Variant; strong fertility and mortality decline until 2025, then constant

3: Slow Fertility Decline; UN fertility decline 25 years delayed, UN medium mortality

4: Rapid Fertility Decline; TFR=1.4 all over the world in 2025, UN medium mortality

5: Immediate Replacement Fertility; assumed TFR=2.1 in 1990; UN medium mortality

6: Constant Mortality; TFR=2.1 all over the world in 2025, constant mortality

7: Slow Mortality Decline; UN mortality decline 25 years delayed, TFR=2.1 in 2025

8: Rapid Mortality Decline; life expectancy of 80/85 yearrs and TFR=2.1 in 2025

9: Third World Crisis; constant fertility and 10% increase in mortality in Africa and Southern Asia; TFR=2.1 in 2025 and UN mortality for the rest of the world

Note: "TFR" is the Total Fertility Rate (= average number of children per woman)

Source: Lutz and Prinz 1991.

(Gilland 1983), 12 billion (Clark, C. 1958), 40 billion (Revelle 1976), to 50 billion (Brown, H. 1954). The underlying problem is, of course, how to establish the appropriate level of kilocalories for each human being (Blaxter 1986). "For humans, a physical definition of needs may be irrelevant. Human needs and aspirations are culturally determined: they can and do grow so as to encompass an increasing amount of 'goods,' well beyond what is necessary for mere survival" (Demeny 1988:215-6).

Other authors point out that the concept of carrying capacity and self-regulating mechanisms applied to animal populations should not be extrapolated to human populations (Sánchez 1989:26), and argue that socioeconomic, technological, or environmental changes are so decisive in altering carrying capacity that the concept itself is of little use (Blaikie and Brookfield 1987). A more appropriate exercise, then, should be to try to develop global accounting systems that relate population, per capita resource use, and wealth distribution.

A different approach is taken by those trying to forecast the socioeconomic and environmental impacts of population projections. Gordon and Suzuki, in their 1991 assessment of environmental change, present a harsh scenario for the year 2040: overpopulation, unbreathable air, high temperatures, desertification, loss of land due to soil erosion and rising seas, food rioting in the Third World, and so on. They insist that the "sacred truths" that the earth is infinite and progress is possible must be discarded.

Another issue is the differential demographic growth of regions and of diverse ethnic and religious groups within countries. Here, population overlaps with political processes. Figure 1.2 shows the difference in the proportion of population between the North and the South based on United Nations "medium" growth assumptions (United Nations Development Programme 1990). By the year 2010, 6 out of every 7 people will live or will have been born in less developed countries of the South (FAO 1990) (Table 1.2). Such differential growth is being seen increasingly as a potential threat to countries in the North (Grant and Tanton 1981).

During the last few years, population growth has increasingly been considered the driving force in environmental degradation, increased pollution, and in fostering international conflicts--for example, over water in the Middle East (Myers 1987). It is interesting to note that in linking population growth to environmental depletion, the disparities in the natural resource base and in the distribution of goods and services in different societies are frequently left out of the picture. It would seem, in fact, that population growth, in some cases, is used to compensate for such existing disparities and inequities. To understand this, population numbers must be analyzed according to human development indicators. We will return to this point later.

FIGURE 1.2 World Population Trend and North South Distribution (billions of people)

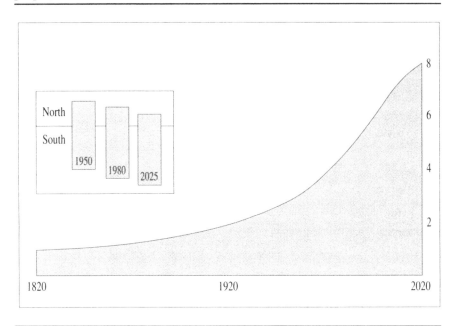

Source: United Nations Development Programme 1990.

The Population Debate

The debate on population has polarized into two major positions. One position holds that an increasing population is the principal driving force that threatens the planet's finite resources. Paul Ehrlich mentions that over-consumption, increasing dependence on ecologically unsound technologies to supply that consumption, and unequal access to resources and poverty play a major role in environmental crises (Ehrlich 1989). He concludes that "the key to understanding overpopulation is not population density but the number of people in an area relative to its resources and the capacity of the environment to sustain human activities" (Ehrlich and Ehrlich 1991:38-39). Much more extreme views on this position have likened human population expansion to a cancerous growth bound to kill its hospitable planet (Hern 1990:30) or to a jar of yeast, which makes alcohol that eventually kills it (Grant and Tanton 1981).

TABLE 1.2   Population Indicators for Major Regions of the World

Region	Population Millions 1990	Av. Rate Growth % 1990-95	IMR* per 1000 1990	% Urban 1990	Urban Growth % 1990-95
World	5,292.2	1.7	63	45	3.0
"North"	1,206.6	0.5	12	73	0.8
"South"	4,084.6	2.1	70	37	4.2
Africa	642.1	3.0	94	34	4.9
North America	275.9	0.7	8	75	1.0
Latin America	448.1	1.9	48	72	2.6
Asia	3,112.7	1.8	64	34	4.2
Europe	498.4	0.2	11	73	0.7
Oceania	26.5	1.4	23	71	1.4
USSR	288.6	0.7	20	66	0.9

* Infant Mortality Rate
Source: Sage and Redclift 1991.

However, case studies have been reported indicating that "there is no linear relation between growing population and density, and such pressures [towards land degradation and desertification]" (Caldwell 1984). In fact, one study found that land degradation can occur under rising pressure of population on resources (PPR), under declining PPR, and without PPR (Blaikie and Brookfield 1987). Therefore, the scientific agenda must look toward more complex, systemic models where the effects of population pressures can be analyzed in relationship with other factors (García 1990). This would allow us to differentiate population as a "proximate" cause of environmental degradation, from the concatenation of effects of population with other factors as the "ultimate" cause of such degradation (Asian Development Bank 1990). Also concerned about population growth, other authors place greater emphasis on improving humanity's lot (Eckholm 1982).

The other position holds that population growth can be dealt with through technological solutions that human creativity will continue to find (Simon 1990; Kasun 1988). An even more optimistic view, originally put forth by Hirschman, saw high rates of population growth as stimulating economic development through inducing technological and organizational changes (1958).

This position, though, ignores the dangers of environmental depletion implicit in unchecked economic growth: consumption increases and rapidly growing populations that can put a very real burden upon the resources of the earth and bring about social and political strife for control of such resources. This position also makes the questionable assumption that technological creativity will have the same outcomes in the South as in the North. Finally, it heavily discounts the importance of the loss of biodiversity--a loss that is irreversible and whose human consequences are as yet unknown.

Neither position, according to some authors, represents the state of the art of scientific understanding (Johnson and Lee 1987; Repetto 1987). Other authors state that "population is not a relevant variable" in terms of resource depletion, and stress income inequality, that is, poverty, as a more important factor (Gallopin 1990; Leff 1990). More specifically, resource consumption, particularly overconsumption by the affluent, is considered by many authors as the key factor to environmental depletion (Hardoy and Satterthwatte 1991; Harrison 1990; Durning 1991). OECD countries represent only 16% of the world's population and 24% of land areas; but their economies account for about 72% of world gross product, 78% of road vehicles, and 50% of global energy use. They generate about 76% of world trade, 73% of chemical products export, and 73% of forest product imports (OECD 1991). The main short-term policy instrument in this case is reducing consumption.

Finally, a "revisionist" position states that "neither alarmism nor total complacency about population growth can be supported by the current evidence" (Kelley 1986). This author cites Kuznets' summary judgment: "we have not tested, or even approximated, empirical coefficients, with which to weigh the various positive and negative aspects of population growth. While we may be able to distinguish the advantages and disadvantages, we rarely know the character of the function that relates them to different magnitudes of population growth."

Importantly, historical demographic studies have demonstrated that no simple correlations can be established between population and environmental transformations. In the volume *The Earth Transformed by Human Action*, researchers found that the time scale of population variability is asynchronous with given environmental transformations and recovery (Whitmore et al. 1990:37). Consistently, the authors stress "the need for caution in using population as a simple surrogate for environmental transformation" (Whitmore et al. 1990:37).

They also found evidence of divergence between global and regional population trends and concluded that "if the experience of past regional population changes and their accompanying environmental transformations has relevance for the future, the projected global scale population 'leveling out' need not diminish the scale and profundity of global environmental change. This is particularly true on the regional or local scale, where global zero population growth (of population or transformation) need not be accompanied by local or regional equilibrium" (Whitmore et al. 1990:37). Furthermore, they warn that "regional population declines are possible, potentially brutal, and even likely, accompaniments to zero population growth on a global scale" (Whitmore et al. 1990:37).

Beyond a bipolar population debate, then, we must develop a clearer understanding of what population numbers mean in different social settings.

What the Population Numbers Mean:
Migration and Urbanization

While intensive ethnographic research methods may point to unexpected and crucial relationships, the rethinking of questions must also extend into new areas of research. While most of the theorizing around the population-environment issue has been directed toward population growth, other demographic processes such as migration and urbanization must be clearly incorporated. Let us for illustrative purposes look at some of the macro issues surrounding these processes.

The demise of agrarian societies the world over, coupled with population increase, has contributed to massive migratory movements in the last few decades. In the early 1980s the number of economic migrants was estimated at around 20 million; adding a similar count for illegal migrants, perhaps 40 to 50 million people have moved in the hope of having a bigger share of the world's development benefits (United Nations Development Programme 1990:28). It is estimated that in the first decades of the next century such movements will increase and diversify, adding to their contingents ecological refugees, who will be moving principally from South to North, but also in regions within the South (United Nations Development Programme 1990:28).

The classic pattern of rural outmigration linked to price changes that led to the breakup of the European peasant economies--which sent out 52 million Central and Western European migrants overseas from 1848 to 1912, even though population growth was slow (Brinley 1961)--continues to be the main driving force in outmigration in many developing countries. At present, it is associated with the fall in the price of agricultural products in the world market: in the 1980s, subsidies in the European Economic Community and other developed economies to protect their own farmers have led to the breakup of peasant economies of many countries of the South, thus leading to massive rural exodus. Such pressures may also push farmers toward environmentally unsound activities: deforestation, monoculture, crop intensification, and overuse of fertilizers or pesticides, all of which mean depletion of natural resources.

Population growth may increase the rate at which such activities are carried out, but the structures that drive such activities are economic and financial, not demographic. If this were not the case, Europe would not have sent out as many migrants overseas during the last century.

The important question is what will happen as market forces continue to integrate agrarian societies into a globalized market, thus continuing pressures for rural outmigration. Measures to protect farmers from the market have failed in most countries. They have often transferred the problem to other countries, as has already been mentioned for the EEC, so that a local Western European solution to an economic and political problem has fueled local problems of impoverishment, outmigration, and resource depletion in many countries of the South. In turn, as in the case of deforestation, that intensifies yet another *global* problem.

Traditional agrarian societies have responded in myriad ways, such as increased self-exploitation in work, decrease of caloric intake, especially among poor rural women, intensification and diversification of income-generating activities, and permanent or recurrent--"swallow" migration as it is called in Mexico--outmigration (Arizpe 1978, 1982).

An important question is whether having more children survive has been used by farmers in this unequal struggle to adapt to a global market. The answer is yes, in many ways; we focus on those related to migration strategies. Since farmers in traditional agrarian societies are not having more offspring but rather are finding that more of them survive, what is being asked of them is that they change an age-old pattern of reproduction. This requires not only a strong economic incentive, which in the case of the more than 1 billion rural poor in the South is not there, but a cultural and political change whereby rural people will feel they have a stake in the new global society that is being created. For small agricultural family producers, it is well known that children do have an economic value in farm labor. Additionally, in recent decades, they have been helpful to their families in relay migration, by sending remittances back to offset the family economic deficit (Arizpe 1982). Such a strategy, though, is possible only when cities offer unlimited opportunities for economic gain, which increasingly is not the case in many mega-cities in developing countries.

Thus, adaptive strategies to disappearing rural livelihoods, given present market conditions, are limited, so it may be assumed that rural outmigration will continue to increase and to be directed toward mega-cities and to international destinations.

Rural-urban disparities continue to increase the attraction of cities: in most countries, urban incomes per person run 50% to 100% higher than rural incomes (United Nations Development Programme 1990:30). In Nigeria the average urban family income in 1978-79 was 4.6 times the rural; in Mexico urban per capita income was 2.6 times the rural (United Nations Development Programme 1990:30).

Projections show that urbanization will become the dominant social process in the next 50 years and most mega-cities will be in low latitudes in tropical regions (Douglas 1991), as can be seen in Table 1.3. Some authors, however, believe that the rate of urbanization in the Third World will not be as high as United Nations figures suggest (Hardoy and Satterthwaite 1991).

The trend towards urbanization fostered by rural push factors rather than urban attraction factors is cause for concern, since "in the places where man's activities are most densely concentrated--his settlements--the environmental impact is greatest and the risks of environmental damage are most acute" (United Nations 1974). Indeed, since the United Nations pointed this out in 1974, urbanization has continued at a rapid pace. Globally, it has been forecast that 24 million hectares of cropland will be transformed to urban-industrial uses by the year 2000; this is only 2 percent of the world total, but it is equivalent to the present-day food supply of some 84 million people (Douglas 1991:8). The loss of agricultural land to urbanization is most severe in the developing countries, where more than 476,000 hectares of land a year will be built up in the remaining years of the twentieth century (World Resources Institute 1988, cited in Douglas 1991).

TABLE 1.3 Projected Increases in Urban Population in Major World Regions, 1985-2000

Region	Urban population (millions)		Absolute Increase (millions)	Percentage Increase
	1985	2000		
Africa	174	361	187	108
Asia	700	1,187	487	70
Latin America	279	417	138	49
Oceania	1.3	2.3	1	77
Developing countries	1,154	1,967	813	70
Industrial countries	844	950	106	13
World	1,998	2,917	919	46

Source: United Nations Development Programme 1990.

While in previous decades cities in developing countries were able to absorb, even under dire poverty conditions, migrants from their rural hinterland, this is no longer the case in the 1990s. Migrants from Eastern Europe and Africa are overflowing into Western Europe, and Peruvians and Bolivians have now joined Mexicans and Central Americans in migrating to the United States (Grant and Tanton 1981). No doubt the numbers of economic and ecological migrants knocking at the North's door would be lower if population growth in the South were decreased, but the trend would still be there. If capital investments do not flow to where the people are, then the people will flow to where capital investments are. One example will suffice: in California alone, it is estimated that 7 million jobs will be created in the 1990s, most of them low-paying, nonskilled jobs Americans will not be able to fill.

Consequently, with both push factors in the South and pull factors in the North at work, the flow of migrants from South to North will continue to grow in the next decades and may become one of the most contentious issues internationally, although an increase of South-South migration can also be foreseen. Thus, the population and environment debate must incorporate these concerns to better model these complex changes. This means, among other things, developing new research priorities and methods to monitor population processes but, in particular, analyzing them in relation to global change phenomena.

Exploring New Research Methods

Research should focus not on population as an isolated variable but on the relationship between population and the use of natural and human-made resources (Demeny 1988:217; Harrison 1990; Durning 1991; Arizpe, Constanza, and Lutz 1992). The methodological and theoretical challenges are considerable.

One research priority must be to explore methods for more precisely estimating the relationship between population and resource use. William Clark suggests that the "Ehrlich identity" (Pollution/Area = People/Area x Economic Production/People x Pollution/Economic Production) can be operationalized as $(CO_2 \text{ Emissions}/km^2 = \text{Population}/km^2 \text{ x GNP/Population x } CO_2 \text{ Emissions/GNP})$ (1991). Clark and his colleagues examined data for 12 countries from 1925 to 1985 and concluded that the same loading of pollution on the environment can come from radically different combinations of population size, consumption, and production. Thus no single factor dominates the changing patterns of environmental loadings across time.

Another research priority is to look at the effect that adding a new person has on resources according to consumption levels and the effect that efficiency has on rising levels of consumption. Gretchen Kolsrud and Barbara Torrey

(1993) examine population growth and energy efficiency in several countries and conclude that the very small population growth forecast for developed countries over the next 40 years will add a burden of CO_2 emissions that will be *equal* to that added by the much larger population growth forecast for the less developed countries. Improving energy efficiency in developed countries could dramatically decrease CO_2 emissions globally (if consumption per person remains constant). It is only under a scenario of severe constraints on emissions in the developed countries that population growth in less developed ones plays a major global role in emissions. If energy efficiency could be improved in the latter as well as the former, then population increase would play a much smaller role. José Goldemberg, a participant at the meetings where these two papers were originally presented, has pointed out that enabling developing countries to "leapfrog" in the adoption of new energy efficient technologies could accomplish this goal.

The need for local studies of causal relations in systemic combinations of populations, consumption, and production is clear, but these local studies need to aim for a general theory that will account for the great variety of local experience. We will illustrate this further below with material from micro-level field research in the Lacandón rainforest.

What the Population Numbers Mean at the Micro Level:
The Lacandón Rainforest

We illustrate several things with the example of the Lacandón rainforest. First, we stress the complexity of the relationship between population and the environment and the number of social, political, historical, and economic factors, *in particular combinations*, that affect their interplay. Second, we introduce the views of the people themselves. Attitudes about the environment and childbearing and their value must be incorporated into models and programs aimed at sustainability.

The colonization and deforestation of the Lacandón rainforest in southeast Mexico began at the turn of the century with the extraction of precious woods by foreign companies until the 1960s and by national government agencies until 1988. Government-sponsored colonization of the rainforest began as part of the Alliance for Progress agricultural settlement programs in the 1960s; it slowed down in the 1970s but faced with the influx of Guatemalan guerrillas and refugees between 1982 and 1988, migration into the rainforest increased along the border with Guatemala. Voluntary migration to the rainforest also occurred as Indians came down from the highlands in the 1970s, pushed both by demographic pressure on lands and by land concentration in the hands of the politically powerful families left intact even after the Mexican Revolution. In 1988 the government of President Carlos Salinas de Gortari decreed a total ban on the cutting of trees in the Lacandón rainforest, which is still being enforced. Programs to develop a sustainable agriculture and agroforestry in the region have been reinforced.

Our research focused on perceptions of deforestation and was based on a survey of 432 households both in the Palenque area and the Marqués de Comillas area deep in the rainforest area at the corner of the border with Guatemala.

Average number of children are displayed in Table 1.4 as well as the significant range of variation in these averages between areas, communities, and by relative wealth.

The population-environment debate as described in the preceding sections would lead us to ask: do the low-income groups have more children because they are poor or are they poor because they have more children? The question is, of course, simplistic and its answer would be misleading. In fact, the data from the fieldwork and from the survey show how much more complex this issue is.

For example, Table 1.4 contains puzzling data on the survival rates of offspring, which do not support a simple correlation as would be predicted by demographic transition theory. Some of the patterns do seem to be predicted by the theory. The highest survival rate is indeed that of urban Palenque children, due to medical services and higher incomes; the high-income women, knowing their children will survive, have lowered the average number of children they have to 2.6. The lower-income group in Palenque, despite having a high survival rate, still has twice the number of children of the affluent group. Perhaps this is because many of them are migrants who brought with them rural fertility rates or because in unemployment situations having more children is useful.

On this second possibility, when asked what the greatest threat is in the world today, the low-income Palenque group had the highest rate of concern about poverty, even higher than the shifting cultivators of the rainforest. Yet, in spite of this perception, they have a lower average number of children. Another slight discrepancy is that the group with the lowest survival rate of children does not have the highest average number of offspring. Other factors, clearly, are at work. We will consider only two here: ethnicity and religion.

Table 1.5 breaks down our sample according to ethnic identity. A startling difference in the survival rate between Indian and non-Indian children is evident, and it appears to correlate with a very high average number of offspring per woman among Indians. Interviews found that family planning has made no inroads among Indian families, for any number of reasons, including lack of access to family planning programs, resistance by men, isolation, and young marriage age.

TABLE 1.4 Average Number of Children per Woman and Survival Rate by Community

Communities	Average	Rate
Pico de Oro/Ref. Agraria. Old ejidos.	5.8	88
Victoria/Nvo. Chihuahua. Poorest ejidos.	6.9	85
Lacandón/La Unión. Deforested ejidos.	5.5	83
Palenque Bajos. Low-income group.	4.2	90
Palenque Altos. High-income group.	2.6	90
Cattle ranchers.	2.6	92
Average	5.1	87

Source: Survey in Lacandón rainforest by Arizpe, Paz and Velázquez

TABLE 1.5 Average Number of Children per Woman and Survival Rate by Ethnic Identity

Identity	Average	Rate
Indians	6.5	76
Non-indians	4.8	87

Source: Survey in Lacandón rainforest by Arizpe, Paz and Velázquez

That the influence of such factors can be overridden is demonstrated, however, by data on other ethnic groups in Mexico who have lowered their fertility rate in the past two decades (Direccíon General de Estadística 1991). This has involved cultural change, women's education and autonomy, and a lessening of geographical and economic isolation.

Similarly, religion also has major effects on fertility levels. Catholic women in the survey have an average of 5.1 children and those belonging to non-Catholic sects, including Jehovah's Witnesses, Seventh-Day Adventists, and Evangelists, average 5.9, in contrast to 2.9 for nonbelievers. But such figures do not fit neatly into patterns. Although child survival is highest among non-Catholic sects it has not led to a lower number of children. However, once again, data show that the religious factor can be overridden, since Catholic women in the Palenque region--with access to schools, clinics, and mass media-- have an average of 5.3 children as compared to those in the Marques de Comillas area, who have an average of 6.6 children.

In our view, for any given group it is the *combination* of factors-- geographical accessibility, schools, income, women's range of activities, ethnicity, religion, access to information, mass media--that is more important than any single factor. We propose an *inverted nested model* for understanding this complexity. While the nested approach is well suited to studies starting from the global-national level and moving to the local level, an inverted nested model could be useful when the research begins at the local level. It would allow identification of the most salient factors affecting population reproduction at the local level, but would help explain why certain factors take on more salience in particular settings. Thus, for example, the same combination of factors leads to a slightly different average number of offspring between Catholic women in Palenque and in Marqués de Comillas because of the relative importance of different factors.

To the question "what is the greatest threat in the world today?" only 3.2% of the total sample referred to overpopulation. Surprisingly, most of them were men and most were cattle ranchers of Palenque. They see the greatest threat to their cattle-ranching business as invasions by landless peasants and are concerned that these peasants should not continue to reproduce at the present rate. A nested approach would allow us to identify population expansion as a concern of both shifting cultivators and cattle ranchers; because of the combination of other factors, these groups end up assigning a different priority to overpopulation as a problem.

The question is no longer why poorer people have more children. Instead it has to be rephrased: what is the nested combination of the most relevant factors in a given local situation that leads to given fertility levels? This implies a much more time-consuming and qualitative approach to data, but the explanatory potential, we believe, is much greater.

Population and Human Development

Notable progress in human development made during this century and especially in recent decades makes it even more difficult to accept our present predicament. On a world scale, life expectancy has risen from 54.9 (1950-70) to 61.5 (1985-90) (World Resources Institute 1990:256). In developing countries, average infant mortality decreased from nearly 200 deaths per 1,000 live births to about 80 in about four decades (1950-88), "a feat that took the industrial countries nearly a century to accomplish" (United Nations Development Programme 1990:2). Primary health care was extended to 61% of the population, and safe drinking water to 55% and despite the addition of 2 billion people in developing countries, the rise in food production exceeded the rise in population by about 20% (United Nations Development Programme 1990:2).

In spite of this progress, in 1985 more than a billion people in developing countries were trapped in absolute poverty, with some groups living in poverty also in developed nations. In 12 of the 23 developing countries where such a comparison is available, the income of the richest groups was 15 times or more that of the poorest group, notably in Latin America (UNDP 1990:22-23). FAO estimates that about 30 million agricultural households have no land and about 138 million are almost landless, two-thirds of them in Asia (UNDP 1990). A major conclusion of research is that some 500 million to 1 billion poor rural women in developing countries suffer the greatest deprivation, "For them, there has been little progress over the past 30 years" (UNDP 1990: 33).

Importantly, inequities in the distribution of financial and human capital did not decrease but actually grew in the 1980s, both within and among nations. This is illustrated in Figure 1.3, which shows the financial flows from developing to developed countries. What recommendations have been given to solve this reversal in the fragile progress in human development? The authors of UNDP's 1990 Report on Human Development conclude that "resumed economic growth is thus essential to allow the expansion of incomes, employment and government spending needed for human development in the long run. Without some end to the continuing debt and foreign exchange crisis in much of Africa and Latin America, the impressive human achievements recorded so far may soon be lost" (UNDP 1990:36).

Clearly, the range of variables impinging on the population question is much more diverse than is usually taken account of in the ongoing population debate in the United States and Europe.

Population and Resource Use in Social Contexts

The gradual decline in human mortality since the end of the last century must be considered one of the greatest achievements of Western civilization, both on scientific and on human management grounds. Rarely in history has such a unanimous, concerted action been successfully undertaken by such a large number of human agents: scientists, doctors, firms producing medicines, voluntary groups, pharmaceutical companies, and governments--even if some may be guided also by their own economic or political interests.

The Hippocratic oath defending human lives over and above any other consideration gave philosophical sustenance to their activities. Many other cultures, though, do not give such preeminence to the struggle against death; for

FIGURE 1.3 Reversing Resource Flows: North to South Net Transfers (US$ billions)

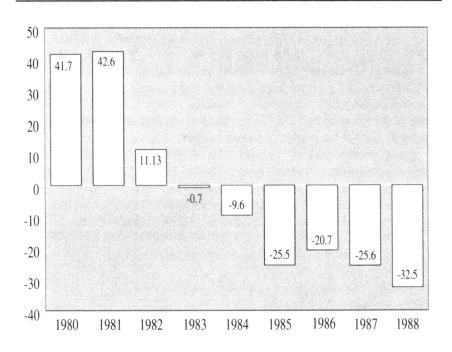

Source: United Nations Development Programme 1990.

some it means coming closer to spiritual liberation, or an opportunity for the soul to transmigrate into other realms or beings. Importantly for our discussion, many cultures, especially those living in inclement natural habitats, did subordinate individual human life to the survival of the group. A vast array of fertility control and abortive practices were, and in some cases still are, present in many non-Western cultures. The point is that for centuries many societies had evolved some kind of accounting system whereby the number of people in their group and their age structure were thought of in relation to available natural resources. This was especially true in hunting and gathering, pastoralist and horticultural societies, less so in agrarian societies, where food cultivation could be expanded and more people fed. Other, more aggressive societies obtained the resources needed to sustain their populations through warfare.

These built-in social and ecological accounting systems in many indigenous cultures placed the responsibility for managing sociodemographic processes on the societies themselves. This has been greatly undermined as a result of four driving forces. First, the centralization of power led to the subordination of rural societies to the needs of the urban systems. Second, the loss of cultural diversity eroded cohesive social mechanisms as traditional societies were pulled into market economies and subjected to uniform educational and media systems. Third, due to the spread of urban culture, people are no longer in direct contact with the natural sources of the things they eat, use, or play with, and so lose their bearings as to the depletion of natural stocks. Indeed, the steel and concrete urban environment gives the impression that goods appear purely out of thin air through technological manipulation. Urbanites, however, are becoming keenly aware of the piling up of human bodies in cities, thereby reinforcing their view that the problem is the number of human bodies occupying space and competing for goods, rather than the overall relationship of the urban pattern of consumption of natural resources and planetary stocks. Fourth, scientific models have tended to leave out the cultural and social matrices in which population processes are embedded, and thus have undermined local and meso-level capabilities of organizing their own sociodemographic processes.

Such socially operated cultural accounting systems, which foster local and regional social management, must be revitalized in societies around the world. In developing regions this would allow communities and local peoples to adjust their reproductive behavior to real expectations of sustainable livelihoods, natural resource availability, and locally defined measures of quality of life. Adapting to the environmental limits to growth in developing countries, however, need not entail accepting the economic limits imposed by continuing subsidies for the wasteful, polluting, affluent life-style of some sectors in industrialized countries of the North.

Curbing the Growth in Population and in Resource Use

A global perspective of the population-resource use issue means that a reduction in population growth in underconsuming nations must go hand in hand with reducing consumption among affluent groups and nations. This can only be achieved by lowering birth and death rates, alleviating poverty, reducing pressures on resources, and improving women's opportunities, employment-generating policies, and health care (Repetto 1986). At present the debate in the North deals more with population policies to implement in the South rather than with curbing overconsumption in the North (Worldwatch Institute 1988).

Population control is known to be insufficient: it has repeatedly been shown that it is not easily achieved in and of itself and that important social and economic transformations, such as reduction of poverty, must accompany it. "Population can only be expected to fall when livelihoods are secure, for only then does it become rational for poor people to limit family size" (Chambers

1988; Sen and Grown 1988). According to a World Bank study of 64 countries, when the income of the poor rises by 1 percent, general fertility rates drop by 3 percent (Lappe and Schurman 1988). Reduction of poverty may be a necessary condition for decreasing fertility; it is not, however, a sufficient condition, as the cases of Kerala, Sri Lanka, and other regions demonstrate (Gordon and Suzuki 1991). Nor do lower population growth rates translate immediately to improved environmental conditions. Even in those cases where population growth has been successfully controlled, as in China, the welfare of the people has not necessarily improved and the environment is not necessarily exposed to lower rates of hazard.

To reduce pressures on resources, research priorities should look at situations where demand, either subsistence or commercial, becomes large relative to the maximum sustainable yield of the resource, where the regenerative capacity of the resource is relatively low, or where the incentives and restraints facing the exploiters of the resource are such as to induce them to value present gains much more highly than future gains (Repetto and Holmes 1983). Natural resource scarcity studies indicate that a transition will have to be made during the next century from cheap, plentiful use of oil to inherently less desirable sources of energy (Mackellar and Vining 1987), although some authors are more optimistic about unlimited availability of energy (Gilland 1986).

As to the problem of food, prudent optimism largely prevails as to the possibilities of increasing agricultural productivity to feed the increase in population through the year 2000 (Srinivasan 1987; Mackellar and Vining 1987). Some authors, however, are not as optimistic (Brown, L. 1983). To analyze such possibilities, the real problem of production of more food must be separated from the economic and political problem of hunger, that is, of food distribution--the food versus feed issue. Biotechnology provides grounds for optimism, although it seems that its commercial applications will not be seen immediately.

Deforestation, on the other hand, presents a rather more pessimistic picture although different sources cannot agree on the rates of deforestation (Mackellar and Vining 1987; FAO 1990; Williams 1991). In 1950, industrialized countries imported 4.2 million square meters of tropical woods; in 1980 they imported 66 million (Myers, R. 1981). The outcome will depend on whether consumption of tropical woods and population pressures on the fringes of tropical rainforests are decreased.

Toward a Global Society

It would seem contradictory to argue in this chapter that one of the driving forces behind the population "bomb" was that population, as a variable, was abstracted from actual societies with highly disparate natural resource and income distribution bases, and yet emphasize that a "global society," another abstract construction, must be built. Indeed, we do agree with the Brundtland Commission that there will be no future if we are unable to build one world. To be more precise, we would say that a global society must begin to be interpreted as such, so that it can be seen as such, and therefore built (Arizpe 1991).

But the answer is that such a global society must be built in the same way that the nation-states have been built. Almost without exception, they are internally plural in ethnic and religious identities, per capita income, economic regionalisms, demographic growth rates, and so on, yet juridically and politically they function as a unit. In other words, almost without exception, the unity of nation-states is not an empirical reality, yet the transactions of national and international life are undertaken on the basis of this unity.

In the same way, one can posit that a global society has become a juridical, political, and even cultural necessity, yet the global empirical reality will always be made up of nations and societies, themselves made up of a plurality of trends, some converging, others diverging, that are still not fully understood or susceptible to being totally controlled. They can, however, through negotiations, be successfully managed and pointed in the right direction--if a direction can be agreed upon. Thus, abstracting population as a single factor in models purporting to represent complex empirical reality is inappropriate, but dealing with population as one of the main issues in building a global society is not only appropriate but necessary.

The deeper issue here is one that underlies debates all the way from the Lacandón rainforest in Mexico to the United Nations General Assembly: who is going to build this new economic and accounting system for the world? This is, indeed, a political issue at an international level. Since nations are still trying to enhance their own "wealth of nations," never having left the harbor of classical economics, each will try to build a system that, minimally, will keep its own interests untouched or, maximally, will increase its benefits.

At a more local level, the question of who is creating the new rules of a global society is perceived in more immediate terms as who is going to bear the cost, actual or potential, of preventing or adapting to new conditions. Whether the debate engages rainforest cattle ranchers and indigenous peoples on deforestation or poor urban dwellers and rich urbanites on urban pollution or corporations and ecologists on economic development or the North and the South on the future of the world, what is at stake is the capacity of human beings to negotiate a common future. And for this purpose, the concept of humanity seems more germane than that of population.

References

Alba, F., and J. Potter. 1986. "Population and Development in Mexico since 1940: An Interpretation." In *Population and Development Review*, 12,1 (March):47-75.

Arizpe, Lourdes. 1978. *Migración, etnicismo y cambio económico*. México: El Colegio de México.

————. 1982. "Relay Migration and the Survival of the Peasant Household." In *Why People Move*, ed. J. Balan Paris: UNESCO.

————. 1991. "The Global Cube: Social Models in a Global Context." In *International Social Science Journal*, 130, Nov.:599-608.

————, R. Costanza, and Wolfgang Lutz. 1992. "Population and Resources." In J. Dooge, G. Goodman, and J. la Riviere (eds.) *An Agenda for Science in Environment and Development*. London: Cambridge University Press.

Asian Development Bank. 1990. "Population Pressure and Natural Resource Management: Key Issues and Possible Actions." Paper No. 6.

Blaikie, M. and B. Brookfield. 1987. *Land Degradation and Society*. London: Methuen.

Blaxter, Kenneth F.R.S.. 1986. *People, Food and Resources*. Cambridge University Press.

Bongaarts, J., W. Mauldin, J. Phillips. 1990. "The Demographic Impact of Family Planning Programs". *Population Council Research Division* Working Paper No. 17.

Brinley, Thomas. 1961. *International Migration and Economic Development*. UNESCO.

Brown, Harrison. 1954. *The Challenge of Man's Future*. New York: The Viking Press.

Brown, Lester. 1983. "Global Food Prospects: Shadow of Malthus". In: Glassner, Martin I, *Global Resources: Challenges of Interdependence*. New York: Praeger.

Caldwell, John. 1984. "Desertification: Demografic Evidence, 1973-83". Australian National University. Occasional Paper No. 37.

Clark, Colin. 1958. "Population Growth and Living Standards" in *The Economics of Underdevelopment*, ed. A.N. Agarwal and S.P.Singh. London: Oxford University Press, 32-5.

Clark, William. 1991. Paper presented at the Annual Meeting of the American Association for the Advancement of Science, Washington, DC.

Chambers, R.. 1988. *Sustainable Livelihoods, Environment and Development: Putting Poor People First*. Brighton, U. K.: Institute of Development Studies, University of Sussex.

Costanza, R. (ed.). 1991. *Ecological Economics: The Science and Management of Sustainability*. New York: Columbia University Press.

Demeny, Paul. 1988. "Demography and the Limits of Growth"; in: *Population and Development Review Supplement*. Vol. 14: 213-244.

————. 1990. "Population" in Turner, B.L. et.al., *The Earth Transformed by Human Action. Global and Regional Changes in the Biosphere over the Past 300 Years*. New York: Cambridge University Press with Clark University, 41-54.

Dirección General de Estadística. 1991. *Population Census 1990*. Mexico: DGE.

Douglas, Ian. 1991. "Human Settlements". Paper presented at the Office for Interdisciplinary Earth Studies Workshop on Global Change held on July 28-August 10, Snowmass, Colorado.

Durning, Alan. 1991. "Asking How Much is Enough"; in: Brown, L.R., et. al. *State of the World 1991*. A Worldwatch Institute Report on Progress Toward a Sustainable Development. 153-169.

Eckholm, R. 1982. *Down to Earth: Environmental and Human Needs*. New York: Norton.

Ehrlich, Paul, et al. 1989. "Global Change and Carrying Capacity: Implications for Life on Earth" in Ruth DeFries and Thomas Malone (eds.) *Global Change and Our Common Future: Papers from a Forum*. Washington: National Academy Press, 19-27.

Ehrlich, Paul R. and Anne H. Ehrlich. 1991. *The Population Explosion*. Touchstone, Simon & Schuster Inc., New York.

FAO (United Nations Food and Agriculture Organization). 1990. *Vital World Statistics*. Rome:FAO.

Gallopin, Gilberto C. 1990. "Global impovershment, sustainable development and the environment". Ecological Analysis Group.

García, Rolando. 1990. "Interdisciplinariedad y sistemas complejos," in Leff, E. (coord.), *Las ciencias sociales y la formación ambiental a nivel universitario*. Mexico.

Gilland, Bernard. 1983. "Considerations on World Population and Food Supply"; in: *Population and Development Review*, 9,2: 203-211.

———————. 1986. "On Resources and Economic Development"; in: *Population and Development Review*, 12,2: 295-305.

Gordon and Suzuki. 1991. *It's a Matter of Survival*. Cambridge: Harvard University Press.

Grant and Tanton. 1981. "Immigration and the American Conscience" in *Progress as if Survival Mattered: A Handbook for a Conserver Society*, Hugh Nash (ed.) (San Francisco: Friends of the Earth).

Hardoy, E and W. Satterthwatte. 1991. "Environmental Problems of the Third World Cities: A Global Issue Ignored?"; in: *Public Administration and Development*, vol. 11.

Harrison, Paul. 1990. "Too Much Life on Earth?" *New Scientist*. May 19.

Hern, Warren M.. 1990. "Why Are There So Many of Us? Description and Diagnosis of a Planetary Ecopathological Process" in *Population and Environment: A Journal of Interdisciplinary Studies*. Volume 12, Number 1, Fall.

Hirschman, Albert. 1958. *The Strategy of Economic Development*. New Haven, Conn.: Yale University Press.

Jacobson, Jodi. 1988. "Planning the Global Family"; in: Brown, L.R., et al. *State of the World 1988*. A Worldwatch Institute Report on Progress Toward a Sustainable Development. 151-16.

Johnson, D. Gale and Lee, D. Ronald (eds.). 1987. *Population Growth and Economic Development: Issues and Evidence*. Madison: University of Wisconsin Press. A Publication of the National Research Council Committee on Population.

Kasun, Jacqueline P.. 1988. *The War Against Population: The Economics and Ideology of World Population Control*. (San Francisco: Ignatius Press).

Kelley, Allen. 1986. "Review of the National Research Council Report Population Growth and Economic Development: Policy Questions" in *Population and Development Review*, 12, 3, Sept.:563-567.

Keyfitz, Nathan. 1991. "Need We Have Confusion on Population and Environment?" International Institute for Applied Systems Analisis, Laxenburg, Austria. August.

--------------. 1991b. "From Malthus to Sustainable Growth". International Institute for Applied Systems Analysis, Laxenburg, Austria. July.

Kolsrud, Gretchen, and Barbara Boyle Torrey. 1993. "The importance of Population Growth in Future Commercial Energy Consumption," in *Global Climate Control*, edited by James White (New York: Plenum Press), 127-141.

Lappe and Schurman. 1988. *Taking Population Seriously*. New York: Earthscan.

Leff, Enrique. 1990. "Población y medio ambiente. Es urgente detener la degradación ambiental". In: *DEMOS*. Carta demográfica sobre México. México.

Little, P. and Horowitz, M. with A. Endre Nyerges (eds.). 1987. *Lands at Risk in the Third World: Local-Level Perspectives*. Westview Press.

Lutz, W. and C. Prinz. 1991. "Scenarios for the World Population in the Next Century: Excessive Growth or Extreme Aging". WP-91-22. International Institute for Applied Systems Analisis, Laxenburg, Austria.

Mackellar, F.L. and Vining Jr., D. R.. 1987. "Natural Resource Scarcity" in *Population Growth and Economic Development*, Johnson and Lee, eds. Madison: University of Wisconsin Press.

Maihold, Gunter and Urquidi, Victor. Compiladores. 1990. *Diálogo con nuestro futuro común*. Perspectivas latinoamericanas del Informe Brundtland. Fundación Frederich Ebert, México. Venezuela: Editorial Nueva Sociedad.

Meadows, Donella. 1988. "Quality of Life". In: *Earth '88: Changing Geographic Perspectives*. National Geographic Society. Washington, D.C., 332-349.

Myers, N.. 1987. *Not Far Afield: US Interests and the Global Environment*. Washington: World Resources Institute.

Myers, R. 1981. "Deforestation in the tropics: who gains, who loses?" in *Where have all the Flowers Gone? Studies in Third World Societies*, ed. Williamsberg.

Organization for Economic Co-operation and Development (OECD). 1991. *The State of the Environment*. Paris:OECD.

Repetto, Robert. 1986. *World Enough and Time*. New Haven: Yale University Press. -

--------------. 1987. *Population, Resources, Environment: An Uncertain Future*. Washington: Population Reference Bureau.

--------------, and Holmes, Thomas. 1983. "The Role of Population in Resource Depletion in Developing Countries"; in: *Population and Development Review*, Vol. 9, No. 4, December.

Revelle, Roger. 1976. "The Resources Available for Agriculture". *Scientific American*, September, 165-178.

Sage, Colin and Redclift, M. 1991. "Population and Income Change: Their role as Driving Forces of Land-Use Change". Document prepared for the Workshop 1991 Global Change Institute on Global Land-Use/Cover Change. Office for Interdisciplinary Earth Studies, July 28-August 10. Snowmass, Colorado.

Sánchez, Vicente, Margarita Castillejos y Leonora Rojas. 1989. *Población, recursos y medio ambiente en México*. Fundación Universo Veintiuno, A.C. México.

Sen, Gita and Caren Grown. 1988. *Development, Crises and Alternative Visions: Third World Women's Perspectives.* (New Delhi, INDIA: DAWN).

Simon, Julian. 1990. *Population Matters: People, Resources, Environment and Inmigration.* New Brunswick: Transaction Publishers.

Srinivasan. 1987. "Population and Food". In: *Population Growth and Economic Development.* Johnson and Lee, eds. Madison, University of Wisconsin Press.

SSRC/ISSC/DAWN. 1991. "Recasting the Population-Environment Debate: A Proposal for a Research Program".

Toledo, Víctor Manuel. 1990. "Modernidad y ecología. La nueva crisis planetaria." Document, April. México.

UNDP (United Nations Development Programme). 1990. *Human Development Report 1990.* New York: Oxford University Press.

United Nations. 1990. *Global Outlook 2000. An Economic, Social and Environmental Perspective.* United Nations Publications.

United Nations. 1974. *Human Settlements: the Environmental Challenge* London: Macmillan.

United Nations Department of International Economic and Social Affairs. 1989. *World Population Prospects 1988.* New York.

United Nations Population Fund. 1991. *The State of the World Population 1991.* New York: Oxford University Press.

Whitmore, Thomas M., et al.. 1990 "Long-Term Population Change" in Turner, B.L. et al. *The Earth Transformed by Human Action. Global and Regional Changes in the Biosphere over the Past 300 Years.* New York: Cambridge University Press with Clark University, 25-39.

Williams, Michael. 1991. "Forest and Tree Cover". Paper presented at the Office for Interdisciplinary Earth Studies Workshop on Global Change held on July28-August 10. Snowmass, Colorado.

World Resources Institute. 1990. *World Resources 1990-91. A Guide to the Global Environment.* New York: Oxford University Press.

Worldwatch Institute. 1988. *State of the World 1988. A Worldwatch Institute Report on Progress Toward a Sustainable Society.* New York.

Worldwatch Institute. 1990. *State of the World 1990. A Worldwatch Institute Report on Progress Toward a Sustainable Society.* New York.

Worldwatch Institute. 1991. *State of the World 1991. A Worldwatch Institute Report on Progress Toward a Sustainable Society.* New York.

2

World Population Trends: Global and Regional Interactions Between Population and Environment

Wolfgang Lutz

When discussing the interactions between changes in population patterns and degradation of the environment, it is useful to distinguish between regional and global environmental changes. Because the determinants and consequences of these environmental changes are different, their interactions with population are also of a different nature.

On the global level it seems to be the accumulation of greenhouse gases and the resulting climate warming together with the depletion of the ozone layer that cause the greatest concern. The increasing rate of extinction of species are also often discussed. Even more immediate global threats to human life, and therefore through mortality to population, are the possibility of global war and shortages in the global food supply, dangers possibly linked to environmental problems but also the result of many other factors. Food supply has a global and a local dimension. Local food shortages can be compensated, at least theoretically--given good political and socioeconomic conditions--by over-production in other regions. Because these distributional mechanisms do not always work, we at present see starvation in many parts of the world although the global food supply should be more than sufficient for the more than 5 billion people living today. But with rapidly increasing population sizes, possible degradation of agricultural land, and continuing infrastructural deficiencies there is the definite possibility that in the not-so-distant future there would even be a global food deficit in which there is no surplus to compensate for local shortages.

On the local and regional level environmental problems are manifold and more difficult to define. Most prominent are the local consequences of global climate change, including erosion of the soils, loss of vegetation, and, closely linked to them, crops and grazing intensities resulting in the local food supply. These aspects seem to be closely interwoven, and local water supply plays a crucial role in all of them. There are also local environmental aspects that more immediately affect human health, namely the toxification of food, water, and air, as well as the general disease environment and the availability of clean drinking water.

In this chapter I will take a closer look at the population side of the population-environment interaction. This paper will demonstrate which future population sizes and patterns are at all possible on global and regional scales, and how they could be influenced. This requires first a closer look at the nature of population dynamics and a distinction between the determinants and the characteristics of changing population patterns.

Dynamics of Changing Population Patterns

Many of the above items are threats to people and therefore have the potential to affect population patterns directly through increasing mortality and more indirectly through changing migration and fertility patterns. Under-nutrition, exposure to toxic chemicals or radiation, and other factors associated with environmental conditions may result in immediate death (e.g., due to starvation or flooding) or in increasing the risk of death in the future (e.g., through cancer), thus affecting the mortality level of the population. Migration, the second fundamental determinant of regional population size, structure, and distribution, is also likely to be dependent on many of the listed environmental aspects. Throughout human history populations have moved from one place to another in search of better living conditions, and with the modern means of transportation available, they are even more likely to do so.

Human fertility, the third and in the long run most influential component of population change, tends to react rather slowly to changes in the outside world. Because reproductive patterns are deeply embedded in cultures and religions, direct reactions to environmental changes are less likely in the shorter run. Under certain conditions fertility patterns may even react to environmental conditions in the "wrong" directions, namely families having more children under worsening conditions because the experience of history tells them that this way they may have more surviving adult children to help them with the basic means of subsistence and to care for them when they are elderly. There is a large body of literature on the determinants of reproductive behavior in less and more developed countries that will not be discussed here. The bottom line of the discussion is that there is no simple monocausal explanation. Neither economic development per se nor the provision of family planning services alone will bring down fertility levels. A comparative analysis of past fertility transitions seems to indicate that social development tends to be more important than mere economic development. But since the two usually go together, their effects are difficult to disentangle. There are, however, some cases, such as Finland at the beginning of this century or Mauritius during the 1960s, where the transition to low fertility happened before any significant economic development but was associated with sociocultural changes. On the other side we find countries such as some oil-exporting countries in the Persian Gulf region that have seen rapid economic development but have not yet experienced the fertility transition because of very traditional social systems. Numerous studies have shown that in most countries female educational status plays a crucial role in lowering fertility levels (Lutz 1989).

Figure 2.1 shows how population dynamics interact with the outside world, which is not only the natural environment but also the socioeconomic and cul-

FIGURE 2.1 Determinants and Basic Characteristics of Changing Population Patterns

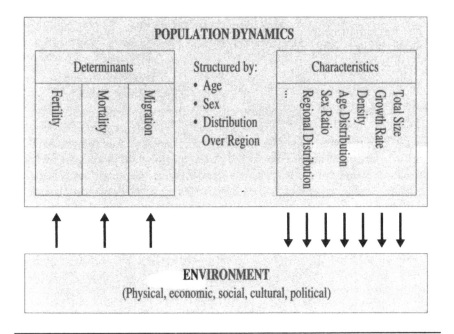

tural surroundings. It is important to understand first that population size and structure can only be influenced through three variables: fertility adding new people, mortality depleting the population, and migration moving people from one region to another. Every change of population patterns can only work through one or more of these variables. Because all three variables vary significantly with age, age is generally regarded as the most important structural characteristic of the population. It is evident that neither population size nor the growth rate can be directly steered, but can only indirectly be influenced by these three basic variables. And there is sometimes quite a substantial time lag between the change in one of the determining variables (e.g., a fertility decline) and the full effect (e.g., stabilization of population size). Even the growth rate, which is usually considered a very sensitive indicator, needs some time to adjust fully to new fertility or mortality levels. This is due to the momentum of population growth, which will be discussed below.

Generally, the following characteristics of population are of importance to the economy, society, and environment in which the population lives: total population size; the growth rate of the population; population density; age distribution of the population; sex ratio in the population; and the regional distribution of the population. The components of population dynamics through which the determining variables of fertility, mortality, and migration affect the characteristics listed above are changes in the size of age groups for both sexes separately and, if regional distribution is considered, the sizes of these groups in the various regions. The structure of the population by age, sex, and region is not formed instantly but results from history that may even go several generations back. Only in the very long run will the effect of an initial age distribution be eliminated.

In the short to medium term the population characteristics always represent a mixture of recent levels in the vital rates of fertility, mortality, and migration, and the past population structure. If, for instance, a population due to traditionally high levels of fertility has a very young age structure, even a decrease in fertility measured by the mean number of children per woman may result in a situation where from year to year the absolute number of births is increasing because larger and larger age groups come into the main reproductive ages 20-30, resulting in more births due to the simple fact that there are more potential mothers. This phenomenon is called the *momentum of population growth*. Generally, momentum refers to the fact that changes in any one of the determining variables need quite a while until they become fully effective.

One good example of the momentum of population growth after a fertility decline is the case of Mauritius, which experienced what was perhaps the world's most rapid fertility decline in the late 1960s. Despite replacement fertility (about 2.1 children per woman) now, the number of births on Mauritius still increases significantly from year to year because the large numbers of children born during the 1960s increase the number of women in the prime childbearing ages. Another case, a thousand times larger than Mauritius, is China. Despite China's rapid fertility decline and its present fertility level that is only slightly above replacement level, the total population of China is still growing by 1.4 percent a year. It is projected that China's present population of about 1.2 billion will increase by at least an additional 300 million people by the year 2025. Most of this increase would be due to the momentum of population growth.

Another, more extreme example of the momentum of population growth can be derived from the following scenario calculations. Since Africa has at present the highest level of fertility (about 6 children per woman on average), its population age structure is also the youngest in the world. Even in the entirely unrealistic case that fertility in Africa were to decline within a year to replacement level, the population of Africa would still increase by about 50 percent. In an even more dramatic projection: if fertility in Africa did not decline to replacement level within 35 years (as assumed in the UN medium variant) but took 25 years longer, this would mean 1.5 billion additional people in Africa. This difference alone is about three times the current population size of Africa.

When talking about population and environment and especially when discussing which variables could most easily and most effectively be influenced, it is important to be aware of this momentum of population growth. Population growth cannot simply be switched on or off. It is hard enough to influence behavioral variables such as fertility. But even if this were entirely successful the age structure would ensure some significant further population growth.

It should also be mentioned that demographers often consider other structural elements of the population in addition to age, sex, and location. Most relevant to demographic processes seem to be marital status, educational status, and labor force participation, as well as urban or rural place of residence. It is also relevant to study the family and household compositions. Since these units are usually the agents of consumption and often even production, their consideration is most relevant to the study of environmental impacts. If these additional structural characteristics are considered in addition to age and sex, then transition rates from one status into another--e.g. from single to married, from childless to with children, from primary education to secondary, or from not in the labor force to in the labor force--become determining variables in addition to fertility, mortality, and migration.

Alternative Population Projections for Major World Regions

It follows from the above description of population dynamics that changes in population size and age structure have great inertia. Unless wars, famines, or epidemics kill significant proportions of the population, or massive migratory streams empty some regions and fill others, the future population of a certain region can be projected with high certainty in the short run. If one also assumes that fertility varies only within a rather narrow range, projections for the next 20-30 years are very reliable and insensitive to minor changes in mortality, migration, and fertility. Most of the change in size and age structure is already preprogrammed in the population's present age structure. In the following scenario calculations we will see that even widely differing assumptions will yield very similar results up to the years 2010-2020. Thereafter, however, the range of possible futures opens widely.

The usual approach to population projections, as practiced by most national statistical offices and international organizations, is to calculate some (usually three) variants based on given combinations of fertility, mortality, and migration assumptions. The assumptions involved are generally not very obvious for the users. Traditionally the high and low variants differ from the medium variant only by the assumed levels of fertility. The time horizon of these projections tends to be 2025-2030. The range appearing between the low and high variant is often mistakenly interpreted as a confidence interval, although the assumptions on which they are based are more or less arbitrary.

Aside from the United Nations (1991), the World Bank regularly prepares population projections on a global scale for all countries and regions (Vu 1985). For the World Bank the time horizon is much longer, but only one variant is given. Generally, it is assumed that in every country fertility will reach replacement level in a certain year (e.g., 2010 in Brazil). After that year all rates are kept constant and projections are done up to the year in which the population will become stationary; that is, cease to grow due to the momentum of growth (2155 in Brazil). That stationary population size is then used as the ultimate size of a country's population. This method was used in the 1985 World Bank projections, resulting in a world population of 10.4 billion in 2100. There are no alternative variants given by the World Bank.

The scenario approach chosen for the calculations in this paper and in related earlier studies (Lutz and Prinz 1991; Arizpe, Costanzo, and Lutz 1992) will demonstrate the implications of several alternative future paths of fertility and mortality that need not necessarily reflect present mainstream thinking. Some scenarios look at the consequences of possible discontinuities in past observed trends; others just caution the general belief that fertility will soon enter a steep decline in many less developed regions, especially in Africa. The value of such a scenario approach lies only to a lesser extent in the possibility for everyone to choose one's own favorite scenario and look at its long-term implications. The main value of such a set of alternative scenario projections, based on controversial but informed guesses about the future, lies in giving a picture of the possible range of future population sizes and structures. This will help to distinguish almost inevitable trends from changes that are very sensitive to slight modifications in the assumption. Given in reasonable regional detail, such data should prove useful for studying the necessary resilience of ecosystems to alternative demographic futures.

For the following scenario calculations six regions of the world were defined and projected separately. They are Africa, Latin America, Eastern Asia (China, etc.), Southern Asia (India, etc.), Other Asia (Arabic-speaking countries, Indonesia, etc.)/Oceania, and Europe/USSR/North America. This definition of regions should not be considered a final one, as it was difficult to find a distribution that matches all reasonable demographic, geographic, environmental, political, and statistical criteria. Age structures of the population for the starting year (1985) were chosen from UN data (UN 1991). A more elaborate and systematic set of scenario projections for 12 world regions is presently being preposed for publication in book form (Lutz, forthcoming).

Table 2.1 gives the present levels of fertility (as measured by the Total Fertility Rate, TFR, which may be interpreted as the mean number of children per woman) and mortality (as measured by life expectancy at birth) as well as total population size in 1990 for the regions defined above. Today, average fertility levels range from 1.97 in the region Europe/USSR/North America to 6.24 in Africa. For individual countries the variance is much larger, e.g., Italy

TABLE 2.1 Fertility, Life Expectancy, and Population Size in 1990 by Region

Region	Total Fertility Rate	*Life Expectancy* Male	Female	Total Population Size (millions)
Africa	5.93	55.7	52.3	640
Latin America	3.30	70.8	65.2	450
Eastern Asia	2.20	73.5	69.8	1,340
Southern Asia	4.32	59.4	58.8	1,200
Other Asia/ Oceania	3.59	66.1	62.4	600
Europe/USSR/ North America	1.96	77.8	70.4	1,060

Source: United Nations (1991). *World Population Prospects 1990*. New York: UN.

at 1.3 children per woman and Rwanda at 8.0. Life expectancy ranges from an average of 50.3 years for men in Africa to 77 years for women in the industrialized countries. In Japan female life expectancy at birth has already reached more than 82 years, whereas in Afghanistan it is only 44 years. These sharp differences are supposed to diminish steadily in most population projections. The UN medium variant, for instance, assumes a reduction in the TFR gap from 4.27 in 1985 to only 1.24 in 2025, and in the maximum regional difference in life expectancies from 26.7 years in 1985 to 18.2 years in 2025.

The 10 scenarios considered in this study are defined as follows. A Constant Rates Scenario (1) keeps fertility and mortality rates constant at their 1985-1990 level. This scenario is very unlikely because in some regions, especially in Africa, it is difficult to imagine how a population many times that of today could survive. If fertility does not decrease significantly one would almost inevitably have to expect an increase in mortality. On the other hand one could argue that Africa still has huge agricultural potential, and that under high standards of agricultural technology and social organization, it could be theoretically possible to provide a livelihood for many more people.

In terms of Malthusian thinking this scenario assumes the absence of both a positive check (through higher mortality) and a preventive check (fertility decline). The only way historical Europe could avoid the positive check (at a population growth rate that was much lower than in many less developed countries, or LDCs, today) was by massive emigration to the New World and through important technological progress. Today the first option hardly exists and it is very doubtful that technological change could be strong enough to feed ten times the Africans of today in some 80 years. The second factor that makes a continuity of constant rates unlikely for the more distant future is the hypothesis of demographic transition, which says that a fertility decline follows the mortality decline with some lag for adjustment of social norms. It is unclear how long this lag will last, but empirical evidence shows that this seems to be a universal tendency.

The second scenario corresponds to the UN Medium Variant projection (2). Since these assumptions only go to the year 2025, rates are kept constant at their 2025 levels for the rest of the next century. The UN Medium Variant assumptions may be regarded as rather optimistic, because--with the exception of Africa--they assume fertility to reach replacement level everywhere within only 35 years. Life expectancy is assumed to increase by around 10 years until 2025, distributed over the regions from only 5 years for women in the industrialized world to more than 15 years for women in Southern Asia, thus resulting in a convergence in life expectancy in the long run. These assumed strong increases in Third World life expectancy do not consider possible new threats, ranging from AIDS in Africa to famines or environmental catastrophes.

More for demonstration than for serious consideration in the long run, the Constant Fertility Scenario (3) assumes an even more extreme case than the

Constant Rates Scenario, with age specific fertility rates constant at their present level but mortality improving according to the UN Median Variant Scenario described above. Such a situation has been observed in parts of Africa in the recent past but is impossible to continue forever for the reasons described above.

In the Slow Fertility Decline Scenario (4) fertility is assumed to decline to the UN Medium Variant level within 60 years instead of 35 years. Hence, this scenario is less optimistic than Scenario 2, since fertility rates close to replacement level are reached in 2050 instead of 2025. Mortality assumptions are identical to Scenario 2.

As another extreme, the Rapid Fertility Decline Scenario (5) investigates the population growth and structure we could expect in the case where current very low Western European fertility levels--with an average number of 1.4 children per woman--would be reached all over the world by 2025. Fertility declines linearly between today and the TFR of 1.4 in 2025, and again the life expectancy of Scenario 2 is assumed. Thus, it corresponds to a situation with completed demographic transition in every region of the globe. This scenario, which does not seem very likely at first sight, would reflect a situation of very high social and economic development in all parts of the world. With constant mortality, a TFR of 1.4 implies that each subsequent generation is diminished by one-third. Hence, in the very long run the world population size would be declining.

For a better understanding of the extent of the momentum of population growth, the Immediate Replacement Fertility Scenario (6) makes the highly unlikely assumption that replacement fertility (a TFR of 2.1) is immediately reached in all parts of the world. This is combined with the mortality assumption of Scenario 2. The extent of the momentum of population growth could also be derived analytically (Keyfitz 1977:155-157). In the case of the initial age distribution being stable, the formula is very simple. In the empirical case of past changes in fertility and mortality patterns, a projection has the advantage of being able to compare population sizes at each point in time.

The following four scenarios which consider different future trends in mortality, all assume replacement-level fertility in all regions by the year 2025. This is done not because it was considered the most likely fertility assumption, but for simplicity. In analogy to Scenario 3, the Constant Mortality Scenario (7) reflects a situation of completed fertility transition in 2025 (i.e., replacement fertility everywhere) together with a stagnation in life expectancy at current levels. This stagnation might have several reasons, ranging from new spreads of infectious diseases to nutritional deficits or environmental degradation. In the Slow Mortality Decline Scenario (8) mortality is assumed to decline as in the UN Medium Variant but with a slower path of improvement. UN mortality levels are reached in 2050 instead of 2025. The other extreme is represented by the Rapid Mortality Decline Scenario (9). Here it is assumed that life expectancy at birth will reach 85 years for women and 80 years for men in

2025. Again, linear paths of improvements and no differences between regions are assumed.

Finally, one scenario was designed where assumptions differ strongly by region. The so-called Third World Crisis Scenario (10) defines the future "Third World" as Africa and Southern Asia only. Hence, the scenario assumes that these two regions will not manage their population growth problem and a positive Malthusian check will increase mortality levels. In those two regions the crisis is defined as an increase in mortality (about 10 percent decline in life expectancy) together with continued high fertility (i.e., constant fertility rates). The rest of the world is assumed to complete the demographic transition by at least 2025, that is, fertility reaches replacement level and mortality improves as in Scenario 2.

Of course, an unlimited number of different scenarios could be defined here especially when one assumes diverging demographic trends in the more distant future, as in the last scenario. Furthermore, the scenario definitions do not consider the possibilities of massive migratory streams from one region to another. In future calculations a larger number of scenarios with more differentiated assumptions will be defined and calculated. The above ten scenario definitions should suffice to give a crude first impression of the consequences of the alternative assumptions.

At the starting year of our projections 1990, the planet accommodated around 5 billion people. Under all scenarios considered over the next 30-40 years, the world population will increase to a size of at least around 8 billion. Even immediate replacement fertility in all parts of the world would result in an additional 2 billion or more people, due only to the momentum of population growth. Under the Rapid Fertility Decline Scenario, assuming only two-thirds replacement by 2025, the total population size would peak in 2040 at around 8 billion and only decline thereafter.

Hence we may conclude on the lower side of the spectrum that, unless completely unexpected major threats to life kill great proportions of the world population over the coming 30-40 years, the world will have to accommodate an extra number of people at least as large as half of the world population today. Under the assumption of sustained sub-replacement fertility in all regions of the world the population might then decline again in the very long run, and possibly by the year 2100 reach a size that is lower than that of today. But still, in the transition period the 8 billion mark will be reached.

On the higher end of the spectrum of possible future population sizes according to the scenarios defined here we have to distinguish between the three scenarios that look at the case of continued growth and those assuming a leveling off.

Obviously, exponential growth cannot continue forever and therefore in the longer run is not only unrealistic but also impossible. Nevertheless, it is instructive to study the results especially in the short- to medium-term future,

TABLE 2.2 Total Population Sizes in Billions in 2050 and 2100 According to Selected Scenarios

Year/ Scenario	Africa	Latin Amer.	East Asia	South Asia	Other Asia/ Oceania	Eur./ N.A./ USSR	World Pop.
1990	0.64	0.45	1.34	1.20	0.60	1.06	5.29
2050							
Scen. 1	3.59	1.38	2.07	4.45	2.04	1.07	14.59
Scen. 2	2.20	0.96	1.72	2.60	1.28	1.13	9.89
Scen. 5	1.32	0.72	1.58	2.09	1.03	0.93	7.68
Scen. 9	1.91	1.00	2.16	2.92	1.43	1.26	10.69
Scen. 10	3.17	0.93	2.05	3.95	1.32	1.21	12.61
2100							
Scen. 1	14.50	3.17	2.47	12.91	5.32	0.94	39.31
Scen. 2	3.46	1.19	1.33	2.60	1.41	0.99	10.96
Scen. 5	0.72	0.42	0.87	1.18	0.61	0.49	4.29
Scen. 9	2.16	1.10	2.24	3.23	1.61	1.26	11.60
Scen. 10	11.19	0.95	2.08	10.14	1.38	1.20	26.93

and compare them to other scenarios. Furthermore, an assumed continuation of currently observed levels is in almost every scientific discipline a standard for comparison unless there is certainty that this level will change in one specific direction. In the case of population growth we have good reason to assume a change in rates, but we are far from any certainty about the extent and the timing of this decline. In any case, because of the inertia, over the coming three decades the Constant Rates Scenario will not result in very different total population sizes than most other scenarios. Around 2015 the 8 billion would be reached and 10 billion only after 2025. Under this scenario a world population of 15 billion would first appear after 2050. Under Scenario 3, assuming constant fertility and improving life expectancy, greater than exponential growth would result in even 17 billion by 2050. For the second half of the next century continued exponential growth would lead to ever-increasing absolute increments, resulting, under Scenario 1, in about 40 billion in the year 2100. Further continuation of this growth would soon result in a "standing room only" situation. By definition, these exponential scenarios do not assume a feedback from population size to fertility or mortality.

For the scenarios assuming a decline of fertility to replacement level at some point in the future, it appears that even relatively small differences in assumptions concerning the timing of fertility decline have a major impact on population size. Projecting UN Medium Variant assumptions up to the year 2100 gives a leveling off in population growth at around 11 billion, the population size in 2050 already being 10 billion. Delaying the fertility decline by 25 years, Scenario 4 gives a population size of more than 14 billion in 2100, and population growth does not seem to stop before having reached 15-16 billion in the 22nd century. Likewise, with a rapid linear fertility decline to a TFR of only 1.4 children per woman in the year 2025 for every region, Scenario 5 gives a totally different picture: after an increase to 8 billion in 2030-2040, population size may decline to a figure below 5 billion in the very long run (2100).

Population size is, to a lesser extent, influenced by assumptions on mortality. In the medium term, a constant mortality level in conjunction with replacement fertility by 2025, Scenario 7 would delay the growth in population size by some ten years as compared to the UN Medium Variant assumptions on mortality improvements. In the very long run, population size in the absence of mortality improvements tends to level off at around 7.5 billion, which is 2 billion below that in a corresponding scenario with increasing life expectancy.

Delaying the assumed improvement in life expectancy by 25 years, Scenario 8 has virtually no impact on total population size. Assuming a rapid increase in life expectancy to 85 years for women and 80 years for men in all regions in the year 2025, Scenario 9 increases population size by 1 billion in 2050 and 2 billion in 2100.

It is safe to say that population growth will occur unevenly across the globe and result in major changes in the global population distribution. Currently, the world population is distributed over the continents as shown in Table 2.3. Somewhat more than 20 percent each live in Europe/North America/USSR, Southern Asia, and Eastern Asia, while around 10 percent each live in Africa, Latin America, and Other Asia/Oceania.

In the future, due to present great differentials in fertility levels resulting in different age structures and therefore a differential momentum of growth, and due to the assumption that fertility changes will only be gradual, the populations of Africa and Southern Asia will grow fast whereas Eastern Asia and Europe will hardly grow. Consequently, all scenarios result in a change of regional population distributions, which in the case of Scenario 2 might read as follows in 2050 (see Table 2.3): only 11% would live in Europe/North America/USSR, only 17% in Eastern Asia, but 22% in Africa. According to this UN Medium Variant Scenario, the total world population will double over the next 70 years:

TABLE 2.3 Regional Population Distribution According to Selected Scenarios

Year/ Scenario	Africa	Latin Amer	East Asia	South Asia	Other Asia/ Oceania	Eur./ USA/ USSR	World Pop. (bil- lions)
1990	12%	9%	25%	23%	11%	20%	5.29
2050							
Scen. 1	25%	9%	14%	31%	14%	7%	14.59
Scen. 2	22%	10%	17%	26%	13%	11%	9.89
Scen. 5	17%	9%	21%	27%	13%	12%	7.68
Scen. 9	18%	9%	20%	27%	13%	12%	10.69
Scen. 10	25%	7%	16%	31%	10%	10%	12.61
2100							
Scen. 1	37%	8%	6%	33%	14%	2%	39.31
Scen. 2	32%	11%	12%	24%	13%	9%	10.96
Scen. 5	17%	10%	20%	28%	14%	12%	4.29
Scen. 9	19%	10%	19%	28%	14%	11%	11.60
Scen. 10	42%	4%	8%	38%	5%	4%	26.93

Africa will grow fourfold, while the industrialized world will stay its current size.

However, Table 2.3 shows that the future population distribution might also be more extreme. The Constant Rates Scenario (1) results in 25% Africans but only 7% Europeans/North Americans/Soviets by the year 2050, and hypothetically 37% and 2%, respectively, in 2100. Aside from the unimaginable population increase in Africa, it is also interesting to look at Asia: while there is about one Southern Asian to each Eastern Asian today, it will be two to one in 2100 according to Scenario 2, or even five to one according to Scenario 1.

A very extreme change in regional distributions is caused by the scenario assuming differential demographic developments between the regions. Scenario 10, assuming a crisis in Africa and Southern Asia resulting in higher mortality and constantly high fertility, but demographic stability in all other parts of the world, gives a distribution that is even more extreme than under Scenario 1:

despite increasing mortality in those regions, 80% of the world population would then live either in Africa or in Southern Asia, as compared to only 35% today.

Scenarios that assume a fast or moderate convergence to replacement or sub-replacement fertility (Scenarios 5-9) by definition result in more moderate distributional changes. Between 10% (Scenario 6) and 19% (Scenario 9) would be Africans, the corresponding percentages for Europe/North America/USSR being 18% and 11%, respectively.

There is no doubt that the world population will be aging significantly in the future. Even the Constant Rates Scenario, which is very unlikely in the long run, gives an increase in the mean ages of all regions over the next 50 years. Scenario 2, based on the UN Medium Variant, will result in much more significant aging in all regions of the world. While in today's industrialized countries the mean age is expected to increase from the present 35 years to more than 43 years by 2070, the extent and pace of aging will be even stronger in Eastern Asia. There, the UN Medium Variant expects an increase in the mean

TABLE 2.4 Mean Age of the Population (in Years) in 1990 and in 2050 According to the 10 Scenarios

Year/ Scenario	Africa	Latin Amer	East Asia	South Asia	Other Asia/ Oceania	Europe/ N.Amer/ USSR	World Total
1990	22.2	25.8	29.8	24.8	24.8	35.8	28.0
2050							
Scen. 1	22.9	29.0	35.7	25.8	26.3	40.2	27.9
Scen. 2	31.6	36.3	43.1	37.4	36.4	42.8	37.5
Scen. 3	23.0	30.0	38.0	26.5	28.1	42.5	28.7
Scen. 4	28.6	35.1	42.4	34.1	34.6	42.8	35.0
Scen. 5	40.6	42.7	45.7	42.2	41.3	47.6	43.2
Scen. 6	40.9	40.0	40.5	41.3	39.0	41.1	40.6
Scen. 7	35.4	36.2	37.5	36.0	33.8	39.2	36.4
Scen. 8	36.2	37.4	39.5	37.3	36.1	41.1	37.9
Scen. 9	37.6	39.4	41.5	39.0	38.6	42.9	39.7
Scen. 10	23.0	37.6	39.9	25.6	36.5	41.5	30.8

age from presently 29 years to 44 years by 2070. Even in Africa the mean age is expected to increase by more than 12 years to about the same level found in Europe and North America today.

As could be expected, the Rapid Fertility Decline Scenario (5), which is the only one that would ultimately bring the world population below its present size, results in the most extreme aging of the world population. Under this scenario, in almost every continent the mean age of the population would reach about 50 years by the end of the next century. What this means in terms of changes to the social and economic structure is hard to imagine, not to mention medical expenses and retirement benefits. The age structure of the population is expected also to have impacts on socioeconomic development and environmental questions because individual behavior and consumption vary significantly with age.

In summary, this comparison of the consequences of various scenarios on total population size and the age structure of the population makes clear the fundamental dilemma of future population trends under low mortality conditions. All scenarios that limit population growth, even at a level that is two to three times the current world population, will result in extreme aging of the population. Only further exponential growth of the population will keep the populations young.

Put simply, either the population explodes in size or it ages to an unprecedented extent. The explosion will sooner or later result in higher mortality levels because it cannot go on forever; the aging makes necessary painful social adjustment processes and a complete remodeling of both family and state support systems for the elderly. Probably the future will bring a combination of both undesirable phenomena. Possible effects of aging and the associated increase in economic dependence ratios, as well as likely changes in the household structure (more single-person households) on the environment, have hardly been studied and deserve much more scientific attention.

Alternative Paths of Population Growth and the Environment

The above-described alternative scenarios of regional population growth are linked to environmental questions in two ways. First, the components of population change--fertility, mortality, and migration--depend to some degree on environmental factors. Especially mortality is closely related to sufficient food supply, which of course is closely related to the quality of soils, rainfall, temperature, and other environmental variables. It is rather unlikely, for instance, that even under ideal conditions Africa would be in the position to sustain constant rates (Scenario 7), which imply a twentyfold population 100 years from now. If fertility does not decline, sooner or later population growth could be checked by increasing mortality.

Aside from the subsistence of the population that is a precondition for survival, population size also has an impact on environmental variables with no immediately perceived or no contemporaneous effect on human living conditions. While soil degradation seems to be more directly linked to food supply, impacts on biodiversity, deforestation, or greenhouse gas emissions may have only longer-term consequences on the population itself. Nevertheless these population impacts on the environment are relevant at many different levels. Without doubt, population growth is one of the driving forces of global environmental change. How important a driving force is a question of heated scientific, ideological and even diplomatic debate. On the one side, scientists from the natural sciences and ecological sciences tend to point to the limited carrying capacity and the fact that twice as many people cause twice as much harm or even more. On the other side, some economists point to human ingenuity for solutions. Still another group considers the wasteful behavior of the rich the major reason for environmental degradation.

It is highly problematic to link figures of absolute numbers of people directly to environmental questions without considering the intermediate behavioral and technological factors. After all it is the behavior of people, what they consume and what they do, that has consequences on the environment. Even a small number of people can do tremendous harm to the environment, whereas one can imagine large numbers of people having environmentally friendly life-styles. Any realistic assessment of the relative importance of these factors requires a careful analysis of specific consumption behaviors and specific environmental conditions. Ideally such assessments should be based on a number of careful case studies.

Despite the lack of an unambiguous direct link between population and environmental change, it is certainly possible to make simple calculations on the environmental impact of alternative population sizes by keeping per capita consumption patterns and technology constant or even making certain specific assumptions on future behavior and technology. To illustrate a potential environmental impact of alternative paths of population growth and also to point to some of the difficulties in attempts to assess the relative effects of population change, some simple calculations on population and carbon emissions will be given below.

Table 2.5 combines some of the results from the scenario calculations above with alternative assumptions on future per capita carbon emissions. In 1990 the 5.3 billion people on earth produced approximately 5.7 gigatons of carbon (not counting deforestation), which results in a per capita carbon emission of 1.07 tons per year. For reasons that will be discussed below, the calculations are made separately for both hemispheres, with "North" referring to Europe (plus the former USSR) and North America and "South" being the rest of the world. This choice of hemispheres made the carbon data compatible to the population data described above. The table shows that at present, two-thirds of total carbon

TABLE 2.5 Population Growth and Carbon Emissions, 1990-2050, Alternative Scenarios (Gigatons and Tons Carbon)

1988-1990	Emissions p.c. (tC)	Population Size (billions)	Total Emissions (Gt C)
World	1.07	5.29	5.66
North*	3.52	1.06	3.73
South*	0.46	4.23	1.94
2050			
I) Population Scenario 1 (Constant Rates)			
A: Constant p.c. emissions in each hemisphere			
North	3.52	1.07	3.76
South	0.46	13.52	6.22
Total			9.98
B: North constant, South triple p.c. emissions by 2050			
North	3.52	1.07	3.76
South	1.38	13.52	18.66
Total			22.42
II) Population Scenario 2 (Extended UN Medium Variant)			
A: Constant p.c. emissions in each hemisphere			
North	3.52	1.13	3.98
South	0.46	8.76	4.03
Total			8.01
B: North constant, South triple p.c. emissions by 2050			
North	3.52	1.13	3.98
South	1.38	8.76	12.08
Total			16.06
III) Population Scenario 6 (Immediate Replacement Fertility)			
A: Constant p.c. emissions in each hemisphere			
North	3.52	1.25	4.40
South	0.46	6.16	2.83
Total			7.23
B: North constant, South triple p.c. emissions by 2050			
North	3.52	1.25	4.40
South	1.38	6.16	8.50
Total			12.90

are emitted by the North, which has one-fifth of the world population. Per capita emissions in the North are 6-7 times higher than in the South.

Table 2.5 also gives some hypothetical figures for 2050 combining population Scenarios 1 (constant rates), 2 (UN medium), and 6 (immediate replacement fertility) with three alternative assumptions on future per capita carbon emissions: A shows constant per capita emissions in each hemisphere; and B, constant per capita emissions in the North but tripling of per capita emissions in the South by 2050.

The table shows at first sight that the variations due to assumptions on alternative per capita emissions are much greater than those due to alternative population assumptions. Under the assumption of constant per capita emissions the population scenario with constant rates would result in an increase of total carbon emissions from the present 5.7 gigatons to about 10 gigatons by 2050. Under the extended UN medium variant it would only increase to 8.01 and under the immediate replacement fertility scenario to 7.23 gigatons. Hence the increase in total carbon emissions between 1990 and 2050 that would be entirely due to population increase (no change in per capita emissions) would be between 28% and 77% according to the two most extreme population scenarios.

In the case of the South reaching Northern consumption levels by 2050, the increase in total carbon emissions would be between 350% and 800% depending on the population scenario. The variant with tripled per capita consumptions in the South has an intermediate position with increases between 130% and 300%. It is immediately clear that the variability due to possible changes in per capita consumption patterns is much larger than that due to alternative population trends.

Certainly one can argue that this conclusion is merely a function of the assumptions made and that there could be different standards applied for population and per capita consumption. But in the case of population, scenarios 1 (constant rates) and 6 (immediate replacement fertility) are about the most extreme that can be made. Both are impossible on practical grounds, although still on the margin of being theoretically possible. In the case of per capita consumption the alternative extreme assumptions certainly are not likely, but it would not be difficult to think of more extreme cases that are theoretically possible. Choosing more extreme consumption alternatives, however, would only increase the importance of per capita consumption versus population growth.

But in any case the above-described decomposition attempts of environmental impacts should be looked at only with extreme caution, because irrespective of the assumptions made the results are highly dependent on the level of aggregation chosen. Had the exercise been performed for the world total without distinguishing between North and South, the result would have been entirely different. Since this is a general problem that is difficult if not devastating in any analysis of this sort, it will be treated more explicitly below. This question is also related to the basic micro-macro question that underlies this volume.

Different Levels of Aggregation Change the Results

It is difficult to make unambiguous statements on the role of population growth on the environment even under a very simplistic approach only linking population size to per capita carbon emissions. This can be seen from the following calculations. Table 2.6 gives some of the calculations performed in Table 2.5 with identical data and assumptions, the only difference being that the world is viewed as one aggregate rather than North and South. For simplicity this is only done for the constant per capita emissions variant (A). Applying the 1.07 tons of carbon for 1990, which is the global per capita emission, to the 14.59 billion people resulting from the constant demographic rates scenario in 2050, this results in 15.61 gigatons of total emissions in 2050. This figure is significantly greater than the 9.98 gigatons in Table 2.5, which distinguished between North and South. The reason for this difference is that the global average is highly influenced by the North whereas the population growth will mostly take place in the South.

Table 2.6 also shows that the difference between the corresponding figures in Tables 2.5 and 2.6 is greater the more rapidly the population grows. This is because the above-described phenomenon is more relevant under rapid population growth in the South.

In other words, the high dependency of the calculated total carbon emissions from 2050 results from a strong negative correlation between population growth and carbon emissions. The higher the emissions, the lower the fertility. And there is no reason to assume that this problem is solved simply by distinguishing between North and South. This negative correlation is likely to be present within each hemisphere, but especially within the South. Even going down to national level does not solve the problem because within each country the rich have fewer children and emit significantly more than the poor. India, for example, has a per capita carbon emission of only 0.21 tons. Although this is one of the lowest in the world there is every reason to assume that the richest 10 percent in India emit at least 10 times more than the bulk of the population and that the expected future population growth of India comes almost entirely from the poor segments of the population. If this is true, the actual impact of population growth on carbon emission will be much less than national averages would imply.

Is there any solution to this problem? The only solution would be to go down to a level of aggregation where the population is sufficiently homogeneous so that there is no more visible correlation between carbon emissions and family size. But since such correlations are likely to exist even within villages it might

TABLE 2.6   Alternative Calculations of Total Carbon Emissions in 2050, Calculated with Global Averages, i.e., Assuming a Homogeneous World

Emissions p.c. (tC)	Population Size (billions)	Total Emission (Gt C)	% of Result in Table 2.5
I.A. Population Scenario 1 / Constant p.c. emissions			
1.07	14.59	15.61	156%
II.A. Population Scenario 2 / Constant p.c. emissions			
1.07	9.89	10.58	132%
III.A. Population Scenario 3 / Constant p.c. emissions			
1.07	7.41	7.92	110%

be necessary to go down to household levels. Certainly this is unfeasible for a macro-level analysis of population and environment.

The only viable solution for treating this problem on a macro level is to stop considering regional or national entities and define a certain number of global groups that are assumed to be homogeneous with respect to growth rates and per capita carbon emissions. This means pooling all the poorest poor of the world into one group, as well as the richest and some intermediate groups. Some preliminary hypothetical calculations with four groups across all continents showed that even an increase in the world population to 14.6 billion as calculated under the constant rates scenario and constant per capita emissions within each group would result only in a total carbon emission of 6.9 gigatons in 2050, as compared to the 5.7 gigatons presently produced. This is significantly less than the 10.0 gigatons resulting from a distinction between North and South, not to speak of the 15.6 gigatons resulting from global averaging.

The calculations above show two things. First, most of the calculations given so far on the effect of population growth on carbon emissions are of rather limited use because they tend to refer to only one specific level of aggregation. The second, more substantive conclusion, is that the "real" or likely effect of population growth on carbon emissions tends to be less than implied by aggregate calculations not considering the heterogeneity within groups. This is a very reassuring conclusion but it cannot be generalized to all other environmental questions. For some environmental aspects there may be no correlation to growth rates and for others there may be even a positive correlation. The latter might be the case for certain forms of overgrazing or deforestation for firewood. If the environmental impact is more severe from groups that have higher rates of population growth, then aggregate calculations tend to underestimate the problem. And this is a worrisome conclusion.

Conclusion: Are There Limits to Population Growth?

This paper showed that there are significant uncertainties concerning the future size of the world population, its regional distribution, and its age distribution. A lot will depend on the future course of fertility in the countries that today still have large family sizes. But this uncertainty looks minor in comparison to the range of possible environmental impacts of human behavior on the planet. Human behavior can change more radically and more quickly than population size and structure. The calculations performed above on possible carbon emissions in 2050 showed clearly that alternative trends in per capita emissions have a much greater effect on future total emissions than even the most extreme alternatives for future population growth. Certainly, this statement on the differential impact of changes does not say which factor is easier to be changed. Is it easier to induce behavioral changes that result in a more moderate population growth or changes that result in lower per capita emissions? In practice both strategies will have to be followed, because they are not in competition with each other. But finding the right mix will be the key question for sustainable development policies over the coming years.

But we are not interested in the population only because it affects the environment. Our primary interest lies in the fact that we are part of the population and are interested in our survival under favorable conditions. The main reason for us to worry about the environment is that environmental destruction may have negative impacts on our quality of life and ultimately on our survival. In this context the key question that comes up again and again is, how many people can live on the earth without undermining their own livelihood. Is there a limit to population growth?

In an era of quickly developing technologies and rapid socioeconomic and cultural changes it is increasingly difficult to speak of limits in an absolute sense. In some fields, such as the speed of computers, limits seem to move away more quickly than anybody had anticipated even a few years ago. In other fields limits have moved away more slowly or have even come closer, such as in the case of global ozone depletion. In both cases limits have started to move, and there seem to be fewer and fewer cases in human life where it is safe to say that limits are fixed and unchangeable. Even concerning the question of our individual life spans on earth, previously unchallenged views about absolute limits to the length of human life have become highly controversial (see, e.g., Menton 1991). But still on an individual level every premature death of a person--and there are many millions each year--may be viewed as having reached and transcended a limit.

Less directly affected by technology but still changing in historically unprecedented speed are the spheres of social infrastructure and organization. This observation of global socioeconomic and cultural change, however, does not necessarily imply that things are becoming better. For instance, higher degrees of inequality and poverty, stagnating or even declining life expectancy in some regions, and pressing environmental problems are partly the consequences of the accelerated speed of change in most areas. Even without judging whether things are changing for the better or the worse, it is safe to say that today there are greater opportunities as well as greater dangers to human life than ever before in history.

When speaking about limits to population growth the question seems to be easier than in many other fields because--at least theoretically--the limit could be quantitatively defined as a discrete number. On the other hand this number depends on a large array of assumptions that must be made, many of which are fuzzy and qualitative in nature. The most basic question of definition in this context is what quality of life one wants to assume for people at any one moment in time. Is it pure survival, is it survival in good health, is it life with a certain material standard of living, is it life with certain cultural values and in a certain intact environment? If one wants to have lots of woods around one's house, then large parts of today's more and less developed countries are unable to provide this quality to a large proportion of its citizens and have already passed the limit. If everyone is satisfied living in a little box eating synthetic nutrient pills, then--given a certain technology and infrastructure--the world could probably be home to more than ten times the present population.

Another important question that needs to be defined when trying to come up with a figure of a maximum population that can live with given resources and in a given environment is the question of heterogeneity--or, more bluntly, inequality--in the society concerned. Will such a society tolerate individuals or groups that consume hundreds of times the raw materials, energy, and space of the lower quartile of the population? Does one want to include the tremendous existing inequalities in the calculations or should one rather assume a homogeneous standard of living that can provide basic life support for more people? Geographic distribution of the population among continents and within regions (urbanization) is another kind of population heterogeneity that has consequences on the environment, the use of raw materials, and the functioning of life-support systems.

Because of all the problems of definition it will not even be attempted in this paper to calculate a specific number of a certain limit to population growth. The alternative population projections given in the previous section show what will happen to world population size under certain trends of fertility and mortality. These trends themselves are the results of highly complex interactions of sociocultural, economic, and even natural changes. If a limit is approached and fertility does not react through a decrease in the number of children born (the preventive Malthusian check), then sooner or later the natural system will reduce population growth through an increase in mortality (the positive Malthusian check). But these feedback mechanisms, and especially their timing under differing conditions, are not well enough understood to come up with any quantitative estimates on the dynamics of these loops. A viable but certainly less instructive alternative is the simple if-then calculations given above.

The major conclusion of this paper is that we know very little about possible effects of population growth on the environment. But it has become somewhat clearer what lines of research could be followed to ultimately better understand the population-environment links and identify some policy priorities to help ensure sustainable development. I will quickly mention four such lines of research in the population-environment field. Of course these priorities are rather subjective.

The first priority for better understanding the interactions between population change and the environment should be in the field of modeling in specific case studies. The emphasis must be on case studies because of the high complexity of the interactions and their dependence on very specific cultural, socioeconomic, geographical, and environmental conditions. Population-environment interactions are clearly very different for different cultures even under identical climate conditions (such as the Sahel and the Australian dry regions) or between primitive subsistence farming in different climatic or geological zones. The emphasis should also be on modeling simply because the human brain is not able to simultaneously consider several interacting effects. These computer models should be as clear and parsimonious as possible and try to avoid all magic. The user should be able to follow each chain of causation and understand why the model produces certain results. Work on such a model on the island of Mauritius has been recently completed (Lutz 1994). Certainly several such case studies under diverse conditions would be needed before one can attempt to draw more general conclusions on the macro level.

The second line of research has to do with specific attention given to the timing of changes. It is increasingly problematic in many fields to speak about fixed absolute limits. An alternative would be to speak about speed limits instead. In that case we focus on the rate of change (the rate of population growth) and study how the environment, but also society and the economy, can accommodate a certain proportional increase in a given time period. If we speak about population growth over the next 20 years we can have a much better picture of what will be the status of technology, food production, and various environmental factors in different regions of the world. Next, one can try to set realistic priorities concerning the standard of living, environmental conservation, etc., and study what the role of different rates of population growth in different regions of the world would be in achieving these priorities. The one big problem is, however, that population growth cannot be simply switched on and off because of the great momentum of population changes described above. For that reason one must have a time horizon of at least a few decades.

The third point to make in this context has to do with the feedback from the environment to population. More specifically it concerns the role of mortality and life expectancy as an indicator of quality of life. Mortality is one of the basic determinants of population size and structure, but survival is also the basic criterion for success or failure in all efforts toward sustainable development and preservation of the environment. The significant increase in the average human life expectancy has probably been the most important and least controversial achievement in human history over the past century. The mere fact of survival is a necessary prerequisite for all other qualities of life. It is one of the few aspirations that all human populations of all cultures seem to have in common. Furthermore, empirical studies show that life expectancy is highly correlated with socioeconomic development. The United Nations Development Programme recently proposed a new development index that includes GNP per capita, literacy and life expectancy at birth (UNDP 1991). Among these three, life expectancy has the highest correlation with the others, indicating that it may be the single best indicator of development.

There is no space here to extensively discuss the pros and cons of life expectancy as an indicator of quality of life. The major point is not to discuss environmental impacts as an ends in themselves but to view them in the light of possible implications on human populations. This can help to set priorities.

A final point has to do with the status of women. It is now well established in demographic research that female educational status is one of the most important, if not the single most important, variables for increasing age at marriage and the transition from having a large number of children fewer children for whom they could assure a higher quality of life. A higher status of women within the family will also help to bring fertility down to the woman's desired family size, which in high-fertility countries tends to be lower than that of men. The question in this context is how important the status of women is in the context of environmental impacts. In short, would there be less environmental destruction if women had a greater say? If the answer to this is even a qualified yes, then enhancing the status of women would have a double effect in the desired direction quite apart from its ethical and societal desirability. It clearly would be the top policy priority.

References

Arizpe, Lourdes, Robert Costanza, and Wolfgang Lutz. (1992). "Population and Nature Resource Use," in *An Agenda of Science for Environment and Development into the 21st Century* J.CI. Dooge et al (eds.), (Cambridge, UK: Cambridge University Press), 61-78.

Keyfitz, Nathan. (1977). *Applied Mathematical Demography.* New York: John Wiley & Sons.

Lutz, Wolfgang.   (1989).   *Distributional Aspects of Human Fertility: A Global Comparative Study.* London: Academic Press.

Lutz, Wolfgang. (1992). "Project status report. Population-development-environment interactions. A case study on Mauritius." *Popnet.* 21 (Spring): 1-12. Laxenburg, Austria: International Institute for Applied Systems Analysis.

Lutz, Wolfgang and Christopher Prinz. (1991). *Scenarios for the World Population in the Next Century: Excessive Growth or Extreme Aging.* WP-91-22. Laxenburg, Austria: International Institute for Applied Systems Analysis.

Menton, Kenneth. (1991). "New biotechnologies and the limits to life expectancy." Pps. 97-116 in Wolfgang Lutz, Ed. *Future Demographic Trends in Europe and North America: What can we assume today?* New York: Academic Press.

UN. (1991). *World Population Prospects 1990.* New York: United Nations.

UNDP. (1991). *World Development Report.* New York: United Nations.

Vu, My T. 1985. *World Population Projections 1985.* Baltimore, Maryland: Johns Hopkins University Press, for the World Bank.

3

Women, Poverty and Population: Issues for the Concerned Environmentalist

Gita Sen

Differences in perceptions regarding the linkages between population and environment became particularly acute during the buildup to the UN Conference on Environment and Development, variously known as the Earth Summit or Rio '92. Disagreement between Southern and Northern countries on the degree of attention to be given to population received considerable publicity during the preparatory meetings to the conference. At the nongovernmental level, too, the issue of population has been of late a subject of considerable debate among environmentalists (especially those from the North), feminists, and population lobbyists.

The basis of these differences often appears baffling; the apparent lack of willingness to compromise or to acknowledge the obvious merits of opposing views seems to indicate a lack of analytical rigor. The debate appears, to some at least, to be based on passionately held but ultimately ephemeral differences. I wish to argue that, although the positions taken in the policy debate have been exaggerated at times, some of the oppositions have deeper roots. They arise from conceptual and possibly paradigmatic differences rather than from disagreements regarding the "truth value" of particular scientific propositions. These shape the protagonists' perceptions of problems, the analytical methods used, and the weight assigned to different linkages and relationships. In particular, varying views regarding development strategies, the linkages between poverty and population growth, and the role of gender relations in shaping those links color the positions taken in the debate.

This essay is an attempt to examine the different perspectives on these issues held by environmental scientists and environmental activists on the one hand, and women's health researchers and feminist activists on the other. More explicitly I mean the dissonance between mainstream environmentalists from the North and women's health researchers and activists from both North and South. My motivation is twofold: first, to identify the positions taken by these two broad groupings within the larger discourses on development and on population; and second, to propose a possible basis for greater mutual understanding. I must state that my own position is that of someone who has come to these debates from a background of working on issues of gender and development, and the paper will perforce tilt heavily toward spelling out the position taken from within the women's movements. I do not claim to be able to explicate how the mainstream of the environmental movement (especially in the North) has come to the particular definitions it has of "the population problem."

At first glance, this may seem an impossible task, since the differences appear to be fundamental. Many environmentalists believe that population growth is a major cause of environmental degradation, while many feminists argue that population growth in itself is simply not a significant contributor to global environmental problems. For an example of the former view, see Consortium for Action to Protect the Earth (1992); of the latter, see Committee on Women, Population and the Environment (1992). The dual aim of this paper is premised on the belief that despite the apparent divergences in their views on population, environmentalists and women's health advocates have much in common. They share similar critiques of the patterns of economic growth; both believe that currently dominant patterns of economic growth are unsustainable whether from an ecologist's standpoint or from a standpoint of human survival and justice. In addition, important sections of both groups derive their knowledge from grassroots and community activism, generating not only a shared skepticism of dominant interests, but also reliance on methods of popular participation in development decisions.

These commonalities provide strong reasons to believe that the two groups can forge closer alliances in the shaping of development policy, provided there is greater clarity not only with regard to the views held, but also how they fit into the evolution of the larger discourses on development and population. On the other hand, a unique opportunity may be missed if, for example, mainstream Northern environmental groups' concerns regarding rapid population growth are simply pursued along a trajectory of traditional demographic concerns, without adequate appreciation of the underlying values and divergent perceptions of experiences and without hard evidence regarding how population, environment, and development actually intersect. While this dangerous process is in stream right now, there is also strong dissent from within the environmental movement itself, as well as in the population field. Both Lohmann (1990) and Erickson (1990), for example, couple environmental thinking and ecological concern with issues of social justice. Ironically, mainstream Northern environmentalism appears to be rediscovering outmoded ways of thinking about population precisely at a time when the population policy field itself has begun to respond to internal challenges, to distance itself from old-line demographic imperatives, and to become more responsive to reproductive health concerns and the lessons learned from four decades of family-planning experiences in the field.

Building on the commonalities between environmentalism and the women's health movement will, therefore, not be easy. But it is still possible. It will require first, clarity of understanding regarding major shifts and controversies in both development and population thinking, and the history of the population movement, and, second, the development of a concrete set of prescriptions regarding how both women's health and population interests and Northern environmental concerns can be mutually advanced. A recognition of the evolution of controversies in the population field and *how these have been influenced by shifts in development thinking* could provide a needed corrective to the belief that population is a simple problem of numbers susceptible of an easy technological fix. The next sections of the paper attempt to provide the building blocks for such an understanding.

Evolution of the Development Debate

Over the five decades since the end of World War II, as the development agenda was shaped, public debate about socioeconomic development has undergone many twists and turns. During the 1950s and 1960s, optimism was high about the possibilities for accelerating the pace of economic growth in the newly decolonized states of Asia and Africa. This, combined with sobering assessments of the vicious circles of poverty and backwardness, gave considerable legitimacy to state-led projects of planning for growth. The principal task was viewed as the need to raise the available stock of physical capital by accelerating investment and mobilizing resources through the use of "surplus" labor domestically, as well as inflows of foreign capital (in different forms) to complement aggregate domestic savings. These early supply-side arguments envisioned a leading role for the state and were buttressed by arguments for import substitution and protection of domestic markets.

While there was an emerging counterargument in favor of free trade and private investment during this phase, there was no real challenge to the primacy of the state's role in mobilizing and allocating economic resources. The major development debate of the early and mid-1960s in Asia was on planning methodologies and techniques. In Latin America there was a growing debate between the structuralists grouped around the UN Economic Commission for Latin America (CEPAL) and the monetarists headed by the International Monetary Fund (IMF) about the causes and cures of inflation and balance of payments problems. While the Latin American debate foreshadowed the neoliberal supply-side arguments propounded by the World Bank in the 1980s, it was not at the time seen as an attack on the state itself. The positions on either side of the free market versus import substitution debate have been viewed sometimes as representing a struggle between different groups of producers, different corporate interests, or different sections of the ruling groups; be that as it may, all of them believed in the importance of the state and in the primary goal of raising the rate of economic growth. This belief also animated the development assistance provided by most donor agencies. During this period the only major international agency propounding stricter controls over the states' allocation of economic resources was the IMF; the World Bank did not, at the time, seriously question the state's role.

By the end of the 1960s there were emergent critiques of this approach as a "trickle-down" strategy, that was not only ineffective in raising general living standards in many countries, but had set in motion processes leading to the expropriation of resources by the powerful both across and within nations. The criticism of the international political economic order by the "dependency" school complemented the newly emerging policy arguments in favor of directly targeting poverty alleviation and basic needs provision.

Within development agencies attention shifted from economic growth per se to the entitlements and needs of the poor. In a 1972 mission report on Kenya, the International Labor Organization (ILO) first defined the concept of basic human needs to include health, education, adequate nutrition, clean water, sanitation, safe housing, etc. The concept was adopted by the World Bank and rapidly gained currency in the development thinking of the decade. Poverty reduction and/or social equity did not appear to follow automatically from high economic growth, as evidenced for example by the case of Brazil, which grew very rapidly but inequitably after the mid-1960s. On the other hand social development, and especially the fulfillment of basic needs such as health, education, housing, sanitation, and a secure livelihood, did not appear to require high rates of economic growth as a prerequisite. The experiences of both socialist countries, such as China, and nonsocialist ones, such as Sri Lanka and Costa Rica, began to be documented as cases where basic needs had been fulfilled in the absence of high economic growth. The growing importance of basic needs in the programmatic directions being shaped at the World Bank and the International Labor Organization lent strength to the argument that social development ought to be tackled directly, and not as an uncertain side effect of economic growth.

In the global political economy, the boom in primary product prices and the spectacular success of OPEC in capturing a greater share of the rents from oil production created an optimism among Third World countries about the possibility of creating a new international economic order. Within Third World countries, the 1970s saw considerable experimentation with alternative development paths and models. The experience of Third World socialist states, such as Cuba, China, and Vietnam, in tackling problems of basic needs with equity exerted an influence on countries such as Jamaica, Nicaragua, and Mozambique. These experiments responded to the growth of social movements, comprising different combinations of industrial workers, peasants, students, the "middle classes'" and others, which had been fueled by the inequitable patterns of growth in the 1960s coupled with oppression by dictatorial regimes. These movements demanded economic change and political participation. An important part of their challenge was the criticism by nongovernmental development practitioners and others of bureaucratic methods of planning and implementation that ignored or excluded the views and needs of people and thereby alienated them.

The 1970s also saw the emergence of an international women's movement. While drawing upon other social movements, it attempted to define its own agendas, in which the possibilities for gender equity were seen as part of new approaches to overall development. Sen and Grown (1987) provide a schematic history of such approaches in the 1970's and 1980's. Women's health groups were a vital part of this movement. Their popular base derived from activist work dealing with the health problems of women in poorer urban and rural communities. The social origins of the women's health groups are different depending on particular country histories and experiences. Many progressive activists came to realize the importance of women's health concerns and of safe reproductive health services only by working with women in base-level popular organizations, whose primary concerns were not initially women's health at all. In the process many of them came to two important conclusions: first, women's health can only successfully be addressed within the context of generally community health as well as basic needs such as education, sanitation, clean water, good nutrition, and secure livelihoods; second, the bureaucratic approach to family planning was not only ineffective, but often created reproductive health problems for women, violated their basic dignity as human beings, and diverted resources away from other primary and preventive health. Strong calls for rethinking population policies and revamping population programs began to crystalize (Hartmann 1987).

The 1970s were thus a period of considerable ferment and change, with the emergence and growth of new ideas, new actors, and new policy approaches. By contrast the 1980s witnessed significant reversals in both development thinking and policy. Externally supported subversion and aggression, together with the oppressiveness of bureaucratic centralism, undermined the capacity and legitimacy of Third World socialist experiments. The slow growth of the world economy and of world trade combined with the debt crisis to bring to the fore arguments favoring austerity and the play of market forces. Structural adjustment programs were implemented in country after country, usually accompanied by popular protest as the cost of living soared and living standards dropped. As in the 1960s, supply-side arguments that focused on laying a basis for economic growth arose. There were, however, key differences between the growth arguments of the 1960s and those of the 1980s. The state was no longer viewed as an engine of growth; its functions were to be minimized. Growth itself was to be based not on domestic market creation but on linking up with the global economy on the latter's terms, i.e., through competitive exports based on cheap labor, which was seen as the Third World's most available "resource" (World Bank 1991).

The dominance of such views and the policies based on them has not gone without challenge. Among the international development policy agencies, the views held at the World Bank and the IMF run counter to the positions held at the ILO, UNICEF, and United Nations Development Programme (UNDP). The latter give greater emphasis to basic needs, to the importance of developing "human resources," and to the problems inherent in promoting structural and institutional changes within inequitable global and national economies.

Internal challenges within countries have also grown, with greater calls for democracy and popular participation. At the same time, social movements within the Third World have linked environmental concerns to the deterioration of the resource base of poor people consequent on inequitable development processes. Many such movements strongly oppose current development patterns (*a fortiori*, their structural adjustment variant) as destroying both the environment and the livelihood base of large groups of people. In addition to large movements such as that against the damming of the Narmada River in India, there are large numbers of more localized settings in which local community organizations or nongovernmental organizations (NGOs) are pitted against "development" interests, which are often simply private commercial interests (Rush 1991; Schmink and Wood 1992; Peluso 1992). Few, if any, such groups would agree that population growth is a major reason for local ecological damage; based on grassroots experience, they perceive government policies serving powerful private interests as a major causal factor. These environmental groups often have strong participation from women who, as those responsible for the household's basic needs, are often very aware of both causes and consequences of local environmental degradation (Agarwal 1991 and chapter 4 of this volume). Along with much of the rest of the Third World women's movements, these organizations criticize current patterns of economic growth and development policy and argue for focusing more on the basic needs and rights of the poor.

These organizations usually acknowledge the importance (in Third World contexts of very low per capita incomes) of developing the potential for economic growth. Nor do they deny the importance of ecological sustainability. But their understanding is that in today's world, the dominant crisis for the majority of the world's population is the crisis of survival occasioned by inequitable global and national economic and political structures. They argue, therefore, that strategies to promote either economic growth or ecological sustainability that run counter to the basic needs, livelihoods, or political inclusion of the less powerful in societies are likely to be inequitable and ultimately counterproductive in their own terms. Thus, for example, a national energy policy that does not take adequate account of the needs of the poor for fuel is unlikely to be sustainable. They believe, on the other hand, that development strategies appropriate to the needs of the majority and politically inclusive in their conception can be environmentally sustainable.

What lessons can be learned from this description of the evolving development debate? First, development policy is not a simple matter of a supply-side fix, but requires paying attention to both supply *and* demand; in particular, policies targeted at improving macroeconomic management or increasing GNP growth while ignoring or worsening the incomes and livelihoods of the majority are not politically or economically sustainable. Second, supply-side, trickle-down economics (whether at the national or international levels) benefit a few disproportionately while marginalizing many. Third, that a direct policy/program focus on poverty alleviation and basic needs not only promotes justice and equity, but also lays a firm basis for human resource development, without which no country can hope to progress. Fourth, top-down, bureaucratic development program methods are ineffective, insensitive, and often coercive.

These challenges to and lessons from development policy experience need to be borne in mind when we turn to the subject of population, because the population issue is really a sub-theme within development, and as such, is continuously being framed in the context of one or another approach to the development discourse. Some of this complex history of the debate about the links between population and development is addressed in the next sections.

Poverty and Population:
Populationists Versus Developmentalists

During the late 1960s and much of the 1970s, the principal debate in the field of population policy centered on the impact of poverty on population growth. Earlier explanations of demographic transitions in different countries stressed the role of per capita income growth in reducing first mortality and then fertility rates. If poverty (in the sense of low per capita incomes) was seen as the main factor behind high death and birth rates, then the solution was to be economic growth, aided by strong family planning programs that would make contraceptives and knowledge about them widely available.

On the other side, unchecked growth in population was seen as a drag on economic growth through a variety of mechanisms, such as reduction in domestic savings rates and diversion of funds away from productive investment. This traditional "populationist" view clearly perceived economic growth as both a necessary *and* a sufficient condition for reduction in population growth, when combined with expanded availability of birth control methods through family planning. Increases in per capita income would generate the demand for contraception, which needed to be matched by an increased supply provided by family planning services.

By the late 1960s and early 1970s, this view came under increasing pressure due to the perceived sluggishness of family planning programs in reducing birth rates. Field evidence of contraceptives lying unused in rural homes (which family planning officials had believed to have been willing recipients of birth control technologies) pointed to the urgent need for a fresh look at both policies and programs. Not only did the micro-social and, in particular, anthropological research base need strengthening, but the macro development strategies framing population policies appeared to need rethinking. Renewed debate on the precise nature of the links between socioeconomic development and population growth culminated in a major revision in thinking that arose during the World Population Conference at Bucharest in 1974.

As we have seen, the mid-1970s was a period of considerable rethinking in development policy internationally. The belief in economic growth as a panacea for all development problems had been largely discredited. In this climate of ferment and challenge to existing orthodoxies, the aphorism coined at Bucharest that "development is the best contraceptive" became the harbinger of fresh thinking about the links between population growth and development. The emphasis shifted from income increases per se to improvements in general health (children's health in particular) and in education (especially women's education) as keys to reducing infant and child mortality rates, thereby laying the basis for reducing the "demand" for children and raising receptiveness to contraceptive technologies. The policy debate of the time counterposed a strategic emphasis on health and education versus an emphasis on increasing the availability, i.e., the *supply* of family planning services. While the former "developmentalist" view also believed in the importance of family planning, it placed greater emphasis on raising the *demand* for family planning through improved health and education.

This debate fueled a significant amount of new research on the micro-foundations of fertility decisions, as well as cross-country analysis of the causes of population change. As a partial result, the 1980s saw the emergence of a more synthetic view of the links between population and development. Improving people's access to secure incomes, rather than high national economic growth per se, began to be seen as linked to improvements in health and education (Krishnan 1992).

The economic realities of the 1980s provided, however, a harsh counterpoint to this emerging synthesis. Many countries in Latin America and sub-Saharan Africa in particular, and even in Asia (where overall economic growth rates tended to be highest), faced significant declines in real government expenditures on the social sectors (Jolly et al. 1991). The primacy of economic growth in laying the basis for poverty alleviation and meeting social needs began to be asserted once again. The *World Development Report 1990* (World Bank 1990) is a good example of this. Even though this report has been lauded because it seemed to indicate a renewed sensitivity at the World Bank to the problem of poverty after nearly a decade of concentration on structural adjustment, one of its basic messages is that poverty is best addressed by raising economic growth rates rather than *direct* antipoverty programs. The latter, the report argues, should be targeted to the destitute.

International decisionmakers in institutions such as the World Bank appeared to believe in the existence of an implicit trade-off rather than complementarity between economic growth and improving the social sectors. This belief tended to ignore the time dimension of the problem. Health and education, like many other social infrastructures pay off on investment in the medium and longer terms; likewise, some of the more damaging effects of social sector disinvestment also tend to be felt in the longer term. Health and education, along with other basic needs, can contribute to raising the quality of a country's labor force, which can become critical in determining its growth potential and competitiveness over the long run. Thus, while one can obviously see a trade-off in a country's current expenditures between economic and so-called social sectors, even the distinction between "economic" and "social" may become fuzzier in the longer term.

More important from the perspective of linkages between population growth and development, real disinvestment in the social sectors, such as occurred during the 1980s and is continuing in a number of countries, might slow down the pace of mortality and fertility declines in the poorest countries. Population policy, as influenced by the "developmentalist" thinking that grew after Bucharest, and the dominant trends of structural economic reforms appear therefore to be at cross-purposes. In the World Bank's approach, this contradiction is addressed by focusing on the need to increase the *efficiency* of social sector expenditures through better targeting. Laudable as this may be in upholding the principle of more efficient management, in practice it has led to reduced per capita real expenditures in countries whose spending on the social sectors was already inadequate.

These issues on the demand side of family planning have been matched by equally serious problems on the supply side, stemming from a concerted attack on family planning institutions by the fundamentalist right wing. The spillover of the internal U.S. political battles over women's reproductive choices and decision-making autonomy into the international arena during and after the United Nations population conference in Mexico City in 1984 meant relative stagnation in U.S. government funding for family planning assistance both bilaterally and to multilateral institutions (Conly et al. 1991). Since the U.S. government until then had been both the major source of funds and the main supporter of the importance of family planning, this occasioned considerable concern within the population community. Attempts to restore U.S. government funds for family planning have become the single most important item on the agenda of many within the family planning "establishment." Arguments have also been made implying that concern about the *quality* of family planning services is an unaffordable luxury during a time of financial stringency.

This single-minded focus by some institutions on increasing funding for family planning has met with considerable opposition. As discussed earlier, a growing international, grassroots-based women's health network has, during the 1970s and 1980s, articulately challenged the quality of traditional family planning programs and policies. This challenge has both a theoretical component rooted in an analysis of gender relations, and an empirical component based on the actual experiences of women with family planning programs.

Perceptions about Women in the Population Field

Policy Approaches

In the history of population policy, women have been viewed typically in one of three ways. The narrowest is the view of women as the principal "targets" of family planning programs, of women's bodies as the site of reproduction, and therefore as the necessary locus of contraceptive technology and reproductive manipulation. The early history of population programs is replete with examples of such views, but even more recently, the "objectification" of women's bodies as fit objects for reproductive re-engineering independent of a recognition of women as social subjects continues apace (Hubbard 1990).

A second view of women which gained currency after the Bucharest Population conference saw women as potential decision makers whose capabilities in managing childcare, and children's health in particular, could be enhanced through greater education. Women began to be viewed as social subjects in this case, but the attention given to women's education has not spun off (in the population policy literature) into a fuller consideration of the conditions under which the education of girls takes hold in a society, and therefore the extent to which education is embedded within larger social processes and structures. While this view represented a step away from objectification, women were still perceived as a means to a demographic end, their own health and reproductive needs incidental to the process.

A third view, which grew in the 1980s focused on maternal mortality as an important health justification for family planning. This view, which was at the core of the Safe Motherhood Initiative, attempted to claim a health justification for family planning on the basis of rates of maternal mortality. In practice, the initiative has received relatively little funding or support.

Conceptual Approaches

Economic theories of fertility are closely associated with the "new" household economics. Premised on the belief that children are a source of both costs and benefits to their parents, such theories argue that parents determine their "optimum" number of children based on a balancing of costs and benefits at the margin. As a description of differences between societies where children are viewed as a source of both present and future streams of income versus those where children are essentially a cost to parents (balanced by a measure of psychological satisfaction but not by a significant flow of money income), the theory has an appealing simplicity. It purports to explain why the former societies may be more pronatalist than the latter. It also suggests that shifting children away from child labor (a source of parental income) toward schooling (a parental cost) might work to reduce fertility.

Such theories have been criticized on a number of grounds (for a critique, see Folbre 1988). The main criticism centers on the assumption that actual fertility is the result of choices made by a homogenous household unit innocent of power and authority relations based on gender and age. Once such relations are acknowledged, and there is enough anthropological and historical evidence for their existence, the basis of decision making within households has to be rethought in terms of differential short-term gains and losses for different members, as well as strategic choices by dominant members that will protect and ensure their continued dominance. For example, if the costs of child-raising increase, *ceteris paribus*, there may be little impact on fertility if the increased costs are largely borne by subordinate members of the household (such as younger women) who do not have much say in household decision making.

Traditionally, in many societies the costs of high fertility in terms of women's health and work burdens are rarely acknowledged as such, as long as the benefits in terms of access to a larger pool of subordinate children's labor or the social prestige inherent in being the father of many sons continue to accrue to men. Such authority relations are further cemented by ideologies that link a woman's personal status within the authoritarian household to her fertility. Newer game-theoretic models of household behavior provide more interesting and complex theories that take better account of the differential distribution of all types of assets as well as gains and losses within the household (Sen 1987). These have not thus far, however, generated adequate explanations of fertility outcomes. A different theoretical approach that takes better account of the shifts in patterns of inter-generational transfers, and therefore of age-based hierarchies, is contained in the work of Caldwell and Caldwell (1987).

Against the Stream:
Gender Relations and Reproductive Rights

Many of the influential approaches to theory and policy within the population field have been less than able or equipped to deal with the complexity and pervasiveness of gender relations in households and the economies and societies within which they function. Both feminist researchers and activists within women's health movements have been attempting to change the terms of the debate and to expand its scope. An important part of this challenge is the critique of population policy and of family planning programs as being biased (in gender, class, and race terms) in their basic objectives and in the methods that they predominantly use.

The definition of a social objective of population limitation (or in many parts of Europe population expansion through increased fertility), without recognizing that there may be costs to limiting family size that are differential across social classes and income groups, has long been criticized (for an early critique, see Mamdani 1974). In particular, such costs are likely to be less than transparent in nondemocratic polities or even within democratic states where the costs are disproportionately visited on groups that are marginal on ethnic and racial bases and therefore do not have sufficient voice (for a look at Norplant in the contemporary United States, see Scott n.d.).

Population policy has also been criticized by some as being a substitute for rather than a complement to economic development strategies that are broad-based in their allocation of both benefits and costs. For example, if impoverished peasants were persuaded or coerced to limit family size on the premise that their poverty is a result of high fertility--independent of the possible causal impact of skewed landholding patterns, commercialization processes, or unequal access to development resources--then it is questionable whether smaller families would make them more or less poor.

The critique becomes more complicated once the gender dimension is introduced. Critics of population policy on class grounds have sometimes been as gender-blind as the policy itself. Having many children may be an economic imperative for a poor family in certain circumstances, but the costs of bearing and rearing children are still borne disproportionately by the women of the household. Gender concerns cannot be subsumed under class concerns, just as the latter cannot be subsumed under a notion of homogenous national or global concerns. Feminist critics of population policy are highly questioning of development strategies that otherwise ignore or exploit poor women while making them the main target of population programs. But they do not believe that the interests of poor women in the area of reproduction are identical to those of poor men.

In general terms the feminist critique agrees with many other critics that population control cannot be made a surrogate for directly addressing the crisis of economic survival that many poor women face. Reducing population growth is not a sufficient condition for raising livelihoods or meeting basic needs. Even rapid fertility decline may sometimes be indicative of a strategy of desperation on the part of the poor who no longer can access the complementary resources needed to put children's labor to use. In particular the critique qualifies the argument that reducing fertility reduces the health risks of poor women and therefore meets an important basic need. This would be true provided the means used to reduce fertility did not themselves increase the health hazards that women face, or were considerably and knowably less than the risks of childbearing. If family planning programs are to do this, critics argue that they will have to function differently in the future than they have in the past.

The most trenchant criticism questions the *objectives* (population control rather than, and often at the expense of, women's health and dignity), the *strategies* (family planning gaining dominance over primary and preventive health care in the budgets and priorities of ministries and departments), the *methods* (use of individual incentives and disincentives for both "target" populations and program personnel, targets and quotas for field personnel, overt coercion, the absence of medical care either before or after, inadequate monitoring of side effects, and the prevalence of opposing camps), and the *birth-control methods* (a narrow range of birth-prevention methods, and technology that has not been adequately tested for safety or that has not passed regulatory controls in Northern countries) advocated and supplied through programs. A now extensive debate surrounding the "quality of care" has focused particularly on the implications of alternative program methods and birth-control techniques for the quality of family program services (Bruce 1989). More broad-ranging evaluations of population policy objectives and strategies have found them guilty of biases of class, race/ethnicity, and gender (Hartmann 1987).

Viewed as a development strategy, the critics see population policies as usually falling within a class of strategies that are top-down in orientation and largely unconcerned with (and often violating) the basic needs or human rights of target populations. Even the developmentalist concern with improving child health and women's education has received little real support from population programs despite the extensive research and policy debate it has generated.

The critical perspective argues that ignoring corequisites such as economic and social justice and women's reproductive health and rights also makes the overt goal of population policy, i.e., a change in birth rates, difficult to achieve. Where birth rates do fall (or rise as the case may be) despite this, the achievement is often predicated on highly coercive methods, and is antithetical to women's health and human dignity. The women's health advocates argue for a different approach to population policy--one that makes women's health and other basic needs more central to policy and program focus, and by doing so increases human welfare, transforms oppressive gender relations, and reduces population growth rates (Germain and Ordway 1989).

Around the world there is a growing emergence of positive statements about what human rights in the area of reproduction might encompass (Petchesky and Weiner 1990). Many of these statements are culturally and contextually specific, but they usually share a common critique of existing population programs and common understandings of alternative principles. Many of them privilege the perspective of poor women, although they recognize that the reproductive rights of all women in most societies are less than satisfactory. Their attempt to recast population policies and programs is also therefore a struggle to redefine development itself to be more responsive to the needs of the majority.

Enter the Environmentalists

Environmentalist concern with population growth predates the public debate sparked by the UN Conference on Environment and Development. Probably some of the most influential early documents were the Club of Rome's *Limits to Growth* and Ehrlich's *The Population Bomb* (1969). The interest in global and local carrying capacity vis-a-vis growing human population sizes and densities spurred a considerable literature, both scientific and popular. Unfortunately, the popular and activist literature has tended to ignore some of the important anthropological debates about carrying capacity (Little 1992; Blaikie 1985), as well as to assume the inconclusiveness of empirical evidence linking environmental change to population growth (Shaw 1989).

But the argument of both developmentalists in the population field and women's health and rights advocates has been precisely that population is *not* just an issue of numbers, but of complex social relationships that govern birth, death, and migration. The interactions of people with their environments can only partially be captured by simple mathematical relationships that do not take into account the distribution of resources, incomes, and consumption; such mathematical relationships by themselves may therefore be inadequate as predictors of outcomes or as guides to policy. An example is the well-known Ehrlich-Holdren identity, $I = PAT$, linking environmental impact (I) with population growth (P), growth in affluence / consumption per capita (A), and technological efficiency (T).

Furthermore, from a policy point of view, more precise modeling of population-environment interactions has not thus far given us much better guidance about appropriate population policies or programs. Ignoring the wide disparities in the growth rates of consumption between rich and poor *within* developing countries, and hence their relative environmental impacts, as well as the critiques of women's health advocates outlined in the previous sections leads to single-minded policy prescriptions directed once more simply to increasing family planning funding and effort. The leap from overly aggregated population-environmental relations to policy prescriptions favoring increased family planning becomes then an implicit choice of politics, of a particular approach to population policy, to environmental policy, and to development. Because it glosses over so many fundamental issues of power, gender, and class relations, and of distribution, and because it ignores the historical experience of population programs, it has come to be viewed by many as a retrograde step in the population-development discourse.

Population Actors

The preceding discussion suggests that the following are important actors in the population field. The first are those population specialists who traditionally have focused on the size and growth of populations, on age structures, migration, and population composition. In general, they enter the development discourse primarily through their concern with the impact that population growth might have on rates of economic growth. In addition, population projections are mapped onto planning needs in areas such as food production, energy, and other infrastructure as well as health, education, etc. These mappings can be said to belong to a class of simple mathematical planning models that usually ignore problems of distribution (based as they tend to be on *per capita* needs and availabilities), as well as the social and institutional aspects of making a plan actually work.

The second group are the developmentalists, who focus less on the impact of demographic change and more on the prerequisites of sustained mortality and fertility declines. In particular they stress the importance of improving health and women's education. They represent, thus, a major revision of traditional population approaches, but all too often stop short of addressing the problem of sustainability or of livelihoods.

A third group, the fundamentalists, has become increasingly important in the population field during the 1980s, gaining legitimacy through their links to mainstream political organizations. Their primary interest is not the size or growth of populations, but rather control over reproduction and a conservative concern to preserve traditional family structures and gender roles. The moral overtones of the U.S. abortion debate notwithstanding, their interest in procreation appears to derive largely from an opposition to changing gender relations in society.

The fourth group are the Northern environmentalists. At the risk of oversimplification, one might argue that many of these individuals and groups focus mainly on the links between economic growth and ecological sustainability on the one hand, and the size and growth of populations on the other.

The fifth important actors are the women's health groups, which have evolved out of either the feminist movement or out of other social movements or popular organizations. Their understanding of the population problem is distinctive in that they define it as primarily a question of reproductive rights and reproductive health, in the context of livelihoods, basic needs, and political participation. They often acknowledge economic growth and ecological sustainability to be concerns, but believe these ought to be viewed in the context of reproductive rights and health. In particular, many of them give priority to the needs and priorities of poor women in defining issues, problems, and strategies.

Each of these sets of population actors has a view of the population question that is consistent with a particular view of development; as such they tend to overlap with particular sets of development actors, and find a niche within a particular set of development ideas. For example, populationists are attracted to problems of economic growth, developmentalists to basic needs issues, and women's health activists to the problems of livelihoods, basic needs, and political empowerment. Many Northern environmentalists, on the other hand, tend to view population solely through the lens of ecological sustainability, and this accounts for a considerable amount of the dissonance between their views and those of grassroots groups in the South.

Towards More Synergy
Between Environmentalists and Feminists

Despite the dissonance provoked by the population-environment debate, there is much in common between feminists and environmentalists in their visions of society and in the methods they use. Both groups (or at least their more progressive wings) have a healthy critical stance toward ecologically profligate and inequitable patterns of economic growth, and have been attempting to change mainstream perceptions in this regard. Both use methods that rely on grassroots mobilization and participation, and are therefore sensitive to the importance of political openness and involvement. As such, both believe in the power of widespread knowledge and in the rights of people to be informed and to participate in decisions affecting their lives and those of nations and the planet. Indeed, there are many feminists within environmental movements, North and South, and environmentalists within feminist movements.

Greater mutual understanding on the population question can result from a greater recognition that the core problem is that of development *within* which population is inextricably meshed. Privileging the perspective of poor women can help ground this recognition in the realities of the lives and livelihoods of many within the South. Viewed through such a lens, some of the major lessons from population programs may be recapitulated as follows:

1. Population policies that are top-down and bureaucratically driven without responsiveness to lessons from the ground (whether with explicitly stated demographic targets or with health and welfare targets) tend to be ineffective on their own terms and haphazard or downright harmful to health and human rights objectives;

2. Women's health (and men's health for that matter) has to be approached in an integrated way because there are many positive and negative synergies between general health and reproductive health. For women in particular, determinants of general health, such as access to health care and services, nutritional status, the amount of hard physical labor and drudgery, the quality of domestic and work environments, and the availability of sanitation, clean water, and safe housing, are difficult to separate in a policy framework from the more proximate determinants of reproductive health, such as access to reproductive health care services, childbearing and birth-related practices, the type and quality of contraception, and sexual marital behavior and practices. This works in both directions--*general and reproductive health improvements are mutually reinforcing*;

3. There are beneficial reinforcements between health improvements and fertility behavior;

4. There are positive relationships between women's empowerment and autonomy, health-seeking behavior and outcomes, and reduced fertility; and

5. Reproductive health is better viewed as a basic human right and thus *as an end in itself*, rather than as an instrument toward fulfilling demographic targets.

In the context of population and development this means that the population issue must be defined as the right to determine and make reproductive decisions in the context of fulfilling *secure livelihoods, basic needs (including reproductive health), and political participation*. Although the reality in most countries in the world may currently be far removed from such an ideal, an affirmation of basic values would provide the needed underpinnings for policy and action.

Such values imply, first of all, that economic growth and ecological sustainability must secure livelihoods, basic needs, political participation, and women's reproductive rights, not work against them. Thus, environmental sustainability must be conceptualized so as to support and sustain livelihoods and basic needs, and not in ways that automatically counterpose "nature" against the survival needs of the most vulnerable in the present. Where trade-offs among these different goals exist or are inevitable, the costs and burdens must not fall on the poorest and most vulnerable, and all people must have a voice in negotiating resolutions through open and genuinely participatory political processes. Furthermore, environmental strategies that enhance livelihoods and fulfill needs can probably help lay the basis for mortality and fertility reductions.

Second, population and family planning programs should be framed in the context of health and livelihoods agendas, should give serious consideration to women's health advocates, and be supportive of women's reproductive health and rights. This has to be more than lip service; it requires reorienting international assistance and national policy, reshaping programs, and rethinking research questions and methodologies. Using the language of welfare, gender equity, or health while continuing advocacy for family planning as usual will not meet the need.

Third, reproductive health strategies are likely to succeed in improving women's health and making it possible for them to make socially viable fertility decisions if they are set in the context of a supportive health and development agenda overall. Where general health and social development are poorly funded or given low priority, as has happened in the development agendas of many major development agencies and countries during the last decade, reproductive rights and health are unlikely to get the funding or attention they need, no matter what lip service they may receive. Reproductive health programs are also likely to be more efficacious when general health and development are served. A poor woman agricultural wage-laborer, ill-nourished and anemic, is likely to respond better to reproductive health care if her nutritional status and overall health improve at the same time.

Fourth, the mainstream Northern environmental movement needs to focus more particularly on gender relations and women's needs in framing its own strategies, as well as on the issues raised by minority groups. These issues (such as those raised by native peoples and African Americans in the U.S.) tend to link environmental issues with livelihoods and basic needs concerns in much the same way as do the people's organizations in the South (from a personal discussion with V. Miller, co-founder of West Harlem Environmental Action in New York City). Greater sensitivity to the one, therefore, might bring greater awareness of the other.

Wide discussion and acknowledgement of these principles could help to bridge some of the current gaps between feminists and environmentalists, and make it possible to build coalitions that can move both agendas forward.

References

Agarwal, B. (1991). "Engendering the environment debate: lessons from the Indian subcontinent," Michigan State University, East Lansing, *CASID Speaker Series No. 8.*

Blaikie, P. (1985). *The Political Economy of Soil Erosion,* London, Longman Publishers.

Bruce, J. (1989). "Fundamental elements of the quality of care: a simple framework," New York, The Population Council, *Programs Division Working Papers No. 1*, May.

Caldwell, J. and Caldwell, P. (1987). "The cultural context of high fertility in sub-Saharan Africa," *Population and Development Review*, 13:3, September, 409-438.

Committee on Women, Population and the Environment (1992). *Statement*, Boston.

Conly, S. et al (1991). "U.S. population assistance: issues for the 1990's," Population Crisis Committee, Washington D.C..

Consortium for Action to Protect the Earth (1992). "Population, environment and development," *Working Paper*, Washington D.C..

Demeny, P. (1992). "Early postwar perspectives on rapid population growth: diagnosis and prescription," paper presented at the Roger Revelle Memorial Symposium, Harvard University, Cambridge, MA.

Ehrlich, P. (1969). *The Population Bomb*, New York, Ballantine Books.

Erickson, B. (ed) (1990). *Call to Action: Handbook for Ecology. Peace and Justice*, San Francisco, Sierra Club Books.

Folbre, N. (1988). "The black four of hearts: towards a new paradigm of household economics," in *A Home Divided: Women and Income in the Third World*, J. Bruce and D.Dwyer (eds.).

Germain, A. and Ordway, J. (1989). *Population Control and Women's Health: Balancing the Scales*, New York, International Women's Health Coalition.

Hartmann, B. (1987). *Reproductive Rights and Wrongs: The Global Politics of Population Control and Contraceptive Choice*, New York, Harper and Row.

Hubbard, R. (1990). *The Politics of Women's Biology*, New Brunswick, Rutgers University Press.

Jolly, R. et al. (1991). "Rethinking Adjustment," *World Development*, 19:12, December, 1801-64.

Krishnan, T.N. (1992). "Population, poverty and employment in India," *Economic and Political Weekly*, XXVII:46, November 14, 2479—98.

Little, P. (1992). "The social causes of land degradation in dry regions," Binghamton, NY, Institute of Development Anthropology, ms.

Lohmann, L. (1990). "Whose common future?" *The Ecologist*, 20:3, 82-4 1974.

Mamdani, M. (1974). *The Myth of Population Control*, New York, Monthly Review Press.

Peluso, N. (1992). *Rich Forests, Poor People: Resource Control and Resistance in Java*, Berkeley, University of California Press.

Petchesky, R. and Weiner, J. (1990). *Global Feminist Perspectives on Reproductive Rights and Reproductive Health*, Report on the Special Sessions at the Fourth International Interdisciplinary Congress on Women, Hunter College, New York City.

Rush, J. (1991). *The Last Tree*, New York, The Asia Society.

Schmink, M. and Wood, C.H. (1992). *Contested Frontiers in Amazonia*, New York, Columbia University Press.

Scott, J. (n.d.). "Norplant: Its impact on poor women and women of color," National Black Women's Health Project Public Policy / Education Office, Washington D.C..

Sen, A.K. (1987). "Gender and cooperative conflicts," *Discussion Paper No. 1342*, Cambridge, MA, Harvard Institute of Economic Research.

Sen, G. and Grown, C. (1987). *Development, Crises and Alternative Visions: Third World Women's Perspectives*, New York, Monthly Review Press.

Shaw, R.P. (1989). "Population growth: is it ruining the environment?" *Populi*, 16:2, 21-29.

World Bank (1990). *World Development Report 1990*, New York, Oxford University Press.

————— (1991). *World Development Report 1991*, New York, Oxford University Press.

4

The Gender and Environment Debate: Lessons from India

Bina Agarwal

What is women's relationship with the environment? Is it distinct from that of men's? The growing literature on ecofeminism in the West, and especially in the United States, conceptualizes the link between gender and the environment primarily in ideological terms. An intensifying struggle for survival in the developing world, however, highlights the material basis for this link and sets the background for an alternative formulation to ecofeminism, which I term "feminist environmentalism."

In this chapter I will argue that women, especially those in poor, rural households in India, on the one hand, are victims of environmental degradation in quite gender-specific ways. On the other hand, they have been active agents in movements of environmental protection and regeneration, often bringing to them a gender-specific perspective and one which needs to inform our view of alternatives. To contextualize the discussion, and to examine the opposing dimensions of women as victims and women as actors in concrete terms, this chapter will focus on India, although the issues are clearly relevant to other parts of the Third World as well. The discussion is divided into five sections. The

This article is reprinted from *Feminist Studies*, Volume 18, number 1 (1992): 119-158, by permission of the publisher, Feminist Studies, Inc., c/o Women's Studies Program, University of Maryland, College Park, MD 20742.

first section outlines the ecofeminist debate in the United States and one prominent Indian variant of it, and suggests an alternative conceptualization. The next three sections respectively trace the nature and causes of environmental degradation in rural India, its class and gender implications, and the responses to it by the state and grass-roots groups. The concluding section argues for an alternative transformative approach to development.

Some Conceptual Issues

Ecofeminism

Ecofeminism embodies within it several different strands of discourse, most of which have yet to be spelled out fully, and which reflect, among other things, different positions within the Western feminist movement (radical, liberal, socialist). As a body of thought ecofeminism is as yet underdeveloped and still evolving, but carries a growing advocacy. My purpose is not to critique ecofeminist discourse in detail, but rather to focus on some of its major elements, especially in order to examine whether and how it might feed into the formulation of a Third World perspective on gender and the environment. Disentangling the various threads in the debate, and focusing on those more clearly articulated, provides us with the following picture of the ecofeminist argument(s):[1] (1) There are important connections between the domination and oppression of women and the domination and exploitation of nature. (2) In patriarchal thought, women are identified as being closer to nature and men as being closer to culture. Nature is seen as inferior to culture: hence, women are seen as inferior to men. (3) Because the domination of women and the domination of nature have occurred together, women have a particular stake in ending the domination of nature, "in healing the alienated human and non-human nature."[2] The feminist movement and the environmental movement both stand for egalitarian, nonhierarchical systems. They thus have a good deal in common and need to work together to evolve a common perspective, theory, and practice.

In the ecofeminist argument, therefore, the connection between the domination of women and that of nature is basically seen as ideological, as rooted in a system of ideas and representations, values and beliefs that places women and the nonhuman world hierarchically below men. And it calls upon women and men to reconceptualize themselves, and their relationships to one another and to the nonhuman world, in nonhierarchical ways.

We might then ask: In what is this connection between nature and women seen to be rooted? The idea that women are seen as closer to nature than men was initially introduced into contemporary feminist discourse by Sherry Ortner who argued that "woman is being identified with, or, if you will, seems to be a symbol of something that every culture devalues, defines as being of a lower order of existence than itself. . . [That something] is 'nature' in the most generalized sense . . . [Women are everywhere] being symbolically associated with nature, as opposed to men, who are identified with culture."[3] In her initial formulation, the connection between women and nature was clearly rooted in the biological processes of reproduction although, even then, Ortner did recognize that women, like men, also *mediate* between nature and culture.

Ortner has since modified her position which was also criticized by others (particularly social anthropologists) on several counts, especially because the nature-culture divide is not universal across all cultures, nor is there uniformity in the meaning attributed to "nature," "culture," "male," and "female."[4] Still, some ecofeminists accept the emphasis on biology uncritically and in different ways reiterate it. An extreme form of this position is that taken by Ariel Kay Salleh who grounds even women's consciousness in biology and in nature. She argues: "women's monthly fertility cycle, the tiring symbiosis of pregnancy, the wrench of childbirth and the pleasure of suckling an infant, these things already ground women's consciousness in the knowledge of being coterminous with nature. However tacit or unconscious this identity may be for many women... it is nevertheless 'a fact of life.'"[5] Others such as Ynestra King and Carolyn Merchant argue that the nature-culture dichotomy is a false one, a patriarchal ideological construct which is then used to maintain gender hierarchy. At the same time they accept the view that women are ideologically constructed as closer to nature because of their biology.[6]

Merchant, however, in an illuminating historical analysis, shows that in premodern Europe the conceptual connection between women and nature rested on two divergent images, coexisting simultaneously, one which constrained the destruction of nature and the other which sanctioned it. Both identified nature with the female sex. The first image, which was the dominant one, identified nature, especially the earth, with the nurturing mother, and culturally restricted "the types of socially and morally sanctioned human actions allowable with respect to the earth. One does not readily slay a mother, dig into her entrails for gold, or mutilate her body."[7] The opposing image was of nature as wild and uncontrollable which could render violence, storms, droughts, and general chaos. This image culturally sanctioned mastery and human dominance over nature.

Between the sixteenth and seventeenth centuries, Merchant suggests, the Scientific Revolution and the growth of a market-oriented culture in Europe undermined the image of an organic cosmos with a living female earth at its center. This image gave way to a mechanistic worldview in which nature was reconceived as something to be mastered and controlled by humans. The twin ideas of mechanism and of dominance over nature supported both the denudation of nature and male dominance over women. Merchant observes:

> The ancient identity of nature as a nurturing mother links women's history with
> the history of the environment and ecological change . . . in investigating the
> roots of our current environmental dilemma and its connections to science,
> technology, and the economy, we must reexamine the formation of a world view
> and a science that, by reconceptualizing reality as a machine rather than a living
> organism, sanctioned the domination of both nature and women.

Today, Merchant proposes, juxtaposing the egalitarian goals of the women's movement and the environmental movement can suggest "new values and social structures, based not on the domination of women and nature as resources but on the full expression of both male and female talent and on the maintenance of environmental integrity."[8]

Ecofeminist discourse, therefore, highlights (a) some of the important conceptual links between the symbolic construction of women and nature and the ways of acting upon them (although Merchant alone goes beyond the level of assertion to trace these links in concrete terms, historically); (b) the underlying commonality between the premises and goals of the women's movement and the environmental movement; and (c) an alterative vision of a more egalitarian and harmonious future society.

At the same time the ecofeminist argument as constructed is problematic on several counts. First, it posits "woman" as a unitary category and fails to differentiate among women by class, race, ethnicity, and so on. It thus ignores forms of domination other than gender which also impinge critically on women's position.[9] Second, it locates the domination of women and of nature almost solely in ideology, neglecting the (interrelated) material sources of this dominance (based on economic advantage and political power). Third, even in the realm of ideological constructs, it says little (with the exception of Merchant's analysis) about the social, economic, and political structures within which these constructs are produced and transformed. Nor does it address the central issue of the means by which certain dominant groups (predicated on gender, class. etc.) are able to bring about ideological shifts in their own favor and how such shifts get entrenched. Fourth, the ecofeminist argument does not take into account women's lived material relationship with nature, as opposed to what others or they themselves might conceive that relationship to be. Fifth, those strands of ecofeminism that trace the connection between women and nature to biology may be seen as adhering to a form of essentialism (some notion of a female "essence" which is unchangeable and irreducible)[10] Such a formulation flies in the face of wide-ranging evidence that concepts of nature, culture, gender and so on, are historically and socially constructed and vary across and within cultures and time periods.[11]

In other words, the debate highlights the significant effect ideological constructs in shaping relations of gender dominance and forms of acting on the nonhuman world, but if these constructs are to be challenged it is necessary to go further. We need a theoretical understanding of what could be termed "the political economy of ideological construction," that is, of the interplay between conflicting discourses, the groups promoting particular discourses, and the means used to entrench views embodied in those discourses. Equally, it is critical to examine the underlying basis of women's relationship with the nonhuman world at levels other than ideology (such as through the work women and men do and the gender division of property and power) and to address how the material realities in which women of different classes (/caste/races) are rooted might affect their responses to environmental degradation. Women in the West, for instance, have responded in specific ways to the threat of environmental destruction, such by organizing the Greenham Commons resistance to nuclear missiles in England and by participating in the Green movement across Europe and the United States. A variety of actions have similarly been taken by women in the Third World, as discussed later. The question then is: Are there gendered aspects to the responses? If so, in what are these responses rooted?

Vandana Shiva's work on India takes us a step forward. Like the ecofeminists, she sees violence against nature and against women as built into the very mode of perceiving both. Like Merchant, she argues that violence against nature is intrinsic to the dominant industrial/developmental model, which she characterizes as a colonial imposition. Associated with the adoption of this developmental model, Shiva argues, was a radical conceptual shift away from the traditional Indian cosmological view of animate and inanimate nature as Prakriti, as "activity and diversity" and as "an expression of Shakti, the feminine and creative principle of the cosmos" which "in conjunction with the masculine principle (Purusha) . . . creates the world." In this shift, the living, nurturing relationship between man and nature as earth mother was replaced by the notion of man as separate from and dominating over inert and passive nature. "Viewed from the perspective of nature, or women embedded in nature," the shift was repressive and violent. "For women . . . the death of Prakriti is simultaneously a beginning of their marginalization, devaluation, displacement, and ultimate dispensability. The ecological crisis is, at its root, the death of the feminine principle."[12]

At the same time, Shiva notes that violence against women and against nature are linked not just ideologically but also materially. For instance, Third World women are dependent on nature "for drawing sustenance for themselves, their families, their societies." The destruction of nature thus becomes the destruction of women's sources for "staying alive." Drawing upon her experience of working with women activists in the Chipko movement--the environmental movement for forest protection and regeneration in the Garhwal hills of northwest India--Shiva argues that "Third World women" have both a special dependence on nature and a special knowledge of nature. This knowledge has been systematically marginalized under the impact of modern science:

"Modern reductionist science, like development, turns out to be a patriarchal project, which has excluded women as experts, and has simultaneously excluded ecology and holistic ways of knowing which understand and respect nature's processes and interconnectedness as *science*."[13]

Shiva takes us further than the Western ecofeminists in exploring the links between ways of thinking about development, the processes of developmental change, and the impact of these on the environment and on the people dependent upon it for their livelihood. These links are of critical significance. Nevertheless her argument has three principal analytical problems. First, her examples relate to rural women primarily from northwest India, but her generalizations conflate all Third World women into one category. Although she distinguishes Third World women from the rest, like the ecofeminists she does not differentiate between women of different classes, castes, races, ecological zones, and so on. Hence, implicitly, a form of essentialism could be read into her work, in that all Third World women, whom she sees as "embedded in nature," *qua women* have a special relationship with the natural environment. This still begs the question: What is the basis of this relationship and how do women acquire this special understanding?

Second, she does not indicate by what concrete processes and institutions ideological constructions of gender and nature have changed in India, nor does she recognize the coexistence of several ideological strands, given India's ethnic and religious diversity. For instance, her emphasis on the feminine principle as the guiding idea in Indian philosophic discourse in fact relates to the Hindu discourse alone and cannot be seen as applicable for Indians of all religious persuasions.[14] Indeed, Hinduism itself is pluralistic, fluid, and contains several coexisting discourses with varying gender implications.[15] But perhaps most importantly, it is not clear how and in which historical period(s) the concept of the feminine principle in practice affected gender relations or relations between people and nature.

Third, Shiva attributes existing forms of destruction of nature and the oppression of women (in both symbolic and real terms) principally to the Third World's history of colonialism and to the imposition of Western science and a Western model of development. Undeniably, the colonial experience and the forms that modern development has taken in Third World countries have been destructive and distorting economically, institutionally, and culturally. However, it cannot be ignored that this process impinged on preexisting bases of economic and social (including gender) inequalities.

Here it is important to distinguish between the particular model of modernization that clearly has been imported/adopted from the West by many Third World countries (with or without a history of colonization) and the socioeconomic base on which this model was imposed. Pre-British India, especially during the Mughal period, was considerably class/caste stratified, although varyingly across regions.[16] This would have affected the patterns of access to and use of natural resources by different classes and social groups. Although much more research is needed on the political economy of natural resource use in the precolonial period, the evidence of differentiated peasant communities at that time cautions against sweeping historical generalizations about the effects of colonial rule.

By locating the "problem" almost entirely in the Third World's experience of the West, Shiva misses out on the very real local forces of power, privilege, and property relations that predate colonialism. What exists today is a complex legacy of colonial and precolonial interactions that defines the constraints and parameters within which and from which present thinking and action on development, resource use, and social change have to proceed. In particular, a strategy for change requires an explicit analysis of the structural causes of environmental degradation, its effects, and responses to it. The outline for an alternative framework, which I term feminist environmentalism, is suggested below.

Feminist Environmentalism

I would like to suggest here that women's and men's relationship with nature needs to be understood as rooted in their material reality, in their specific forms of interaction with the environment. Hence, insofar as there is a gender and class (/caste/race)-based division of labor and distribution of property and power, gender and class (/caste/race) structure people's interactions with nature and so structure the effects of environmental change on people and their responses to it. And where knowledge about nature is experiential in its basis, the divisions of labor, property, and power which shape experience also shape the knowledge based on that experience.

For instance, poor peasant and tribal women have typically been responsible for fetching fuel and fodder and in hill and tribal communities have also often been the main cultivators. They are thus likely to be affected adversely in quite specific ways by environmental degradation. At the same time, in the course of their everyday interactions with nature, they acquire a special knowledge of species varieties and the processes of natural regeneration. (This would include knowledge passed on to them by, for example, their mothers.) They could thus be seen as both victims of the destruction of nature and as repositories of knowledge about nature, in ways distinct from the men of their class. The former aspect would provide the gendered impulse for their resistance and response to environmental destruction. The latter would condition their perceptions and choices of what should be done. Indeed, on the basis of their experiential understanding and knowledge, they could provide a special perspective on the processes of environmental regeneration, one that needs to inform our view of alternative approaches to development. (By extension, women who are no longer actively using this knowledge for their daily sustenance, and are no longer in contact with the natural environment in the same way, are likely to lose this knowledge over time and with it the possibility of its transmission to others.)

In this conceptualization, therefore, the link between women and the environment can be seen as structured by a given gender and class (/caste/race) organization of production, reproduction, and distribution. Ideological constructions such as of gender, of nature, and of the relationship between the two, may be seen as (interactively) a part of this structuring but not the whole of it. This perspective I term "feminist environmentalism."

In terms of action such a perspective would call for struggles over both resources and meanings. It would imply grappling with the dominant groups who have the property, power, and privilege to control resources, and these or other groups who control ways of thinking about them, via educational, media, religious, and legal institutions. On the feminist front there would be a need to challenge and transform both notions about gender and the actual division of work and resources between the genders. On the environmental front there would be a need to challenge and transform not only notions about the relationship between people and nature but also the actual methods of appropriation of nature's resources by a few. Feminist environmentalism underlines the necessity of addressing these dimensions from both fronts. To concretize the discussion, consider India's experience in the sections below. The focus throughout is on the rural environment.

Environmental Degradation and Forms of Appropriation

In India (as in much of Asia and Africa) a wide variety of essential items are gathered by rural households from the village commons and forests for everyday personal use and sale, such as food, fuel, fodder, fiber, small timber, manure, bamboo, medicinal herbs, oils, materials for housebuilding and handicrafts, resin, gum, honey, and spices.[17] Although all rural households use the village commons in some degree, for the poor they are of critical significance given the skewedness of privatized land distribution in the subcontinent.[18] Data for the early 1980s from twelve semiarid districts in seven Indian states indicate that for poor rural households (the landless and those with less than two hectares dryland equivalent) village commons account for at least 9 percent of total income, and in most cases 20 percent or more, but contribute only 1 to 4 percent of the incomes of the nonpoor (Table 4.1). The dependence of the poor is especially high for fuel and fodder: village commons supply more than 91 percent of firewood and more than 69 percent of their grazing needs, compared with the relative self-sufficiency of the larger landed households. Access to village commons reduces income inequalities in the village between poor and nonpoor households. Also there is a close link between the viability of small farmers'

TABLE 4.1   Average Annual Income from Village Commons in Selected Districts of India

| State[a] and District | Per Household Annual Average Income from Village Commons | | | |
| | Poor Households[b] | | Other Households[c] | |
	Value (Rs)	Percent of Total Household Income	Value (Rs)	Percent of Total Household Income
Andhra Pradesh				
Mahbubnager	534	17	171	1
Gujarat				
Mehsana	730	16	162	1
Sabarkantha	818	21	298	1
Karnataka				
Mysore	649	20	170	3
Madhya Pradesh				
Mandsaur	685	18	303	1
Raisen	780	26	468	4
Maharashtra				
Akola	447	9	134	1
Aurangabad	584	13	163	1
Sholapur	641	20	235	2
Rajastan				
Jalore	709	21	387	2
Nagsur	831	23	438	3
Tamil Nadu				
Dharmapuri	738	22	164	2

Source: N.S. Jodha, "Common Property Resources and Rural Poor," *Economic and Political Weekly*, 5 July 1986, 1176.

  [a]   "State" here refers to administrative divisions within India and is not used in the political economy sense of the word as used in the text.
  [b]   Landless households and those owning < 2 hectares (hs) dryland equivalent.
  [c]   Those owning > 2 hs dryland equivalent. 1ha = 2.47 acres.

private property resources and their access to the commons for grazing draft as well as milch animals.[19]

Similarly, forests have always been significant sources of livelihood, especially for tribal populations, and have provided the basis of swidden cultivation, hunting, and the gathering of nontimber forest produce. In India, an estimated 30 million or more people in the country depend wholly or substantially on such forest produce for a livelihood.[20] These sources are especially critical during lean agricultural seasons and during drought and famine.

The health of forests, in turn, has an impact on the health of soils (especially in the hills) and the availability of ground and surface water for irrigation and drinking. For a large percentage of rural households, the water for irrigation, drinking, and various domestic uses comes directly from rivers and streams in the hills and plains. Again there are class differences in the nature of their dependency and access. The richer households are better able to tap the (relatively cleaner) groundwater for drinking and irrigation by sinking more and deeper wells and tubewells, but the poor are mainly dependent on surface sources.

However, the availability of the country's natural resources to the poor is being severely eroded by two parallel, and interrelated, trends--first, their growing degradation both in quantity and quality; second, their increasing statization (appropriation by the state) and privatization (appropriation by a minority of individuals), with an associated decline in what was earlier communal. These two trends, both independently and interactively, underlie many of the differential class-gender effects of environmental degradation outlined later. Independently, the former trend is reducing overall availability, and the latter is increasing inequalities in the distribution of what is available. Interactively, an altered distribution in favor of the state and some individuals and away from community control can contribute to environmental degradation insofar as community resource management systems may be more effective in environmental protection and regeneration than are the state or individuals. These two trends I call the primary factors, underlying the class gender effects of environmental change. Several intermediary factors impinge on these primary ones, the most important of which, in my view, are the following: the erosion of community resource management systems resulting from the shift in "control rights" over natural resources away from community hands,[22] population growth, and technological choices in agriculture and their associated effect on local knowledge systems. These also need to be seen in interactive terms. Consider each in turn.

Forms of Environmental Degradation

Although there is as yet an inadequate data base to indicate the exact extent of environmental degradation in India and its cross-regional variations, available macro-information provides sufficient pointers to warrant considerable concern and possibly alarm. Degradation in India's natural resource base is manifest in disappearing forests, deteriorating soil conditions, and depleting water resources. Satellite data from India reveal that in 1985-87, 19.5 percent of the country's geoarea was forested and declining at an estimated rate of 1.3 million hectares a year.[23] Again, by official estimates, in 1980, 56.6 percent of India's land was suffering from environmental problems, especially water and wind erosion. Unofficial estimates are even higher. In some canal projects, one-half the area that could have been irrigated and cultivated has been lost due to waterlogging,[24] creating what the local people aptly call "wet deserts." The area under periodic floods doubled between 1971 and 1981, and soil fertility is declining due to the excessive use of chemical fertilizers. Similarly, the availability of both ground and surface water is falling. Groundwater levels have fallen permanently in several regions, including in northern India with its high water tables, due to the indiscriminate sinking of tubewells--the leading input in the Green Revolution technology.[25] As a result, many drinking water wells have dried up or otherwise been rendered unusable. In addition, fertilizer and pesticide runoffs into natural water sources have destroyed fish life and polluted water for human use in several areas.[26]

The Process of Statization

In India, both under colonial rule and continuing in the postcolonial period, state control over forests and village commons has grown, with selective access being granted to a favored few. To begin with, several aspects of British colonial policy have had long-lasting effects.[27] First, the British established state monopoly over forests, reserving large tracts for timber extraction. Second, associated with this was a severe curtailment in the customary rights of local populations to these resources, rights of access being granted only under highly restricted conditions, with a total prohibition on the barter or sale of forest produce by such right holders. At the same time, the forest settlement officer could give considerable concessions to those he chose to so privilege. Third, the colonial state promoted the notion of "scientific" forest management which essentially cloaked the practice of encouraging commercially profitable species, often at the cost of species used by the local population. Fourth, there was virtually indiscriminate forest exploitation by European and Indian private contractors, especially for building railways, ships, and bridges. Tree clearing was also encouraged for establishing tea and coffee plantations and expanding the area under agriculture to increase the government's land revenue base. In effect these policies (a) severely eroded local systems of forest management; (b) legally cut off an important source of sustenance for people, even though illegal entries continued; (c) created a continuing source of tension between the forestry officials and the local people; and (d) oriented forest management to commercial needs.

Postindependence policies show little shift from the colonial view of forests as primarily a source of commercial use and gain. State monopoly over forests has persisted, with all the attendant tensions, as has the practice of scientific forestry in the interests of commercial profit. Restrictions on local people's access to nontimber forest produce have actually increased, and the harassment and exploitation of forest dwellers by the government's forest guards is widespread.[28]

The Process of Privatization

A growing privatization of community resources in individual (essentially male) hands has paralleled the process of statization. Customarily, large parts of village common lands, especially in northwest India, were what could be termed "community-private," that is, they were private insofar as use rights to them were usually limited to members of the community and therefore exclusionary; at the same time they were communal in that such rights were often administered by a group rather than by an individual.[29] Table 4.2 reveals a decline in village commons ranging between 26 and 63 percentage points across different regions, between 1950 and 1984. This is attributable mainly to state policy acting to benefit selected groups over others, including illegal encroachments by farmers, made legal over time; the auctioning of parts of commons by the government to private contractors for commercial exploitation; and government distribution of common land to individuals under various schemes which were, in theory, initiated for benefiting the poor but in practice benefited the well-off farmers.[30] For sixteen of the nineteen districts covered, the share of the poor was less than that of the nonpoor (Table 4.2). Hence the poor lost out collectively while gaining little individually.

Similarly, in the tapping of groundwater through tubewells, there are dramatic inequalities in the distribution of what is effectively an underground commons. Tubewells are concentrated in the hands of the rich and the noted associated fall in water tables has, in many areas, dried up many shallow irrigation and drinking water wells used by the poor. In some regions, they have also depleted soil moisture from land used by poor households.[31]

Now consider the intermediary factors mentioned earlier: the erosion of community management systems, population growth, and choice of agricultural technology and local knowledge systems.

The Erosion of Community Resource Management Systems

The statization and privatization of communal resources have, in turn, systematically undermined traditional institutional arrangements of resource use and fodder, and practices of shifting agriculture which were typically not destructive of nature.[32] Some traditional religious and folk beliefs also (as noted) contributed to the preservation of nature, especially trees or orchards deemed sacred.[33]

Of course, much more empirical documentation is needed on how regionally widespread these traditional systems of management were and the contexts in which they were successful in ensuring community cooperation. However, the basic point is that where traditional community management existed, as it did in many areas, *responsibility for resource management was linked to resource use* via local community institutions. Where control over these resources passed from the hands of the community to those of the state or of individuals, this link was effectively broken.

In turn, the shift from community control and management of common property, to state or individual ownership and control has increased environmental degradation.[34] As Daniel W. Bromley and Michael M. Cernea note "the *appearance* of environmental management created through the establishment of government agencies, and the aura of coherent policy by issuance of decree prohibiting entry to--and harvesting from--State property, has led to continued degradation of resources under the tolerant eye of government agencies."[35]

Property rights vested in individuals are also no guarantee for environmental regeneration. Indeed, as will be discussed at greater length later, individual farmers attempting tree planting for short-term profits have tended to plant quick-growing commercial trees such as eucalyptus, which can prove environmentally costly.

Population Growth

Excessive population growth has often been identified as the primary culprit of environmental degradation. And undoubtedly, a rapidly growing population impinging over time on a limited land/water/forest base is likely to degrade the environment. However, political economy dimensions clearly underlie the *pace* at which this process occurs and *how the costs of it are distributed*. The continuing (legal and illegal) exploitation of forests, and the increasing appropriation of village commons and groundwater resources by a few, leave the vast majority to subsist on a shrinking natural resource base. Added to this is the noted erosion of community resource management systems which had enforced limitations on what people could and did take from communal resources, and which could perhaps have ensured their protection, despite population pressure.[36]

Population growth can thus be seen as exacerbating a given situation but not necessarily as its primary cause. It is questionable that interventions to control population growth can, in themselves, stem environmental degradation, although clearly, as Paul Shaw argues, they can "buy crucial time until we figure out how to dismantle more ultimate causes."[37]

What adds complexity to even this possibility is that in the link between environmental degradation and population growth, the causality can also run in the opposite direction. For instance, poverty associated with environmental degradation could induce a range of fertility-increasing responses--reduced education for young girls as they devote more time to collecting fuel, fodder,

TABLE 4.2 Distribution of Privatized Village Commons in Selected Districts of India

State & Districts	VCs as % of Village Area, 1982-84	% Decline in VC Area, 1950-84	% of Land to		% of Recipients Among		Per Household Area Owned (ha)			
			Poor	Others	Poor	Others	Poor		Others	
							Before[a]	After[b]	Before	After
Andhra Pradesh										
Mahbubnagar	9	43	50	50	76	24	0.3	0.9	3.0	5.1
Medak	11	45	51	49	59	41	1.0	2.2	3.1	4.6
Gujarat										
Banaskantha	9	49	18	82	38	62	0.8	2.0	5.4	8.8
Mehsana	11	37	20	80	36	64	1.0	1.7	9.0	9.8
Sabarkantha	12	46	28	72	55	45	0.5	1.1	7.0	9.8
Karnataka										
Bidar	12	41	39	61	64	36	1.0	2.0	6.4	9.2
Gulbarga	9	43	43	57	60	40	0.8	2.4	4.5	7.7
Mysore	18	32	44	56	67	33	0.9	1.9	4.1	11.6
Madhya Pradesh										
Mansaur	22	34	45	55	75	25	1.2	2.5	7.7	12.4
Raisen	23	47	42	58	68	32	1.3	2.2	6.2	9.0
Vidisha	28	32	38	62	48	52	1.3	2.5	4.9	6.8
Maharashtra										
Akola	11	42	39	61	58	42	1.0	1.6	3.1	4.6

(continues)

TABLE 4.2 (continued)

State & Districts	VCs as % of Village Area, 1982-84	% Decline in VC Area, 1950-84	% of Land to		% of Recipients Among		Per Household Area Owned (ha)			
			Poor	Others	Poor	Others	Poor Before[a]	After[b]	Others Before	After
Aurangabad	15	30	30	70	42	58	1.1	2.2	6.4	6.3
Sholapur	19	26	42	58	53	47	0.7	2.2	3.4	5.6
Rajastan										
Jalore	18	37	14	86	37	63	0.3	1.7	7.2	12.5
Jodhpur	16	58	24	76	35	65	0.4	1.3	2.3	3.8
Nagaur	15	63	21	79	41	59	1.3	2.5	2.4	5.2
Tamil Nadu										
Coimbatore	9	47	50	50	75	25	0.8	2.5	3.8	5.8
Dharmapuri	12	52	49	51	55	45	1.0	1.9	4.6	7.5

Source: N.S. Jodha, "Common Property Resources and Rural Poor," Economic and Political Weekly, 5 July 1986, 1177-78.
[a] Before the distribution of VC land.
[b] After the distribution of VC land.

and so on, leading to higher fertility in the long term, given the negative correlation between female education and fertility; higher infant mortality rates inducing higher fertility to ensure a given completed family size; and people having more children to enable the family to diversify incomes as a risk-reducing mechanism in environmentally high-risk areas.[38] These links are another reminder that it is critical to focus on women's status when formulating policies for environmental protection.

Choice of Agricultural Technology and Erosion of Local Knowledge Systems

Many of the noted forms of environmental degradation are associated with the Green Revolution technology adopted to increase crop output. Although dramatically successful in the latter objective in the short run, it has had high environmental costs, such as falling water tables due to tubewells, waterlogged and saline soils from most large irrigation schemes, declining soil fertility with excessive chemical fertilizer use, and water pollution with pesticides. Moreover, the long-term sustainability of the output increases achieved so far, itself appears doubtful. Deteriorating soil and water conditions are already being reflected in declining crop yields.[39] Genetic variety has also shrunk, and many of the indigenously developed crop varieties (long-tested and adapted to local conditions) have been replaced by improved seeds which are more susceptible to pest attacks. The long-term annual growth rate of agricultural production in India over 1968-85 was 2.6 percent, that is, slightly *lower* than the pre-Green Revolution, 1950-65, rate of 3.08. Crop yields are also more unstable.[40] All this raises questions about the long-term sustainability of agricultural growth, and more generally of rural production systems, under present forms of technology and resource management in India, and indeed in south Asia.

The choice of agricultural technology and production systems cannot be separated from the dominant view of what constitutes scientific agriculture. The Green Revolution embodies a technological mix which gives primacy to laboratory-based research and manufactured inputs and treats agriculture as an isolated production system. Indeed, indiscriminate agricultural expansion, with little attempt to maintain a balance between forests, fields, and grazing lands, assumes that the relationship between agriculture, forests, and village commons is an antagonistic, rather than a complementary, one. By contrast, organic farming systems (now rapidly being eclipsed) are dependent on maintaining just such a balance. More generally, over the years, there has been a systematic devaluation and marginalization of indigenous knowledge about species varieties, nature's processes (how forests, soils, and water are formed and sustained interrelatedly), and sustainable forms of interaction between people and nature. These trends are not confined to countries operating within the capitalist mode.

Similar problems of deforestation, desertification, salination, recurrent secondary pest attacks on crops, and pesticide contamination are emerging in China.[41]

What is at issue here is not modern science in itself but the process by which what is regarded as "scientific knowledge" is generated and applied and how the fruits of that application are distributed. Within the hierarchy of knowledge, that acquired via traditional forms of interacting with nature tends to be deemed less valuable.[42] And the people who use this knowledge in their daily lives--farmers and forest dwellers and especially women of these communities--tend to be excluded from the institutions which create what is seen as scientific knowledge. These boundaries are not inevitable. In Meiji Japan, the farmer's knowledge and innovative skills were incorporated in the broader body of scientific knowledge by a systematized interaction between the farmer, the village extension worker, and the scientist. This enabled a two way flow of information from the farmer to the scientist and vice-versa: "Intimate knowledge of the best of traditional farming methods was thus the starting point for agricultural research and extension activities.[43]

Such attempts contrast sharply with the more typical top-down flow of information from those deemed experts (the scientists/professionals) to those deemed ignorant (the village users). The problem here is only part!y one of class differences. Underlying the divide between the scientists/professionals (usually urban-based) and the rural users of innovations (including user-innovators) whose knowledge comes more from field experience than from formal education, are also usually the divides between intellectual and physical labor, between city and countryside, and between women and men.

Class-Gender Effects

We come then to the class-gender effects of the processes of degradation, statization and privatization of nature's resources and the erosion of traditional systems of knowledge and resource management. These processes have had particularly adverse effects on poor households because of the noted greater dependency of such households on communal resources. However, focusing on the class significance of communal resources provides only a partial picture-- there is also a critical gender dimension, for women and female children are the ones most adversely affected by environmental degradation. The reasons for this are primarily threefold. First, there is a preexisting gender division of labor. It is women in poor peasant and tribal households who do much of the gathering and fetching from the forests, village commons, rivers, and wells. In addition, women of such households are burdened with a significant responsibility for family subsistence and they are often the primary, and in many female-headed households the sole, economic providers.

Second, there are systematic gender differences in the distribution of subsistence resources (including food and health care) within rural households, as revealed by a range of indicators: anthropometric indices, morbidity and mortality rates, hospital admissions data, and the sex ratio (which is 93 females per 100 males for all-India).[44] These differences, especially in health care, are widespread in India (and indeed in south Asia).[45]

Third, there are significant inequalities in women's and men's access to the most critical productive resource in rural economies, agricultural land, and associated production technology.[46] Women also have a systematically disadvantaged position in the labor market. They have fewer employment opportunities, less occupational mobility, lower levels of training, and lower payments for same or similar work.[47] Due to the greater task specificity of their work, they also face much greater seasonal fluctuations in employment and earnings than do men, with sharper peaks and longer slack periods in many regions and less chance of finding employment in the slack seasons.[48]

Given their limited rights in private property resources such as agricultural land, rights to communal resources such as the village commons have always provided rural women and children (especially those of tribal, landless, or marginal peasant households) a source of subsistence, *unmediated by dependency relationships* on adult males. For instance, access to village commons is usually linked to membership in the village community and therefore women are not excluded in the way they may be in a system of individualized private land rights. This acquires additional importance in regions with strong norms of female seclusion (as in northwest India) where women's access to the cash economy, to markets, and to the marketplace itself is constrained and dependent on the mediation of male relatives.[49]

It is against this analytical backdrop that we need to examine what I term the "class gender effects" (the gender effects mediated by class) of the processes of environmental degradation, statization and privatization. These effects relate to at least six critical aspects: time, income, nutrition, health, social-survival-networks, and indigenous knowledge. Each of these effects is important across rural India. However, their intensity and interlinkages would differ cross-regionally, with variations in ecology, agricultural technology, land distribution, and social structures, associated with which are variations in the gender division of labor, social relations, livelihood possibilities, and kinship systems.[50] Although a systematic regional decomposition of effects is not attempted below, all the illustrative examples are regionally contextualized.

On Time

Because women are the main gatherers of fuel, fodder, and water, it is primarily their working day (already averaging ten to twelve hours) that is lengthened with the depletion of and reduced access to forests, waters, and soils.

Firewood, for instance, is the single most important source of domestic energy in India (providing more than 65 percent of domestic energy in the hills and deserts of the north). Much of this is gathered and not purchased, especially by the poor. In recent years, there has been a sevenfold increase in firewood collection time (see Table 4.3). In some villages of Gujarat, in western India, even a four-to-five-hour search yields little apart from shrubs, weeds, and tree roots, which do not provide adequate heat.

TABLE 4.3 Time Taken and Distance Travelled for Firewood Collection

Country/ Regions	Year of Data	Firewood Collection[a]		Data Source
		Time Taken	Distance Travelled	
India				
Chamoli (hills)				
(a) Dwing	1982	5 hr/day[b]	over 5km	Swaminathan
(b) Pakhi		4 hr/day		(1984)
Gujarat (plains)				
(a) Forested		1x every 4 days	n.a.	
(b) Depleted	1980	1x every 2 days	4-5 km	Nagbrahman &
(c) Severely				Sambrani (1983)
depleted		4-5 hr/day	n.a.	
Madhya Pradesh				
(plains)	1980	1-2 times/wk	5 km	Chand & Bezborah (1980)
Kumson (hills)	1982	3 days/week	5-7 km	Folger & Dewan
Karnataka (plains)	n.a.	1 hr/day	5.4 km/trip	Batliwala (1983)
Garwhal (hills)	n.a.	5 hr/day	10 km	Agarwal (1983)
Bihar (plains)	c1972	n.a.	1-2 km/day	Bhaduri & Surin
	1980	n.a.	8-10 km/day	
Rajasthan (plains)	1988	5hr/day	4 km	personal observ.

(*continues*)

TABLE 4.3   (*continued*)

| Country/ Regions | Year of Data | Firewood Collection[a] | | Data Source |
		Time Taken	Distance Travelled	
Nepal				
Tinan (hills)	1978	3 hr/day	n.a.	Stone (1982)
Pangua (hills)	late 1970s	4-5hr/bundle	n.a.	Bajracharya (1983)
WDA[c] (lowlands)				
(a) low deforesta- tion	1982- 1983	1.5 hr/day	n.a.	Kumar & Hotchkiss (1988)
(b) high deforestation		3 hr/day	n.a.	

Sources: Madhura Swaminathan, "Eight Hours a Day for Fuel Collection," *Manushi* (March - April1984); D. Nagbrahman and S. Sambrani, "Women's Drudgery in Firewood Collection," *Economic and Political Weekly*, 1-8 Jan. 1983; Malini Chand and Rita Bezboruah, "Employment Opportunities for Women in Forestry" in *Community Forestry and People's Participation, Seminar Report*, Ranchi Consortium for Community Forestry, 20-22 Nov. 1980; Bonnie Folger and Meera Dewan, "Kumson Hills Reclamation: End of Year Site Visit," (Delhi, OXFAM America, 1983); "Women and Coooking Energy," *Economic and Political Weekly*, 24-31 Dec. 1983; Anil Agarwal, "The Cooking Energy Systems - Problems and Opportunities," (Center for Science and Environment, Delhi); T. Bhadhuri and V. Surin, "Community Forestry and Women Headloaders," in *Community Forestry and People's Participation, Seminar Report*; Linda Stone, "Women and Natural Resources: Perspectives from Nepal," in *Women in Natural Resources: An International Perspective*, ed. Molly Stock, Jo Ellen Force, and Dixie Ehrenreich (Moscow: University of Idaho Press, 1982); Deepak Bajracharya, "Deforestation in the Food/Fuel Context: Historical and Political Perspectives from Nepal," *Mountain Research and Development* 3, no. 3 (1983); Shubh Kumar and David Hotchkiss, "Consequences of Deforestation for Women's Time Allocation, Agricultural Production, and Nutrition in Hill Areas of Nepal," Research Report, no. 69 (Washington, D.C.: International Food Policy Research Institute, 1988).

[a] Firewood collected mainly by women and children.
[b] Average computed from information given in the study.
[c] Western Development Area.
n.a. Information not available.

Similarly, fodder collection takes longer with a decline in the village commons. As a woman in the hills of Uttar Pradesh (northwest India) puts it:

> When we were young, we used to go to the forest early in the morning without eating anything. There we would eat plenty of berries and wild fruits . . . drink the cold sweet (water) of the Banj (oak) roots . . . In a short while we would gather all the fodder and firewood we needed, rest under the shade of some huge tree and then go home. Now, with the going of the trees, everything else has gone too.[51]

The shortage of drinking water has exacerbated the burden of time and energy on women and young girls. Where low-caste women often have access to only one well, its drying up could mean an endless wait for their vessels to be filled by upper-caste women, as was noted to have happened in Orissa.[52] A similar problem arises when drinking water wells go saline near irrigation works.

In Uttar Pradesh, according to a woman grassroots activist, the growing hardship of young women's lives with ecological degradation has led to an increased number of suicides among them in recent years. Their inability to obtain adequate quantities of water, fodder, and fuel causes tensions with their mothers-in-law (in whose youth forests were plentiful), and soil erosion has compounded the difficulty of producing enough grain for subsistence in a region of high male outmigration.[54]

On Income

The decline in gathered items from forests and village commons has reduced incomes directly. In addition, the extra time needed for gathering reduces time available to women for crop production and can adversely affect crop incomes, especially in hill communities where women are the primary cultivators due to high male outmigration. For instance, a recent study in Nepal found that the substantial increase in firewood collection time due to deforestation has significantly reduced women's crop cultivation time, leading to an associated fall in the production of maize, wheat, and mustard which are primarily dependent on female labor in the region. These are all crops grown in the dry season when there is increased need for collecting fuel and other items.[55] The same is likely to be happening in the hills of India.

Similar implications for women's income arise with the decline in common grazing land and associated fodder shortage. Many landless widows I spoke to in Rajasthan (northwest India) in 1988 said they could not venture to apply for a loan to purchase a buffalo under the government's anti-poverty program as they had nowhere to graze the animal and no cash to buy fodder.

As other sources of livelihood are eroded, selling firewood is becoming increasingly common, especially in eastern and central India. Most "headloaders," as they are called, are women, earning a meager 5.50 rupees a day for twenty kilograms of wood.[56] Deforestation directly impinges on this source of livelihood as well.

On Nutrition

As the area and productivity of village commons and forests fall, so does the contribution of gathered food in the diets of poor households. The declining availability of fuelwood has additional nutritional effects. Efforts to economize induce people to shift to less nutritious foods which need less fuel to cook or which can be eaten raw, or force them to eat partially cooked food which could be toxic, or eat leftovers that could rot in a tropical climate, or to miss meals altogether. Although as yet there are no systematic studies on India, some studies on rural Bangladesh are strongly indicative and show that the total number of meals eaten daily as well as the number of cooked meals eaten in poor households is already declining.[57] The fact that malnutrition can be caused as much by shortages of fuel as of food has long been part of the conventional wisdom of rural women who observe: "It's not what's in the pot that worries you, but what's under it." A tradeoff between the time spent in fuel gathering versus cooking can also adversely affect the meal's nutritional quality.

Although these adverse nutritional effects impinge on the whole household, women and female children bear an additional burden because of the noted gender biases in intrafamily distribution of food and health care. There is also little likelihood of poor women being able to afford the extra calories for the additional energy expended in fuel collection.

On Health

Apart from the health consequences of nutritional inadequacies, poor rural women are also more directly exposed than are men to waterborne diseases and to the pollution of rivers and ponds with fertilizer and pesticide runoffs, because of the nature of the tasks they perform, such as fetching water for various domestic uses and animal care, and washing clothes near ponds, canals, and streams.[58] The burden of family ill-health associated with water pollution also falls largely on women who take care of the sick. An additional source of vulnerability is the agricultural tasks women perform. For instance, rice transplanting, which is usually a woman's task in most parts of Asia, is associated with a range of diseases, including arthritis and gynecological ailments.[59] Cottonpicking and other tasks done mainly by women in cotton cultivation expose them to pesticides which are widely used for this crop. In China, several times the acceptable levels of DDT and BHC residues have been found in the milk of nursing mothers, among women agricultural workers.[60] In India, pesticides are associated with limb and visual disabilities.[61]

On Social Support Networks

The considerable displacement of people that results from the submersion of villages in the building of major irrigation and hydroelectric works, or from large-scale deforestation in itself, has another (little recognized) class and gender implication--the disruption of social support networks. Social relationships with kin, and with villagers outside the kin network, provide economic and social support that is important to all rural households but especially to poor households and to the women.[62] This includes reciprocal labor-sharing arrangements during peak agricultural seasons; loans taken in cash or kind during severe crises such as droughts; and the borrowing of small amounts of food stuffs, fuel, fodder, and so on, even in normal times. Women typically depend a great deal on such informal support networks, which they also help to build through daily social interaction, marriage alliances that they are frequently instrumental in arranging, and complex gift exchanges.[63] Also the social and economic support this represents for women in terms of strengthening their bargaining power within families needs to be recognized, even if it is not easy to quantify.[64] These networks, spread over a range of nearby villages, cannot be reconstituted easily, an aspect ignored by rehabilitation planners.

Moreover for forest dwellers, the relationship with forests is not just functional or economic but also symbolic, suffused with cultural meanings and nuances, and woven into their songs and legends of origin. Large-scale deforestation, whether or not due to irrigation schemes, has eroded a whole way of living and thinking. Two close observers of life among the tribal people of Orissa in eastern India note that "the earlier sense of sharing has disappeared . . . Earlier women would rely on their neighbors in times of need. Today this has been replaced with a sense of alienation and helplessness . . . The trend is to leave each family to its own fate."[65] Widows and the aged are the most neglected.

On Women's Indigenous Knowledge

The gathering of food alone demands an elaborate knowledge of the nutritional and medicinal properties of plants, roots, and trees, including a wide reserve knowledge of edible plants not normally used but critical for coping with prolonged shortages during climatic disasters. An examination of household coping mechanisms during drought and famine reveals a significant dependence on famine foods gathered mainly by women and children for survival. Also among hill communities it is usually women who do the seed selection work and have the most detailed knowledge about crop varieties.[66] This knowledge about nature and agriculture, acquired by poor rural women in the process of their everyday contact with and dependence on nature's resources, has a class and gender specificity and is linked to the class specificity and gendering of the division of labor.

The impact of existing forms of development on this knowledge has been twofold. First, the process of devaluation and marginalization of indigenous knowledge and skills, discussed earlier, impinges especially on the knowledge that poor peasant and tribal women usually possess. Existing development strategies have made little attempt to tap or enhance this knowledge and understanding. At the same time, women have been excluded from the institutions through which modern scientific knowledge is created and transmitted. Second, the degradation of natural resources and their appropriation by a minority results in the destruction of the material basis on which women's knowledge of natural resources and processes is founded and kept alive, leading to its gradual eclipse.

Responses: State and Grassroots

Both the state and the people most immediately affected by environmental degradation have responded to these processes, but in different ways. The state's recognition that environmental degradation may be acquiring crisis proportions is recent and as yet partial; and, as we have seen, state developmental policies are themselves a significant cause of the crisis. Not surprisingly, therefore, the state's response has been piecemeal rather than comprehensive. For instance, the problem of deforestation and fuelwood shortage has been addressed mainly by initiating treeplanting schemes either directly or by encouraging village communities and individual farmers to do so.

However, most state ventures[67] in the form of direct planting have had high failure rates in terms of both tree planting and survival, attributable to several causes--a preoccupation with monocultural plantations principally for commercial use, which at times have even replaced mixed forests; the takeover of land used for various other purposes by the local population; and top-down implementation. Hence, in many cases, far from benefiting the poor, these schemes have even taken away existing rights and resources, leading to widespread local resistance. Also, women either do not figure at all in such schemes or, at best, tend to be allotted the role of caretakers in tree nurseries, with little say in the choice of species or in any other aspect of the project. Community forestry schemes, on the other hand, are often obstructed by economic inequalities in the village community and the associated mistrust among the poor of a system that cannot ensure equitable access to the products of the trees planted.

Ironically, the real "success" stories, with plantings far exceeding targets, relate to the better-off farmers who, in many regions, have sought to reap quick profits by allotting fertile cropland to commercial trees. As a result, employment, crop output, and crop residues for fuel have declined, often dramatically, and the trees planted, such as eucalyptus, provide no fodder and poor fuel.[68] The recent government policy in West Bengal (eastern India) of leasing sections of degraded forest land to local communities for collectively planting, managing, and monitoring tree plantations for local use, holds promise. But in several other parts of the country large tracts of such land have also been given to paper manufacturers for planting commercial species.

As some environmentalists have rightly argued, this predominantly commercial approach to forestry, promoted as "scientific forestry," is reductionist--it is nature seen as individual parts rather than as an interconnected system of vegetation, soil, and water; the forest is reduced to trees, the trees to biomass. For instance, Shiva notes that in the reductionist worldview only those properties of a resource system are taken into account which generate profits, whereas those that stabilize ecological processes, but are commercially nonexploitable, are ignored and eventually destroyed.[69] Indeed, the noted effects of development policies on the environment--be they policies relating to agriculture or more directly to forests and water use--point to a strategy which has been extractive/destructive of nature rather than conserving/regenerative. The strategy does not explicitly take account of the long-term complementarity between agriculture and natural resource preservation and therefore raises serious questions about the ability of the system both to sustain long-term increases in agricultural productivity and to provide sustenance for the people.

But should we see people in general and women in particular solely as victims of environmental degradation and of ill-conceived topdown state policies? The emergence of grassroots ecology movements across the subcontinent (and especially India) suggests otherwise. These movements indicate that although poor peasant and tribal communities in general, and women among them in particular, are being severely affected by environmental degradation and appropriation, they are today also critical agents of change. Further, embodied in their traditional intersection with the environment are practices and perspectives which can prove important for defining alternatives.

The past decade, in particular, has seen an increasing resistance to ecological destruction in India, whether caused by direct deforestation (which is being resisted through nonviolent movements such as Chipko in the Himalayan foothills and Appiko in Karnataka) or by large irrigation and hydroelectric works, such as the Narmada Valley Project covering three regions in central India, the Koel-Karo in Bihar, the Silent Valley Project in Kerala (which was shelved through central government intervention and local protests in 1983), the Inchampalli and Bhopalpatnam dams in Andhra Pradesh (against which 5,000 tribal people, with women in the vanguard, protested in 1984), and the controversial Tehri dam in Garwal. Women have been active participants in most of these protests.

Although fueled by differing ideological streams, which Ramachandra Guha identifies as Crusading Gandhian, Appropriate Technology, and Ecological Marxism, these resistance movements suggest that those affected can also be critical agents of change. Common to these streams is the recognition that the present model of development has not succeeded either in providing sustenance or in ensuring sustainability. However, the points from which the differing ideologies initiate this critique are widely dispersed. In particular, they differ in their attitudes to modern science and to socioeconomic inequalities. As Guha puts it, under the crusading Gandhian approach, "modern science is seen as re-sponsible for industrial society's worst excesses,"[70] and socioeconomic inequalities within village communities tend to get glossed over. Ecological Marxism sees modern science and the "scientific temper" as indispensable for constructing a new social order, and there is a clear recognition of and attack on class and caste inequalities (although the position on gender is ambiguous). Appropriate Technology thinking, which falls within these two strands, is not as well-worked-out a philosophic and theoretical position as Gandhism and Marxism. It is pragmatic in its approach to modern science and emphasizes the need to synthesize traditional and modern technological traditions. Although problems relating to socioeconomic hierarchies are recognized, there is no clear program for tackling them. Over the past decade there has been some cross-fertilization of thinking across these different ideological streams.

However, it is important to distinguish here between the perspectives revealed by an examination of *practice* within the environmental movement and the explicit *theoretical* formulation of an environmental perspective. Although dialectically interlinked the two do not entirely overlap. The three ideological streams, as identified by Guha, relate to different ways in which groups adhering to preexisting ideological and philosophic positions (Marxist, Gandhian) have incorporated environmental concerns in their practice. In a sense environment has been added on to their other concerns by these groups. This does not as yet represent the formulation of a new theoretical perspective (that an environmental approach to development needs) by any of these groups.

In terms of practice within the movement, women have been a visible part of most rural grassroots ecological initiatives (as they have of peasant movements in general). This visibility is most apparent in the Chipko movement described below. However, women's participation in a movement does not in *itself* represent an explicit incorporation of a gender perspective, in either theory or practice, within that movement. Yet such a formulation is clearly needed. Feminist environmentalism as spelled out earlier in this paper is an attempt in this direction.

To restate in this context, in feminist environmentalism I have sought to provide a theoretical perspective that locates both the symbolic and material links between people and the environment in their specific forms of interaction with it, and traces gender and class differentiation in these links to a given gender and class division of labor, property, and power. Unlike Gandhism and Marxism, feminist environmentalism is not a perspective that is consciously subscribed to by an identifiable set of individuals or groups. However, insofar as tribal and poor peasant women's special concern with environmental degradation is rooted in this material reality, their responses to it, which have been articulated both in complimentary and oppositional terms to the other ideological streams, could be seen as consistent with the feminist environmentalist framework.

The Chipko movement is an interesting example in this respect. Although it emerged from the Gandhian tradition, in the course of its growth it has brought to light some of the limitations of an approach that does not explicitly take account of class and gender concerns. More generally too it is a movement of considerable historical significance whose importance goes beyond locational specificity, and is a noteworthy expression of hill women's specific understanding of forest protection and environmental regeneration.[71]

The movement was sparked off in 1972-73 when the people of Chamoli district in northwest India protested the auctioning of 300 ash trees to a sports good manufacturer, while the local labor cooperative was refused permission by the government to cut even a few trees to make agricultural implements for the community. Since then the movement has spread not only within the region but its methods and message have also reached other parts of the country (Appiko in Karnataka is an offshoot).[72] Further, the context of local resistance has widened. Tree felling is being resisted also to prevent disasters such as landslides, and there has been protest against limestone mining in the hills for which the villagers had to face violence from contractors and their hired thugs.

Women's active involvement in the Chipko movement has several noteworthy features that need highlighting here. First, their protest against the commercial exploitation of the Himalayan forests has been not only jointly with the men of their community when they were confronting nonlocal contractors but also, in several subsequent instances, even in opposition to village men due to differences in priorities about resource use. Time and again, women have clear-sightedly opted for saving forests and the environment over the short-term gains of development projects with high environmental costs. In one instance, a potato-seed farm was to be established by cutting down a tract of oak forest in Dongri Paintoii village. The men supported the scheme because it would bring in cash income. The women protested because it would take away their only local source of fuel and fodder and add five kilometers to their fuel-collecting journeys, but cash in the men's hands would not necessarily benefit them or their children.[73] The protest was successful.

Second, women have been active and frequently successful in protecting the trees, stopping tree auctions, and keeping a vigil against illegal felling. In Gopeshwar town, a local women's group has appointed watchwomen who receive a wage in kind to guard the surrounding forest, and to regulate the extraction of forest produce by villagers. Twigs can be collected freely, but any harm to the trees is liable to punishment.

Third, replanting is a significant component of the movement. But in their choice of trees the priorities of women and men don't always coincide--women typically prefer trees that provide fuel, fodder, and daily needs, the men prefer commercially profitable ones.[74] Once again this points to the association between gendered responsibility for providing a family's subsistence needs and gendered responses to threats against the resources that fulfill those needs.

Fourth, Chipko today is more than an ecology movement and has the potential for becoming a wider movement against gender-related inequalities. For instance, there has been large-scale mobilization against male alcoholism and associated domestic violence and wasteful expenditure. There is also a shift in self-perception. I have seen women stand up in public meetings of the movement and forcefully address the gathering. Many of them are also asking: Why aren't we members of the village councils?

Fifth, implicit in the movement is a holistic understanding of the environment in general and forests in particular. The women, for instance, have constructed a poetic dialogue illustrating the difference between their own perspective and that of the foresters.[75]

Foresters:	What do the forests bear?
	Profits, resin and timber.
Women (Chorus):	What do the forests bear?
	Soil, water and pure air.
	Soil, water and pure air.
	Sustain the earth and all she bears.

In other words, the women recognize that forests cannot be reduced merely to trees and the trees to wood for commercial use, that vegetation, soil, and water form part of a complex and interrelated ecosystem. This recognition of the interrelatedness and interdependence between the various material components of nature, and between nature and human sustenance, is critical for evolving a strategy of sustainable environmental protection and regeneration.

Although the movement draws upon, indeed is rooted in, the region's Gandhian tradition which predates Chipko, women's responses go beyond the framework of that tradition and come close to feminist environmentalism in their perspective. This is suggested by their beginning to confront gender and class issues in a number of small but significant ways. For instance, gender relations are called into question in their taking oppositional stands to the village men on several occasions, in asking to be members of village councils, and in resisting male alcoholism and domestic violence. Similarly, there is clearly a class confrontation involved in their resistance (together with the men of their community) to the contractors holding licenses for mining and felling in the area.

At the same time, ecology movements such as Chipko need to be contextualized. Although localized resistance to the processes of natural resource appropriation and degradation in India has taken many different forms, and arisen in diverse regional contexts, resistances in which entire communities and villages have participated to constitute a movement (such as Chipko, Appiko, and Jharkhand) have emerged primarily in hill or tribal communities. This may be attributable particularly to two factors: the immediacy of the threat from these processes to people's survival, and these communities being marked by relatively low levels of the class and social differentiation that usually splinter village communities in south Asia. They therefore have a greater potential for wider community participation than is possible in more economically and socially stratified contexts. Further, in these communities, women's role in agricultural production has always been visibly substantial and often primary--an aspect more conducive to their public participation than in many other communities of northern India practicing female seclusion.

In emphasizing the role of poor peasant and tribal women in ecology movements, I am not arguing, as do some feminist scholars, that women possess a specifically feminine sensibility or cognitive temperament, or that women *gua women* have certain traits that predispose them to attend to particulars, to be interactive rather than individualist, and to understand the true character of complex natural processes in holistic terms.[76] Rather, I locate the perspectives and responses of poor peasant and tribal women (perspectives which are indeed often interactive and holistic) in their material reality--in their dependence on and actual use of natural resources for survival, the knowledge of nature gained in that process, and the broader cultural parameters which define people's activities and modes of thinking in these communities. By this count, the perspectives and responses of men belonging to hill or tribal communities would also be more conducive to environmental protection and regeneration than those of men elsewhere, but not more than those of the women of such communities. This is because hill and tribal women, perhaps more than any other group, still maintain a reciprocal link with nature's resources--a link that stems from a given organization of production, reproduction, and distribution, including a given gender division of labor.

At the same time, the positive aspects of this link should not serve as an argument for the continued entrenchment of women within a given division of labor. Rather, they should serve as an argument for creating the conditions that would help universalize this link with nature, for instance, by *declassing* and *degendering* the ways in which productive and reproductive activities are organized (within and outside the home) and how property, resources, knowledge, and power are distributed.

Conclusion

The Indian experience offers several insights and lessons. First, the processes of environmental degradation and appropriation of natural resources by a few have specific class gender as well as locational implications--it is women of poor, rural households who are most adversely affected and who have participated actively in ecology movements. "Women" therefore cannot be posited (as the ecofeminist discourse has typically done) as a unitary category, even within a country, let alone across the Third World or globally. Second, the adverse class gender effects of these processes are manifest in the erosion of both the livelihood systems and the knowledge systems on which poor rural women depend. Third, the nature and impact of these processes are rooted interactively, on the one hand, in ideology--(in notions about development, scientific knowledge, the appropriate gender division of labor, and so on) and, on the other hand, in the economic advantage and political power predicated especially, but by no means only, on property differentials between households and between women and men. Fourth, there is a spreading grassroots resistance to such inequality and environmental destruction--to the processes, products, people, property, power, and profit-orientation that underlie them. Although the voices of this resistance are yet scattered and localized, their message is a vital one, even from a purely growth and productivity concern and more so if our concern is with people's sustenance and survival.

In particular, the experiences of women's initiatives within the environmental movements suggests that women's militancy is much more closely linked to family survival issues than is men's. Implicit in these struggles is the attempt to carve out a space for an alternative existence that is based on equality, not dominance over people, and on cooperation with and not dominance over nature. Indeed what is (implicitly or explicitly) being called into question in various ways by the movements is the existing development paradigm--with its particular product and technological mix, its forms of exploitation of natural and human resources, and its conceptualization of relationships among people and between people and nature. However, a mere recognition that there are deep inequalities and destructiveness inherent in present processes of development is not enough. There is a need for policy to shift away from its present relief-oriented approach toward nature's ills and people's welfare in which the solution to nutrient-depleted soils is seen to lie entirely in externally added chemical nutrients, to depleting forests in monoculture plantations, to drought starvation in food-for-work programs, to gender inequalities in ad hoc income-generating schemes for women, and so on. These solutions reflect an aspirin approach to development--they are neither curative nor preventive, they merely suppress the symptoms for a while.

The realistic posing of an alternative (quite apart from its implementation) is of course not easy, nor is it the purpose of this paper to provide a blueprint.

What is clear so far are the broad contours. An alternative approach, suggested by feminist environmentalism, needs to be *transformational* rather than welfarist--where development, redistribution, and ecology link in mutually regenerative ways. This would necessitate complex and interrelated changes such as in the *composition* of what is produced, the *technologies* used to produce it, the *processes* by which decisions on products and technologies are arrived at, the *knowledge systems* on which such choices are based, and the class and gender *distribution* of products and tasks.

For instance, in the context of forestry programs, a different composition of the product may imply a shift from the currently favored monocultural and commercial tree species to mixed species critical for local subsistence. An alternative agricultural technology may entail shifting from mainly chemical-based farming to more organic methods, from monocultural high-yielding variety seeds to mixed cropping with indigenously produced varieties, from the emphasis on large irrigation schemes to a plurality of water-provisioning systems, and from a preoccupation with irrigated crops to a greater focus on dryland crops. A change in decision-making processes would imply a shift from the present top-down approach to one that ensures the broad-based democratic participation of disadvantaged groups. Indeed, insofar as the success stories of reforestation today relate to localized communities taking charge of their environmental base, a viable solution would need decentralized planning and control and institutional arrangements that ensure the involvement of the rural poor, and especially women, in decisions about what trees are planted and how the associated benefits are shared. Similarly, to encourage the continued use and growth of local knowledge about plants and species in the process of environmental regeneration, we would require new forms of interaction between local people and trained scientists and a widening of the definition of "scientific" to include plural sources of knowledge and innovations, rather than merely those generated in universities and laboratories. This last is not without precedent, as is apparent from the earlier discussion on Meiji Japan's interactive teams which allowed a flow of information not only from the agricultural scientist to the farmer but also the reverse. The most complex, difficult, and necessary to transform is of course the class and gender division of labor and resources and the associated social relations. Here it is the emergence of new social movements in India around issues of gender, environment, and democratic rights, and especially the formation of joint fronts between these movements on a number of recent occasions, that point the direction for change and provide the points of hope.

Indeed, environmental and gender concerns taken together open up both the need for reexamining, and the possibility of throwing new light on, many long-standing issues relating to development, redistribution, and institutional change. That these concerns preclude easy policy solutions underlines the deep entrenchment (both ideological and material) of interests in existing structures and models of development. It also underlines the critical importance of grassroots political organization of the poor and of women as a necessary condition for their voices to be heeded and for the entrenched interests to be undermined. Most of all it stresses the need for a shared alternative vision that can channel dispersed rivulets of resistance into a creative, tumultuous flow.

In short, an alternative, transformational approach to development would involve both ways of *thinking* about things and ways of *acting* on them. In the present context it would concern both how gender relations and relations between people and the nonhuman world are conceptualized, and how they are concretized in terms of the distribution of property, power, and knowledge, and in the formulation of development policies and programs.

It is in its failure to explicitly confront these political economy issues that the ecofeminist analysis remains a critique without threat to the established order.

Notes

This is a substantially revised and abridged version of a paper presented at a conference on "The Environment and Emerging Development Issues" at the World Institute of Development Economics Research, Helsinki, 3-7 Sept. 1990. A longer version is also available as Discussion Paper No. 8: Engendering the Environment Debate: Lessons from the Indian Subcontinent," CASID Distinguished Speaker Series (Michigan State University, 1991).

I am grateful to several people for comments on the earlier versions: Janet Seiz, Gillian Hart, Nancy Folbre, Jean Dreze, Lourdes Beneria, Gail Hershatter, Pauline Peters, Tariq Banuri, Myra Buvinic, and *Feminist Studies'* editors and anonymous reviewers. I also gained from some lively discussions following seminar presentations of the paper at the Center for Population and Development Studies, Harvard University, February 1991; the Center for Advanced Study in International Development, Michigan State University, April 1990, the Hubert Humphrey Institute of Public Affairs, University of Minnesota, April 1990, and the departments of City and Regional Planning and Rural Sociology, Cornell University, May 1990.

1. See especially Ynestra King, "Feminism and the Revolt," *Heresies*, no. 13, "Special Issue on Feminism and Ecology" (1981): 12-16, "The Ecology of Feminism and the Feminism of Ecology," in *Healing the Wounds: The Promise of Ecofeminism*, ed. Judith Plant (Philadelphia: New Society Publishers, 1989), 18-28, Healing the Wounds: Feminism, Ecology, and the Nature/Culture Dualism," in *Reweaving the World: The Emergence of Ecofeminism*, ed. Irene Diamond and Gloria Orenstein (San Francisco: Sierra Club Books, 1990), 98-112; Ariel Kay Salleh, "Deeper than Deep Ecology; the Eco-Feminist Connection," *Environmental Ethics* 16 (Winter 1984): 339-45; Carolyn Merchant, *The Death of Nature: Women, Ecology, and the Scientific Revolution* (San Francisco: Harper and Row, 1980); and Susan Griffin, *Women and Nature: The Roaring Within Her* (New York, Harper & Row, 1978). Also see discussions and critiques by

Michael E. Zimmerman, "Feminism, Deep Ecology, and Environmental Ethics" (pp. 21-44) and Karen J. Warren, "Feminism and Ecology: Making Connections" (pp. 3-20), both in *Environmental Ethics* 9 (Spring 1987); Jim Cheney, "Ecofeminism and Deep Ecology," *Environmental Ethics 9* (Summer 1987): 115-45; and Helen E. Longino's review of Merchant in *Environmental Ethics 3* (Winter 1981): 365-69.

2. King, "Ecology of Feminism," 18.

3. Sherry Ortner, "Is Male to Female as Nature is to Culture?" in *Women, Culture and Society*, ed. Michelle Z. Rosaldo and Louise Lamphere (Stanford: Stanford University Press, 1974), quotes on pp. 72, 73.

4. See the case studies, and especially Carol P. MacCormack's introductory essay in *Nature, Culture, and Gender*, ed. Carol P. MacCormack and Marilyn Strathern (Cambridge: Cambridge University Press, 1980), 13. Also see Henrietta L. Moore, *Feminism and Anthropology* (Minneapolis: University of Minnesota Press, 1989).

5. Salleh, 340.

6. See Merchant, 144.

7. Ibid., 2,3.

8. For this and the previous quote see ibid., xx-xxi, xix.

9. King in "Feminism and the Revolt" (unlike in her earlier work) does mention the necessity of such a differentiation but does not discuss how a recognition of this difference would affect her basic analysis.

10. For an illuminating discussion of the debate on essentialism and constructionism within feminist theory, see Diane Fuss, *Essentially Speaking* (New York, Routledge, 1989).

11. See case studies in *Nature, Culture and Gender*.

12. Vandana Shiva, Staying Alive: Women, Ecology, and Survival (London: Zed Books, 1988), quotes on pp. 39, 42.

13. Ibid., 14-15.

14. Also see the discussion by Gabrielle Dietrich, "Plea for Survival: Book Review," Economic and Political Weekly, 18 Feb. 1989, 353-54. Apart from the religion-specificity of the discourse on the feminine principle, an interesting example of the relationship between different religious traditions and the environment is that of sacred groves. These groves, dedicated to local deities and sometimes spread over 100 acres, were traditionally preserved by local Hindu and tribal communities and could be found in several parts of the country. Entry into them was severely restricted and tree cutting usually forbidden. (See Madhav Gadgil and V.D. Vartak, "Sacred Groves of India: A Plea for Continued Conservation," *Journal of the Bombay Natural History Society* 72, no. 2 [1975]) These groves are now disappearing. Among the Khaki tribe of northeast India, elderly non-Christian Khasis I spoke to identified the main cause of this destruction to be the large-scale conversion of Khasis to Christianity which undermined traditional beliefs in deities and so removed the main obstacle to the exploitation of these groves for personal gain.

15. For instance, the Rig-Veda the collection of sacred Sansikirt hymns preserved orally for over 3,000 years, which constitutes the roots of Brahmanic Hinduism, is said to have been traditionally inaccessible to women and untouchable castes, both of whom were forbidden to recite the hymns on the ground that they would defile the magic power of the words (for elaboration see Weedy O. Fisherty's *Other People's Myths* [New York

and London: Macmillan, 1990]). In contrast, the *Bhakti* tradition, which began around the sixth century, sought to establish a direct relationship between God and the individual (without the mediation of Brahmin priests) irrespective of sex or caste and gave rise to numerous devotional songs and poems in the vernacular languages. Many women are associated with the movement, one of the best-known being the sixteenth-century poet-saint, Mirabai. Today the Bhakti tradition coexists with the more ritualistic and rigid Brahmanic tradition. In fact a significant dimension of the growing Hindu fundamentalism in India in recent years is precisely the attempt by some to give prominence to one interpretation of Hinduism over others--visible, contemporary struggle over meanings.

Similarly, several versions of the great epic, Ramayana have existed historically, including versions where the central female character, Seta, displays none of the subservience to her husband that is emphasized in the popular version (treated as sacred text) and which has molded the image of the ideal Indian woman in the modern mass media. Feminist resistance to such gender constructions has taken various forms, including challenging popular interpretations of female characters in the epics and drawing attention to alternative interpretations. See for instance, Uma Chakravarty's essay "The Seta Myth," *Samya Skakti* 1 (July 1983); and Bina Agarwal's poem, "Seta Speak," *Indian Express*, 17 Nov. 1985.

16. See Irfan Habib, "Peasant and Artisan Resistance in Mughal India," *McGill Studies in International Development* no. 34 (McGill University, Center for Developing Area Studies, 1964), and his essay in *Cambridge Economic History of India*, ed. Tapan Ray Chaudhuri and Irfan Habib (Cambridge: Cambridge University Press, 1982).

17. See especially Kersala Forestry Research Institute, *Studies in the Changing Patterns of Man-Forest Interaction and its Implications for Ecology and Management* (Trivandrum, 1980), 235.

18. It is estimated that in 1981-82, 66.6 percent of landowning households in rural India owned 1 hectare or less and accounted for only 12.2 percent of all land owned by rural households (National Sample Survey Organisation, *Thirty-seventh Round Report on Land Holdings--I, Some Aspects of Household Ownership Holdings* (Department of Statistics, Government of India, 1987). The distribution of operational holdings is almost as skewed.

19. See N.S. Jodha, "Common Property Resources and Rural Poor," Economic and Political Weekly, 5 July 1986, 1169-81; and Piers Blaikie, *The Political Economy of Soil Erosion in Developing Countries* (London and New York, Longman, 1985).

20. Sharad Kulkarni, "Towards a Social Forestry Policy," *Economic and Political Weekly*, 5 Feb. 1983, 191-96.

21. See V. Pingle, "Some Studies of Two Tribal Groups of Central India, pt. 2: The Importance of Food Consumed in Two Different Seasons," *Plant Food for Man* 1(1975); and Bina Agarwal, "Social Security and the Family: Coping with Seasonality and Calamity in Rural India," *Journal of Peasant Studies* 17 (April 1990): 341-412.

22. I prefer to use the term "control rights" here, rather than the commonly used term "property rights," because what appears critical in this context is less who owns the resources than who has control over them. Hence, for instance, the control of state-owned resources could effectively rest with the village community.

23. *Forest Survey of India* (New Delhi: Ministry of Environment and Forests, Government of India, 1990).

24. P.K. Josh and A.K. Agnihotri, "An Assessment of the Adverse Effects of Canal Irrigation in India," *Indian Journal of Agricultural Economics* 39 (July-September 1984): 528-36.25.

25. See, for instance, Jayanta Bandyopadhyay, "A Case Study of Environmental Degradation in Karnataka," (Paper presented at a workshop on Drought and Desertification, India International Center, 17-18 May 1986); and B.D. Dhawan, *Development of Tubewell Irrigation in India* (Delhi, Agricole Publishing Academy, 1982.).

26. Center for Science and Environment, *The State of India's Environment: A Citizen's Report*, 1985-86 (Delhi: Agricole Publishing Academy, 1982).

27. See especially, Ramachandra Guha, "Forestry in British and Post-British India: A Historical Analysis," *Economic and Political Weekly*, 29 Oct.1983, 1882-96.

28. See Malini Chand and Rita Bezboruah, "Employment Opportunities for Women in Forestry," in *Community Forestry and People's Participation—Seminar Report* (Ranchi Consortium for Community Forestry, 20-22 Nov. 1980); and Srilata Swaminathan, "Environment: Tree versus Man," *India International Center Quarterly* 9, nos. 3 and 4 (1982).

29. However, the degree to which the village community acted as a cohesive group and the extent of control it exerted over communal lands varied across undivided India: it was much greater in the northwest than elsewhere (see B.H. Baden-Powell, *The Indian Village Community* {New Haven, Conn.: HRAF Press, 1957})

30. For a detailed discussion on these causes, see Jodha.

31. Bandyopadhahy.

32. On traditional systems of community water management see Nirmal Sengupta, "Irrigation: Traditional vs. Modern," *Economic and Political Weekly*, Special Number (November 1985): 1919-38; Edmund R. Leach, *Pul Eliya - A Village in Ceylon: A Study of Land Tenure and Kinship* (Cambridge: Cambridge University Press, 1967); and David Seklar, "The New Era of Irrigation Management in India" (photocopy, Ford Foundation, Delhi, 1981). On communal management of forests and village commons, see Ramachandra Guha, "Scientific Forestry and Social Change in Uttarakhand" (pp. 1939-52), and Madhav Gadgil, "Towards an Ecological History of India" (pp. 1909-38), both in *Economic and Political Weekly*, Special Number (November 1985); and M. Moench, "Turf and Forest Management in a Garhwal Hill Village," in *Whose Trees? Proprietary Dimensions of Forestry*, ed. Louise Fortmann and John W. Bruce (Boulder, Colo.: Westview Press, 1988). On firewood gathering practices, see Bina Agarwal, "Under the Cooking Pot: The Political Economy of the Domestic Fuel Crisis in Rural South Asia," *IDS Bulletin* 18, no.1 (1987): 11-22. Firewood for domestic use in rural households was customarily collected in the form of twigs and fallen branches, which did not destroy the trees. Even today, 75 percent of firewood used as domestic fuel in northern India (and 100 percent in some other areas) is in this form.

33. The preservation of sacred groves described in note 14 is one such example.

34. Also see discussion in Partha Dasgupta and Karl-Goran Maler, "The Environment and Emerging Development Issues." (Paper presented at a conference on Environment and Development, Wider, Helsinki, September 1990).

122 *Bina Agarwal*

35. Daniel W. Bromley and Michael M. Cernea, "The Management of Common Property Natural Resources," *World Bank Discussion Paper* no. 57 (Washington, D.C.: World Bank, 1989), 25.

36. Ibid.

37. Paul Shaw, "Population, Environment, and Women: An Analytical Framework." (Paper prepared for the United Nations Fund for Population Activities (UNFPA), Inter-Agency Consultative Meeting, New York, 6 Mar. 1989), 7.

38. Mark Rosenzwieg and Kenneth I. Wolpin, "Specific Experience, Household Structure, and Intergenerational Transfers: Farm Family Land and Labor Arrangements in Developing Countries," *Quarterly Journal of Economics* 100, supp. (1985): 961-87.

39. Under some large-scale irrigations works, crop yields are *lower* than in the period immediately prior to the project (Joshi and Agnihotri).

40. C.H. Hanumantha Rao, S.K. Ray, and K. Subbarao, *Unstable Agriculture and Drought* (Delhi: Vikas Publishing, 1988).

41. Bernhard Glaeser, ed. *Learning From China? Development and Environment in Third World Countries* (London: Allen & Unwin, 1987, 1987).

42. Also see Stephen A. Marglin, "Losing Touch: The Cultural Conditions of Worker Accommodation and Resistance," in *Knowledge and Power*, ed. Frederique A. Marglin and Stephen A. Marglin (Oxford: Oxford University Press, 1988).

43. See Bruce F. Johnston, "The Japanese Model of Agricultural Development: Its Relevance to Developing nations," in *Agriculture and Economic Growth--Japan's Experience*, ed. Kazushi Ohkawa, Bruce F. Johnston, and Hiromitsu Kaneda (Princeton: Princeton University Press, 1969), 61.

44. For a review of issues and literature on this question, see Bina Agarwal, "Women, Poverty, and Agricultural Growth in India," *Journal of Peasant Studies* 13 (July 1986): 165-220.

45. These sex ratios are particularly female-adverse in the agriculturally prosperous northwestern regious of Punjab and Haryana where these figures are, respectively, 88 and 87 females per 100 males. For a discussion on the causes of this regional variation see Agarwal, "Women, Poverty, and Agricultural Growth in India" and Barbara Miller, *The Endangered Sex: Neglect of Female Children in North-West India* (Ithaca: Cornell University Press, 1981).

46. Women in India rarely own land, and in most areas also have limited access to personal assets such as cash and jewelry. See Bina Agarwal, "Who Sows? Who Reaps? Women and Land Rights in India," *Journal of Peasant Studies* 15 (July 1988): 531-81.

47. See discussions in Agarwal, "Women, Poverty, and Agricultural Growth in India"; Bina Agarwal, "Rural Women and the High Yielding Rice Technology in India," *Economic and Political Weekly*, 31 Mar. 1984, A39-A52; and Kalpana Bardhan, "Rural Employment, Welfare, and Status: Forces of Tradition and Change in India," *Economic and Political Weekly*, 25 June 1977, A34-A48: 2 July 1977, 1062-74; 9 July 1977, 1101-18.

48. See Agarwal, "Rural Women and the High Yielding Rice Technology in India;" and James G. Ryan and R.D. Ghodake, "Labour Market Behaviour in Rural Villages in South India: Effects of Season, Sex, and Socio-Economic Status," Progress Report, Economic Programme 14, International Crop Research Institute for Semi-Arid Tropics (ICRISAT), Hyderbad (1980).

49. See Bina Agarwal, "Women, Land, and Ideology in India," in *Women, Poverty, and Ideology: Contradictory Pressures, Uneasy Resolutions*, ed. Haleh Afsbar and Bina Agarwal (London: Macmillan, 1989); and Ursula Sharma, Women, Work, and Property in North-West India (London: Tavistock, 1980).

50. For a detailed cross-regional mapping of some of these variables in the context of women's land rights in South Asia, see Bina Agarwal, *Who Sows? Who Reaps? Gender and Land Rights in South Asia*, forthcoming (Cambridge: Cambridge University Press).

51. Quoted in Sundarial Bahuguna, "Women's Non-Violent Power in the Chipko Movement," in *In Search of Answers: Indian Women's Voices in "Manushi,"* ed. Madhu Kishwar and Ruth Vanita (London: Zed Books, 1984), 132.

52. Personal communication, Chitra Sundaram, Danish International Development Agency (DANIDA), Delhi, 1981.

53. Bina Agarwal, "Women and Water Resource Development," photocopy, Institute of Economic Growth, Delhi, 1981.

54. Bahuguna.

55. Shubh Kumar and David Hotchkiss, "Consequences of Deforestation for Women's Time Allocation, Agricultural Production, and Nutrition in Hill Areas of Nepal," *Research Report* 69 (Washington, D.C.: International Food Policy Research Institute, 1988).

56. See T. Bhaduri and V. Surin, "Community Forestry and Women Headloaders," in *Community Forestry and People's Participation Seminar Report* (Ranchi Consortium for Community Forestry, 20-22 Nov. 1980).

57. Michael Howes and M.A. Jabbar, "Rural Fuel Shortages in Bangladesh: The Evidence from Four Villages," Discussion Paper 213 (Sussex, England: Institute of Developent Studies, 1986).

58. Agarwal, "Women and Water Resource Development."

59. Joan P. Mencher and K. Saradamoni, "Muddy Feet and Dirty Hands: Rice Production and Female Agricultural Labour," *Economic and Political Weekly*, 25 Dec. 1982, A149-A167; and United Nations Development Program, Rural Women's Participation in Development, *Evaluation Study*, no. 3, UNDP, New York (June 1979).

60. Rudolf G. Wagner, "Agriculture and Environmental Protection in China," in *Learning from China?*

61. Dinesh Mohan, "Food vs. Limbs: Pesticides and Physical Disability in India," *Economic and Political Weekly*, 28 Mar. 1987, A23-A29.

62. These are apart from the widely documented patron-client types of relationships.

63. See Sharma, *Women, Work, and Property in North-West India*; and Sylvia Vatuk, "Sharing, Giving, and Exchanging of Foods in South Asian Societies," (University of Illinois at Chicago Circle, October 1981).

64. See Amartya Sen, "Gender and Cooperative-Conflict," in *Persistent Inequalities*, ed. Irene Tinker (New York: Oxford University Press, 1990) for a discussion on the bargaining approach to conceptualizing intrahousehold gender relations, and Agarwal, "Social Security and the Family," for a discussion on the factors that affect intrahousehold bargaining power.

65. Walter Fernandes and Geeta Menon, *Tribal Women and Forest Economy: Deforestation, Exploitation, and Status Change* (Delhi: Indian Social Institute, 1987), 115.

66. Among the Garo tribes of northeast India in the early 1960's, Burling found that the men always deferred on this count to the women, who knew of aproximately 300 indigenously cultivated rice varieties. See Robbins Burling, *Rensanggri: Family and Kinship in a Garo Village* (Philadelphia: Pennsylvania University Press, 1963). In Nepal even today it is women who do the seed selection work among virtually all communities. Sae Meena Acharya and Lynn Bennet, "Women and the Subsistence Sector in Nepal," *World Bank Staff Working Paper* no. 526 (Washington, D.C.: World Bank, 1981).

67. For a detailed discussion on these schemes and their shortcomings, see Bina Agarwal, *Cold Hearths and Barren Slopes: The Woodfuel Crisis in the Third World* (London: Zed Books, 1986).

68. D.M. Chandrashekar, B.V. Krishna Murti, and S.R. Ramaswamy, "Social Forestry in Karnataka: An Impact Analysis," *Economic and Political Weekly*, 13 June 1987, 935-41; and Shiva.

69. Vandana Shiva, "Ecology Movements in India," *Alternatives* 11 (1987): 255-73.

70. Ramachandra Guha, "Ideological Trends in Indian Environmentalism," *Economic and Political Weekly*, 3 Dec. 1988, 2578-81.

71. Among the many writings on the Chipko movement, see especially Jayanta Bandyopadhyay and Vandana Shiva, "Chipko," *Seminar*, no. 330 (February 1987); Shiva: Shobhita Jain, "Women and People's Ecological Movement: A Case Study of Role in the Chipko Movement in Uttar Pradesh," *Economic and Political Weekly*, 13 Oct. 1984, 1788-94; and Bharat Dogra, *Forests and People*, published by the author (Delhi, 1984).

72. I understand there have also been cases of people hugging trees to protect them from loggers in the United States, although they appear to have no apparent link with Chipko.

73. There is a growing literature indicating significant gender differences in cash-spending patterns, with a considerable percentage (at times up to 40 percent) of what men earn in poor rural households often going towards the purchase of items they alone consume, such as liquor, tobacco, and clothes, and much of what the women earn going toward the family's basic needs. See especially Joan Mencher, "Women's Work and Poverty: Women's Contribution to Household Maintenance in Two Regions of South India," in *A Home Divided: Women and Income in the Third World*, ed. Daisy Dwyer and Judith Bruce (Stanford: Stanford University Press, 1988).

74. This gender divergence has also been noted elsewhere. See Rita Brara, "Commons Policy as Process: The Case of Rajasthan, 1955-85," *Economic and Political Weekly*, 7 Oct. 1987, 2247-54.

75. Quoted in Shiva.

76. For a critique of these lines of argument, see Helen E. Longino, "Can There Be a Feminist Science?" *Hypatia* 2 (Fall 1987): 51-64.

5

The Relation Between Population and Deforestation: Methods for Drawing Causal Inferences from Macro and Micro Studies

Alberto Palloni

Objectives

Does population growth have identifiable effects on the rate of deforestation at global and local levels? As occurs with analogous queries probing the relation between population pressure and other environmental outcomes, this is a

This chapter was prepared while the author was a Fellow at the Center for Advanced Studies in the Behavioral Sciences, Stanford, California. I am grateful for the intellectual and material support that I received at the Center. I am grateful for the financial support provided by the National Science Foundation #BNS-870084. I am also grateful for the support provided by the Social Science Research Council throughout the life of this project. I would like to thank Guido Pinto for his very able assistance in collecting the materials. Paul Holland read an early version of this manuscript and provided helpful statistical and editorial suggestions. Last but not least, Richard Berk was a severe but fair critic of an early version of this paper. Although due to time pressures and my own limitations I have not included all the changes he suggested, the paper is a better informed piece than it would have been without his contribution. Of course, neither he nor anyone else who came into contact with early incarnations of this work is to blame for the inaccuracies and shortcomings that remain in it.

question that has been intensely debated in the past but left without a satisfactory answer. And just as in other areas where population is invoked as a cause (or effect) of other environmental outcomes, this is not surprising since, as I show later, the conceptual and empirical difficulties to be surmounted are formidable. Somewhat more surprising, however, is the fact that we seem to have only a primitive idea about the nature and diversity of findings uncovered by past research. Indeed, a great deal of the discussion is dominated by grand-theorizing, overarching frameworks rarely if ever based on solid findings, and simulation exercises that impose and assume rather than verify relations between strategic factors. Not infrequently the controversy has been fueled by exegesis based on very selected evidence that intentionally or unintentionally discards and marginalizes alternative results. Yet, a cursory examination of the literature shows that there is an abundance of research findings that directly or indirectly relate to the main problem. Why, then, are we accomplices of a debate unable to marshall if not flawless at least solid evidence?

One must admit that, as a rule, the research findings alluded to above follow the trail of very diverse problems, are obtained with research designs that are hardly comparable at all, and are frequently based on the analyses of empirical cases or data sets that resist even the most ingenious attempts to find a reduction to common baselines. But perhaps a more important factor is the absence of systematic procedures to process these findings. I am not claiming that we lack adequate review studies; we have plenty, spanning the whole gamut of quality levels. Rather the point is that we have failed to develop and apply rigorous techniques for the retrieval of estimates of the relation between population and deforestation or, for that matter, other relevant environmental outcomes.

What I set out to do in this paper is to describe and utilize two alternative procedures to derive inferences about the magnitude and direction of effects of population growth on deforestation. I do this by examining a large sample of studies that were designed to answer the question stated at the outset. The first procedure involves statistical techniques that have evolved rapidly in the last 10 years within the field of what in the social sciences is referred to as meta-analysis. Meta-analysis, however, is part of the rapidly developing field of statistical methods for combining information. The second technique requires the application of very simple principles of qualitative algebra to a set of case studies where the relation(s) of interest are investigated. As will become clear in the discussion that follows, these two procedures are neither situated at the same level of abstraction nor are they in rife competition with each other. Even though one relies on premises about quantification of relations that the other purposely avoids and replaces, they can be seen as complementary. Since the empirical data required by these procedures are different, the inferences that we draw from each of them could be conflictive and may be even difficult to compare. Thus, while we should hail agreements and use them to reinforce conclusions, inconsistencies may be more a matter of incompatible data bases than the outcomes of relations that are unequally detected by the two procedures. I will argue, however, that meta-analysis is more limited not because of inherent shortcomings but due to the fact that its object of reduction, quantitative assessments of pertinent causal relations, cannot produce sufficient information about the range of causal conditions that lead to deforestation.

In the first section below, I describe the nature of the problem, establish the boundaries of what we seek to understand, and highlight the most important analytic problems. In the second, I describe and apply a few procedures from meta-analysis to a handful of quantitative studies, and in the following section I explore the application of qualitative algebra to a sample of about 55 case studies. The final section presents a summary and conclusions.

Consumption of Forests and Population Pressure

The Extent of Modern Deforestation

Clearing of forests is hardly an activity that emerged only recently. Forest clearing of large areas of Central Europe and the Mediterranean had been accomplished by 1300, after which it subsided for an interval of about 300 years (Hosier 1988; Williams 1990). Although the contribution of natural causes cannot be dismissed offhand (Blaikie and Brookfield 1987a), past episodes of deforestation appear to consistently emerge and peak after the onset of favorable economic conditions and wane during periods of economic stagnation. There is little if any evidence suggesting that population growth and removal of tree cover went in tandem, except perhaps to the extent that both were responsive to sharp swings in economic activity (Williams 1989). In the developing world too the onset of deforestation precedes by some time the beginnings of the postwar population explosion. Thus, ecosystems in wretched conditions, such as the hills of Nepal, underwent important degradation long before modern population pressure appeared on the scene (Ives 1987). In the Philippines the loss of forest land can be traced back to the early part of the century, when the country was under American colonialism (Bautista 1990). And while in the Amazon region of Brazil the sharp increases in forest clearing did not occur until after 1975 (Barbier, Burquess and Markandaya 1990), in the south of Brazil (Rio Grande do Sul, Parana) and most of Central America (particularly El Salvador and Honduras), large territories covered with *Araucaria* and some rainforest had been devastated before 1970 (McNeill 1988; Barrows 1991). And yet, in both regions, Brazil and Central America, rapid population growth did not start in earnest until after 1950.

What is novel in the post-war cycle of deforestation is the scale and speed of the phenomenon. According to recent data, the global annual (percent) rate of deforestation measured in units of surface cleared between 1968 and 1978 is

.51 (Allen and Barnes 1985), a rate that is equivalent to a half-life of the forest cover of about 136 years. During the short interval between 1981 and 1985 the global annual rate of deforestation slightly increased to about .58 and the half-life of the forest decreased to 118 years (International Institute for Environment and Development and World Resources Institute 1986). But the rate of deforestation from 1650 until 1850 can be roughly estimated to be at most one-fourth of that (Williams 1990;Table 11.1). Furthermore, unlike most historical episodes of deforestation, current trends affect dense and massive tropical forests that comprise a substantial fraction of all existing forests. Unless arrested or slowed down, the sheer speed of the process and the magnitude of the area exposed to it rules out natural regeneration and charts a course of ecological degradation that may well be irreversible.

In addition to the size of the exposed area and the pace of clearing, current trends have an additional peculiar characteristic; by and large, the areas affected are located in the core of the developing world, in countries and regions experiencing the highest rates of population growth ever recorded. Undoubtedly it is this partially accidental feature--for clearing of dense forest can only proceed in areas where there are dense forests to be cleared--that provides access to an easy imagery sustaining claims about causal connections between population and deforestation. Causal inferences of this sort are further muddied by measurement complications that are conveniently swept aside. A brief summary of the most important among these complications is presented below.

"Bundling" the Diversity of Forests and Deforestation

There is a curious contrast between what this "study of studies" will show and the nature of claims that are based on those studies, most of which identify population growth or high population density as the driving force of deforestation. As I will show below and as others have also shown before, the relations are far more complex than they appear on first scrutiny. Except in extreme cases population growth is only a remote cause of the problem, so removed in fact as to be somewhat useless not only for policy formulations but also for theory construction. And yet after examining the conclusions of nearly 70 studies on deforestation carried out in disparate regions during a period of approximately 15 years, one is left with the strong impression that there is no question about the existence of a direct causal connection between deforestation and population growth or, more generally, population pressure. With slight differences of emphasis but sharper differences in perspectives and methodologies, in all these studies population growth is eventually invoked as a factor contributing to the rate of forest clearing. Some conclusions are blunt and leave scarcely any room for alternative explanations:

> The slow and almost imperceptible increase in population during premodern times and its rapid rise from roughly 1600 onwards has led to a steady decrease of the world's forests as humankind has needed more land for growing food, timber for construction and shelter, and fuelwood to keep warm, cook food and smelt metals. (Williams 1989:181)

or, less directly but not more subtly:

> Given a population growth rate of 3.7 percent per annum, there is no doubt about the need to increase agricultural production. However, this should not be achieved by destroying the last remaining natural forests which are needed to carry out important ecological regulatory functions. (Kleinert 1987:103)

Other authors are more cautious and suggest the presence of complexities that, upon further investigation, could conceivably overshadow the role of population without necessarily displacing it from its prominent position. Here, the importance of population is toned down, inserted in a more conservative causal discourse that reflects some reticence to dispose of the nagging issue altogether, even if that is precisely what the evidence calls for:

> Although the rapid increase in population growth in the region [Central America] is a matter for serious concern, population growth per se cannot adequately explain the destructive land-use patterns that have emerged in the south. (Stonich 1989:290)

Occasionally population will appear as an afterthought, another item in a lengthy and heterogeneous "shopping list" of causes, perhaps more as a concession to an entrenched tradition than as a reflection of relations and mechanisms disentangled in empirical research:

> To summarize, it can be said that deforestation is caused by the combination of a number of factors. It is difficult to identify a single cause and account for its share. The combined influence of economic pressure, population pressure and Government policy have together precipitated the present alarming situation. (Chattopadhyay 1985:225)

or, when the list is longer,

> In other words, deforestation is a consequence of low per capita income, non-availability of fossil fuels, high population density, and poverty; hence, it is intricately linked to socio-economic process. (Bowonder 1985:179)

Diversity in Deforestation. Clearing of semi-arid or moist tropical forest, rainforest, humid temperate forest, and savanna woodland is conceived as an essentially negative outcome because, in addition to seriously undermining the aggregate value of amenities, it leads to loss of biodiversity, laterization, erosion, soil degradation, siltation of streams and reservoirs, and exacerbation of conditions that could produce global warming. The discounted value of the economic benefits derived from forest clearing, so go some persuasive arguments, simply does not stand up to the societal costs that will burden current and future generations. Others argue that under some conditions, deforestation is a perfectly rational activity at least insofar as the calculus of cost and benefits is concerned (Hosier 1988). This raises a first complication related to the homogeneity of the outcome: can one lump in one single bundle all clearing and tree-cutting activity and manipulate one single outcome-measure? Or, alternatively, should one make allowances for heterogeneities that reflect at least partially short- and long-term environmental outcomes? Plainly, the thinning-out of a tropical forest as a result of heavy-handed, high-tech logging activities cannot lead to the same environmental spill-overs as would, for example, the clearing of an equivalent area of dry forest for fuelwood harvesting. It is not just the type of human action that matters--although clearly this contributes to the differences--but the quality of the stock of tree cover being sacrificed and the residual ecological resiliency or fragility. The choice is far from obvious, for while the strategy of "bundling indicators" facilitates measurement but introduces confusion, the other strategy complicates the empirical investigations (and may even make them impossible when the information is simply unavailable) but may preserve significant distinctions.

Though related to the first, the second complication is quite different. Forest clearing involves a wide range of human activities with very different consequences and, one would think, very different determinants: it stretches from treecutting for fuelwood and charcoal production to indiscriminate clearing to extraction of selected and highly valued roundwood. Caught inbetween is the clearing that occurs as a result of swidden agriculture, as shifting cultivators move from one plot to the next in an endless but losing battle to prevent precipitous falls in yields or, alternatively, the establishment of extensive grazing fields to raise cattle in ranching operations geared to export markets.

A third complication is somewhat more pedestrian in inspiration but could, in all likelihood, produce a host of unanticipated and nasty consequences. If one decided that the magnitude of deforestation is to be measured by a single indicator that bundles all tree losses or area cleared, with no distinctions made about the nature of forest cover or the human actions involved, how are we to construct the indicator? Absolute area cleared or volume harvested is not a good choice since decrements will be roughly in proportion to time elapsed and areas (or volume) exposed. Since current "stumpage" stocks are largely concentrated in countries with high rates of population growth, a graph displaying absolute decrements per unit of time against population will inevitably tend to show a positive relation. But this involves no causation, not even an informative relation. Obviously, a similar diagnostic applies when the units of analysis are villages and their contiguous territories or districts and provinces. A far better measure is the monthly or annual rate of clearing or the log of the ratio of forested area at time t+n to forested area at time t. This, of course, presupposes that we have good means of assessing the magnitude of the area cleared and that we have the means to set up equivalence classes of lost forest as a function of potential natural regeneration. With a few exceptions, most literature on the subject proceeds as if this latter complication were trivially solved. We found no good analysis of the potential pitfalls or of the degree of correspondence between observed area cleared and secondary environmental damage.

Finally, the nature of the product being cleared in combination with the methods used to do it surely results in different probabilities of and waiting times for regeneration of the tree cover. When this heterogeneity is considerable, a single indicator of deforestation is likely to hide so much diversity regarding potential environmental outcomes and prospective carrying capacity as to be totally meaningless as a proxy for the ultimate environmental influence that deforestation will exert.

Dimensions of Population Growth. Although elementary, the distinction between population growth and population density is frequently overlooked. Somehow the proposition "high rural population density leads to deforestation" is used interchangeably with "rapid rural population growth leads to deforestation" and no care is taken to establish their inherently different implications. Whereas over time high population density inevitably results from population growth within fixed geographic boundaries, there is no necessary correlation between one and the other in a short interval of time and over diverse geographic locations. The mechanisms linking deforestation to population pressure that results from rapid growth but not high density, on the one hand, and to pressure derived from high density but not rapid growth, on the other, are bound to be different. The differences may exert nontrivial influences on research results particularly when, instead of following the more comfortable strategy of using the "single-bundle" approach, one chooses to distinguish different types of deforestation.

A related but distinct conceptual shortcoming is immediately erected as soon as population is conceived simply as the absolute number of people or its relative growth. There are other dimensions of the structure of a population that may have nontrivial effects on the uses and misuses of forests, such as age distribution, settlement patterns, and distribution of individuals by types of households or familial arrangements. Just as population density and population growth need to be kept separate, so should the various dimensions or the distributional aspects of population.

As with other environmental outcomes, the study of the relation between population and deforestation is usually done on well-defined legal and territorial entities, usually countries or, more rarely, villages and districts. While in all these cases the precise locus of deforestation is well defined, the locus of population growth or density may not be altogether obvious. Indeed, it could depend on the type of activity leading to deforestation. Thus, when deforestation results from land clearing to accommodate an oversupply of agricultural labor force, the approximate unit of reference or measurement is the country or village or district that generates the excess population growth choking labor markets. But, when deforestation occurs as a result of intensive logging, the unit of reference is no longer obvious and largely depends on the exact mechanism linking demand for timber, social distribution of returns from exports, and/or population pressure, if any, that sustains the demand for the product. It also depends on how population pressure is alleviated by allocating the returns from logging.

This brings up a final issue. Unequal regional population distribution--which is, in part at least, a reflection of differential rates of natural increase--may lead to deceivingly simple imageries. Thus, for example, government-sponsored resettlement policies and programs--either directly enforced or indirectly stimulated by subsidies and incentives--reshuffle the spatial and sectorial distribution of labor and, in the process, may alter the demand for land occupied by pristine forests. While the argument that population growth is the ultimate culprit is indeed suggestive, it is also uninformative and misleading for the relations involve many layers that need to be peeled away before the causal mechanism is exposed: unequal population distribution, unequal regional rates of labor-demand formation, and resettlement policies are important pieces in the puzzle and need to be examined before simple statements of relations can begin to be considered seriously.

The Nature of the Relation

Conventionally, population growth and selected economic outcomes (income per capita, for example) have been linked together through generalized "production functions" that translate factor densities into output. The translation is subject to some constraints, usually associated with technology and elasticities of response of output to factor inputs. In regimes with a positive rate of population increase, the long-run behavior of the system largely depends on whether there are *a priori* defined mechanisms to escape excessive population growth. This occurs if returns to scale are more than marginal and figure prominently to arrest slumping per capita output or, alternatively (but more commonly, complementarily), when technological improvements that trigger an upward displacement of output at fixed values of input factors are themselves responsive to population pressure. When no such mechanisms are provided, the entire system is doomed at least in the long run. This, in a nutshell, is the difference between a "Malthusian" view and a "Simonian/Boserupian" approach.

Arguably, a similar representation could be used to describe the process of deforestation. The "product" is a measure of the extent of forest cleared in an interval of time while the factor inputs include population (or labor force), arable land, and technology with associated elasticities:

$$D(t) = g_{\{\alpha,\beta,\delta\}}(P,L,T) \qquad (1)$$

where g is a suitable functional form, P is population, L is arable land, T is technology, and α, β, δ are elasticities. Manipulations of (1) lead to expressions for the rate of deforestation as a function, among others, of the rate of population growth (or population density). As I will show later, this is the type of reasoning that leads to linear models tailored to the estimation of elasticities that are sometimes construed as measures of "causal effects." Most of the quantitative studies that have been carried out in the past 10 years or so rely on formulations analogous to (1). This approach, however, has two main problems.

Outcome Heterogeneity. While (1) may be a sensible representation for some types of deforestation, it may be inaccurate for others. Deforestation that occurs largely as shifting cultivators nibble away progressively large patches at the edges of forests, could in principle be represented by expressions analogous to (1). However, deforestation that occurs as an outcome of heightened demand for cattle pastures or for timber, is likely to fall outside the boundaries of (1). This is a very important and vexing issue that is regularly obscured in almost all quantitative research on the subject though, admittedly, much less so in case studies. The point is not just that the input factors may be different or that the elasticities are variable across outcomes (types of deforestation) but that the relation between them and at least one of these factors, namely population, is far from being reducible to a common expression. It is simply nonsense to argue that population pressure acts homogeneously since it may indeed be the case that it plays no role at all in at least some of the outcomes or, more radically, it may exert influences moving in opposite directions. If, as is very commonly done, one neglects this heterogeneity, estimation of effects will lead to biased and uninterpretable estimated effects.

If most deforestation occurred as a function of one mode of action alone, these caveats would be purely academic. But available data confirm the conjectures that forest clearing involves substantially heterogenous activity and that the degree of heterogeneity varies across regions. Although the data bases are not ideal, they do show that the proportion of forest area cleared attributable to fuelwood demands, timber extraction, and cultivation changes over time and across regions (Williams 1989, 1990; Allen and Barnes 1985; Biswas 1986; Bowonder 1985).

134

FIGURE 5.1 Time Series of Fuelwood and Charcoal Production

Total Roundwood

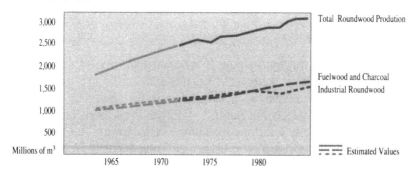

Fuelwood and Charcoal

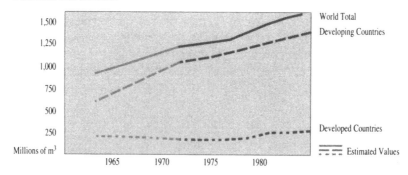

Industrial Roundwood

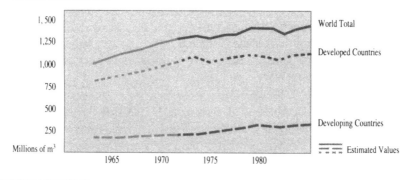

Source: M. Williams "Forests" in *The Earth as Transformed by Human Action*, B.L. Turner, W.C. Clark, R.W. Kates, et al., eds. Cambridge: Cambridge University Press, 1989.

Contingencies and Mediating Factors. But the single most important charge that one can level against the representation in (1) is that it operates in an institutional vacuum. This means that it neglects first, that the effects of population on deforestation are *mediated* by a chain of social processes and, second, that their magnitude and their direction is *contingent* on the presence (absence) of social institutions. Two examples show the importance of mediating factors and contingencies.

The first example is a simple one. It refers to the extent of deforestation produced by clearing for fuelwood. Apparently the connection between area deforested (A) and population (P) is quite straightforward: per capita needs of energy for cooking and heating (e) when multiplied by total population results in the total amount of energy needed. Parameters translating energy requirement into volume of wood (λ_1) and volume of wood into area cleared (λ_2) complete the equation relating population and area of the forest cleared:

$$A = P (e \; \lambda_1 \; \lambda_2) \qquad (2)$$

A straightforward algebraic consequence of the expression is that an increase of x percent in the population will imply an increase of x percent in the area cleared. With some cosmetic changes this type of argument is consistently made in many studies, particularly those that focus on Africa and some parts of Asia; the connection is transparent and immediate and there can be no doubt about what the role of population pressure is. Furthermore, the evidence available does seem to confirm it: the postulated relation is such that if population grows exponentially, so will the area of forest cleared. Figure 5.1, taken from the excellent review by Williams (1990), does indeed indicate that consumption of forests to satisfy fuelwood demands follows a roughly exponential trajectory. More than 90 percent of the total wood extraction in low-income, rapidly growing countries and half of the wood extracted from the forests of the world is for fuelwood usage (Bowonder 1985; Williams 1989). It is not surprising then that total correlations between extent of deforestation and population growth are more than trivial.

But expression (2) is artificially simple to capture the complexities of any empirical case and, in all likelihood, will not lead to correct results if the time horizon is expanded beyond a few years. There is ample evidence that relentless use of the forest to satisfy fuelwood needs does indeed take place *under certain conditions*, e.g., those that allow the unregulated use of common property resources when alternative fuels are prohibitively priced. The outcome could be different if property rights were established so that benefits and costs were spread across all economic agents involved in extraction, exchange and consumption. Under such conditions, price increases would inevitably derail the process and the connection between population and area deforested could be lost or attenuated. But lack of efficacious mechanisms to internalize externalities is only partly responsible for the inaccuracy of the representation. Another element that favors unrestricted forest clearing is excess supply of labor force. As recognized by Whitney in an analysis of Sudan:

> In subsistency fuel gathering, exemplified by much of rural Sudan, location utility is expressed not in monetary terms but in terms of the effort and time expended to collect and transport the wood. If the opportunity cost is zero, or the demand for wood is inelastic, a much greater effort will be expended on fuel collection, with resultant deforestation, than in a situation where the opportunity costs of labor are rising or where the demand for fuel is elastic. In the former situation, fuel gatherers will continue to exploit one area until the perceived marginal effort of producing the next unit of fuel exceeds the effort of moving to a more distant area where wood is more plentiful. (Whitney 1987:126).

Low prices of fuel for the urban population, the source of a large part of the demand for fuelwood in many countries, are indeed desirable to central governments. Import-substitution relying on the more expensive fossil fuels or regulations that decelerate the process of clearing but would increase pressure on prices are not politically viable. Thus, the relation represented by expression (2) rests not only on conditions dominated by "Hardin-type" of transfers (McNicoll 1990) but requires also an acquiescent central authority unwilling to shake up their foundation and, finally, a plentiful supply of labor. The effects of population growth are then a function of, or are contingent upon, the joint occurrence of a peculiar set of property rights, an urban-biased central authority, and a production structure with a low rate of labor absorption. Conceivably, changes in some or all these three conditions could lead to a partial or complete dissociation between population growth and deforestation. If these conditions approximate reality well, we would observe that a function defined in a high dimensional plane relating deforestation, population growth, property rights, central authority, and production system would be highly nonlinear, and that the partial derivatives of deforestation with respect to population growth would be themselves highly nonlinear. This more nuanced representation completely negates the practical value of the commonly used "average" effect of population growth on deforestation.

The second example illustrates the operation of mediating factors and contingencies. At least during the initial stages of a demographic transition, population growth leads to increases in the rate of increase of families or households, not just to increases in the absolute size of the population or population density. In the absence of other changes, this will accelerate land fragmentation and exacerbate the inequality of land distribution. This, in turn, could affect the returns to labor, the number of hours worked, and the kind and quantity of activities devoted to preservation of land quality and conservation of complementary natural resources. Intensification and extensification are both plausible reactions and, in particular, indiscriminate land clearing could end up being the dominant response to the inevitable onset of declining yields. The future turn of events will depend on social and political conditions supporting (undermining) economic transfers. This is hinted at in an interesting paper by Thiesenhusen who suggests that some societies under high population pressure and undergoing rapid fragmentation of property may be highly resilient and could delay or prevent altogether the release of excess labor force into marginal areas (1991). The feat is achieved through the use of networks of social exchanges and systems of obligations. Within limits to be empirically explored, a system of social and economic exchanges and obligations could deflect, in significant ways, the pressure on resources generated by sheer increases in the labor force. Within those limits, the linkage between the growth in the labor force and deforestation or other environmental outcomes will be only a weak one.

This example illustrates two points. First, population growth operates *through* land inequality and both factors are part of a rather lengthy causal chain: the more remote the connection (the higher the number of intermediate stages), the less relevant will population growth be and the higher the likelihood that its effects will be twisted, bent and even dissipated by a sequel of contingencies punctuating the chain. This does not mean that population growth does not have any effects; it only implies that its effects are felt insofar as they are efficiently transmitted by mediating factors. Second, the effects of population growth can be attenuated and delayed within social and political contexts that favor the reallocation of wealth and cushion the fall in per capita yields. These social and political conditions are thus contingencies that alter the responses to changes in population growth.

A review of the literature reveals that the most important mediations and contingencies involve five factors: the fragility of the ecosystem; the stock of active (or passive) production technology; management techniques and knowledge; social and political institutions; and cultural frameworks. Moving the debate beyond polarized views requires one not to abandon altogether the idea that population growth may have an effect on deforestation but to carefully identify those conditions that determine "when" and "where" the effect will be weak or strong. As in the case of another environmental outcome such as land degradation examined by Blaikie and Brookfield, deforestation "can occur under rising population pressure, under declining population pressure and without population pressure. . . Population is certainly one factor in the situation, and the present rapid growth of rural populations in many parts of the world makes it, in association with other causes, a critical factor. But 'in association with other causes' is the essential part of that statement, for the other causes themselves can be sufficient" (1987b:34). The analyses presented later will illustrate well the idea of a complex web of causality and justify the cautious position adopted by Blaikie and Brookfield. Indeed, at least four of the five factors identified above appear repeatedly in a large sample of cases of deforestation.

In an effort to establish a solid if provisional base for theorizing, I now turn to the analysis of available studies. I will attempt to make inferences about the strength, direction and significance of the effects of population growth on deforestation and about the type of mediating mechanisms and contingencies (the "other causes") that modify those effects. I will show that the effects of population growth, if not trivial, are somewhat weak and, more important, cannot be easily interpreted in the absence of theorizing about mediations and contingencies. I begin with a meta-analysis of quantitative studies.

Meta-Analysis of the Effects of Population Growth on Deforestation

There are three types of studies to verify, disprove, or inform the relations being analyzed. The first are simulation models. Through complicated relations they enable us to calculate an estimated ultimate effect of either rate of population growth or density on any specified environmental outcome. There are two types of simulation models: those that produce negative results and those that produce positive results. The main problem with these models is that the relations that drive the simulation models are for the most part assumed and not verified. And since there are so many of them, it is difficult to know to what extent the results are consistent with the actual course of events in a geographic area during a well-defined period of time. The fact that slight modifications to the representation of the main relations lead to diametrically opposite results should alert us to the perils of this type of models. Although they are not well suited for the production of sound inferences, simulation models are important tools to understand the implication of causal representations and to guide the collection of relevant information.

The second type of studies are those that use aggregate relations and employ a variable-oriented approach (Ragin 1987). Here the aim is to examine and quantify the relation between variables, some of which are called independent whereas others are dependent. Techniques for combining information or meta-analysis, are tailored to derive statistically sound inferences about the magnitude and significance of the effects of independent variables on dependent variables from a collection of studies.

The third type involves case studies that deal with a complete causal configuration. It has been conjectured that since the case study approach is holistic one can learn more about mediating mechanisms and contingencies, assess better the degree of causal remoteness between two factors, and identify more efficiently the various causal configurations that lead to some outcomes of interest. The problem is that there may be many case studies but no single standardized procedure to synthesize the results. Later, I will apply a procedure based on qualitative algebra to draw causal inferences from samples of case studies.

A Brief Introduction to Meta-Analysis

Techniques for combining information have been around for a long time. In a recent and timely National Academy of Sciences report summarizing some of the most recent developments and the most pressing difficulties, meta-analysis is identified as the name given in social and behavioral sciences to a set of techniques for combining information to summarize the results of a collection of studies (National Research Council 1992:1).

The aim of meta-analysis is to provide "a single set of numbers that describe and summarize the results of independent pieces of research" (Mullen 1989). It consists of a series of statistical techniques designed to convert disparate statistics used in primary level studies to a common metric, to use the transformed statistics to assess the magnitude and direction of effects, and, finally, to test hypotheses about the relation between results at the primary level and selected characteristics of the studies. The unit of analysis in meta-analysis is a primary level hypothesis test rather than a "subject" (individual, community, country). The hypotheses test might be of the form "X exerts a positive effect on Y" and primary level studies designed to verify it will generally yield statistics representing the magnitude and statistical significance of the effects of X on Y. Procedures in meta-analysis are tailored to generate inferences to verify the hypothesis, and the statistical tools are constructed to gauge typical responses (of Y to changes in X), their degree of dispersion, and the association of variability of responses with characteristics of the primary level studies.

In an ideal primary level study of the relation between population growth and rate of deforestation one would seek to estimate the magnitude of the change in deforestation associated with a change in the rate of increase of the population. For simplicity, assume that I have two groups of populations, one with "high" rates of population growth and the other with "low" rates of population growth. A reasonable measure of the effect of the rate of increase would be D, or the difference in the rate of deforestation between the two groups divided by the standard deviation of rates of deforestation. A simple formula can transform this statistic into a coefficient of correlation between the two variables, r, a unit normal variate or a Z_{Fisher} measure. Similarly, if the study yields t statistics, simple expressions can be applied to convert them into the common metric Z_{Fisher}, D, or r. If one were to identify K studies each producing a statistic transformable into the Z_{Fisher}, D, or r metric, one could calculate a summary or combined statistic measuring the magnitude of the effect across all studies as well as a normal variate statistic to test the null hypothesis of no effects. The combined value is then back-transformed into a more interpretable average value, D^* or, alternatively, into an average level of significance, p^*. The latter is the likelihood associated with D^* that the combined results of the K studies would have been obtained under the null hypotheses (of no effects of population growth on the rate of deforestation).

But not all these techniques are equally acceptable. In particular, unlike those relying on measures of effects, those based on pooled z or p values can be misleading since they do not use all the information contained in the separate data sets (National Research Council 1992). For the most part their use continues--and the applications in this article are no exception--since these techniques can be applied with the minimal information that is normally published with the studies. Meta-analysis can be pushed further to answer the following question: is the inter-study variability in the transformed statistics statistically significant? And if so, what are the factors that could explain the variability? This is an extremely useful feature of meta-analysis. In fact, the previous discussion suggests that deforestation is a heterogeneous outcome responding to sharply different activities each of which may be influenced by disparate determinants. Furthermore, population pressure may exert very different effects on these activities. If the K studies do not differentiate type of deforestation but, instead, use the "bundle" indicator described earlier, it is likely that the estimated (transformed) effects of the rate of population growth will differ significantly from each other. Suppose furthermore that population pressure is only relevant for deforestation that involves "extensification" due to clearing by shifting cultivators. It should then be the case that the (transformed) effects differ significantly across studies and, furthermore, that they are systematically associated with the fraction of deforestation attributable to land clearing for swidden agriculture characteristic in each study.

Meta-Analysis of Quantitative Studies

A bibliographic search extending back to 1975 produced eight eligible studies. Of these I kept only four that satisfied the following conditions: (i) they explicitly formulate a model relating deforestation to population pressure with clearly defined variables, (ii) estimate the magnitude and direction of the effects (possibly in the presence of other variables), (iii) provide information on estimated effects, levels of significance, t or p statistics and sample size information.

The search was based on two different principles. I first located several studies where the evidence for/against deforestation was reviewed, then used an ascendancy approach to trace references to published papers that summarized quantitative findings. I complemented this by tracing papers that cited the earlier studies. Second, I identified a set of major journals that devoted published space to related topics and scanned their indices for the period after 1975.

Three remarks about the nature of these studies are necessary. First, some of the studies grouped observations according to geographic location and used different definitions of population pressure. One study estimated the effects of population density as well as of population growth. In yet another, estimates were obtained separately for Latin America, on the one hand, and Africa and Asia on the other. Finally, rural and urban population growth were treated as alternative indicators of population pressure. Although doing so weakens the validity of the independence assumption on which the meta-statistical tests rest, I treated separately the various sets of estimates, as if they were distinct primary hypothesis tests. This implies that there are not four but eight different (but possibly dependent) tests of hypotheses or observations for the meta-analysis. Had I had access to the original data, I could have produced statistics preserving independence while simultaneously accounting for the diversity of indicators for the independent variable. But this was not possible, and it is rarely possible in other examples of meta-analysis. The lack of independence is one of the two conditions--the other will be identified shortly--that leads to underestimates of the uncertainty of the magnitude and direction of the relation between population growth and deforestation.

Second, all studies use a different model specification, e.g., the set of control variables are not identical. I initially ignore these discrepancies but then make full use of them to seek explanations for patterns found in the studies' findings.

Finally, while it is impossible to standardize the procedures of estimation used in each study, one can fully investigate their potential effects on hypotheses verification. For example, if one were concerned with the effects of outliers on the results, one could gauge their effects provided there were studies that dealt with outliers differently. If in all studies the existence and treatment of outliers is simply not mentioned, it is impossible to discern if and to what extent the inclusion of outliers is relevant for inferences.

The main results are displayed in Tables 5.1 and 5.2. In Table 5.1 I summarize the information retrieved from the studies, their characteristics relative to some strategic factors, the transformed statistics, and some basic statistics for comparison. Note that with one exception, in all studies population pressure increases deforestation (the relation is positive) and that in three of them these effects are statistically significant. The statistics for the weighted (by sample sizes) combination of studies show that the effects are statistically significant (combined Z equal to 5.20 with p equal to .00012). This indicates that the results in Table 5.1 are unlikely to obtain if the null hypothesis (of no relation between deforestation and population pressure) were true.

A good metric for the size of the effects is D (weighted average of the D_i for each study with weights equal to sample sizes, N_i): this statistic is equivalent to the difference in the rates of deforestation (expressed in units of standard deviation) that would be observed if the sample were divided into a group with

TABLE 5.1 Basic Results of Meta-Analysis[a]

Study	Statistic	N_i	Sig_i	C_i	P_i	Z_i	p_i	Z_{Fisher}	r^2	D_i	S_i	S_i^2	w_i	w_i^*
1	t=1.48(34)	39	+	1	0	1.45	.074	.25	.07	.51	.34	.12	8.33	.73
2	t=1.79(20)	25	+	1	0	1.70	.044	.39	.14	.80	.45	.20	5.00	.81
3	t=.53(34)	39	+	1	1	.520	.300	.10	.01	.18	.34	.03	33.33	.94
4	t=.04(20)	25	-	1	1	-.04	.510	-.01	-.00	-.02	.50	.25	4.00	.78
5	t=6.20(8)	11	+	0	1	3.63	.000	1.53	.83	4.40	.71	.50	2.00	.65
6	t=5.14(28)	36	+	1	0	4.27	.000	.87	.49	1.94	.38	.14	7.14	.86
7	t=3.98(28)	36	+	1	0	3.51	.000	.70	.36	1.50	.38	.14	7.14	.86
8	t=.00(7)	10	+	0	1	.000	.50	.00	.00	.000[b]	-	-	-	-

Results of Combined Analysis (weighted)[c]:

$Z = 5.20$, $p = .000$ $r^2 = .41$

Fail Safe Number $= 75$ $D = .89$

$Z_{Fisher} = .43$

Results of Fixed and Random Effects Analysis

$d_1 = .71$ $d_2 = .88$ $Q = 61.49$ (6 degrees of freedom)

$\alpha^2_1 = .015$ $\alpha^2_2 = .18$ $\tau^2 = 1.03$

Key to symbols:

i is an index for the study
The numbers in parentheses are the degrees of freedom.
N_i is the number of observations.
Sig_i is the direction of the effects. A " + " indicates that population pressure increases deforestation.

C_i equals 1 if the model uses appropriate controls.
P_i equals 1 if the model uses population growth rather than density.
Z_i is the Z value corresponding to the observed statistic (first column)
p_i is the level of significance for the observed statistic.
r_i^2 is the square of the correlation coefficient.
D_i is the measure of intergroup distance or effect associated with the observed statistic.
Z_{Fisher} is equal to $.5(\ln (1+r)/(1-r))$.
S_i is the estimated standard deviation of the estimated effect.
S_i^2 is the estimated variance of the estimated effect.
$w_i = 1/S_i^2$
$w_i^* = 1/(S_i^2 + \tau^2)$
τ^2 is the estimated between study variances $\quad \max \{0, \dfrac{Q-(n-1)}{\Sigma w_i - \dfrac{\Sigma w_i^2}{\Sigma w_i}}\}$
$d_1 = \Sigma w_i D_i / \Sigma w_i$
$d_2 = \Sigma w_i^* D_i / \Sigma w_i^*$

[a]References for the quantitative studies can be found in the list of references. The identifying numbers are the following: Allen and Barnes 1985; Arcia, Bustamente, and Paguay 1991; Schram 1987; Stonich 1989.
[b]In case 8, the report does not present sufficient statistical information. I have adopted a conservative strategy and imputed at value of .00.
[c]The weights used are the respective sample sizes.

high population pressure and one with low population pressure. Thus, a (weighted) value of $D =. 89$ (see Table 5.1) indicates that the difference in the rates of deforestation between low and high population pressure environments is equivalent to about .89 of a standard deviation, a relatively modest impact indeed. An alternative interpretation is that if two settings (countries, villages, etc.) were randomly selected from the population (of settings) with rapid population growth and the other from the population (of settings) with slow population growth, the former would show deforestation about 75 percent of the time (this is the probability that the a standard normal variate is less than $.89/\sqrt{2}$) (National Research Council 1992).

Table 5.1 also displays d_1, an alternative statistic that like D, is a weighted average of D_i. Unlike the weights applicable for calculating D, the weights to calculate d_1 are the reciprocal of the variances of each estimate standardized by the sum of the reciprocal of the variances (these appear in the columns labelled s_i, s_i^2 and w_i). The value of d_1 in the sample of studies that excludes the last study, which does not include estimates of standard errors, is .71 and its standard error, σ_1, is .123. The corresponding z value is 5.77.

There are two considerations that weaken the inferences one could draw from these findings. First, it can be argued that the validity of these results is highly dependent on the degree of selectivity associated with the studies that were successfully retrieved and that different results would obtain had I selected *all* the quantitative studies carried out during the period of time considered, not just those published. Indeed, it could occur that the studies that were published in more visible journals (or published at all) are precisely drawn from among those that erroneously reject the null hypothesis of no effect. This is the so-called "file drawer problem" (Rosenthal 1979; see also Begg and Berlin 1988). Although there is no straightforward solution for it, one can gauge its extent by calculating the "fail-safe number" or the minimum number of nonselected studies (unpublished or appearing in marginal publications with a low probability of being traced) showing null results that would be needed to overturn the alternative hypothesis of existence of effects. For this study the number is 75, which, though not comfortably large, is enough to provide some confidence in the results of the meta-analysis. Another way to explore the possibility of publication bias is to examine the existence of general patterns of association. Figure 5.2 displays a plot of the estimated size-effects (D) associated with each study and the sample sizes (N). One would expect this figure to resemble an inverted funnel with its center located approximately over the value of D that represents the true effect (Mullen 1989). The plot in the figure does not replicate an inverted funnel shape and instead does suggest the existence of bias against studies with $D < 0$, precisely the type of expectation harbored by a selectivity argument.

The fail-safe number or the funnel plot are neither the only nor the best tools available to determine the existence, magnitude and direction of selectivity. A more compelling strategy requires that I model the actual selection process so that its mechanisms are reflected in the procedure to calculate the statistic(s) to make inferences. A likelihood approach can then be used to estimate parameters and standard errors (National Research Council 1992). The difficulty with this strategy is that it requires that I have a good idea of the selection mechanism in operation. Alternatively, one could test the sensitivity of results to *competing* selection mechanisms. But this requires the utilization of generalized maximum likelihood procedures that are well beyond the scope of this article.

Second, the tests performed above are consistent with the assumptions of a *fixed* effects model. This means that I have assumed that the estimated effects in each study are deviations from a single population parameter and that the differences among them are attributable to random error. If the studies differ systematically in terms of some characteristic inherent to each study, a fixed effects model would systematically understate the degree of uncertainty in the relations between the variables. In this case one must estimate a *random* effects model, which posits two sources of variance for the estimates of effects in each study: a within-study variance component and a between-study variance component. To assess whether or not the observed between-study heterogeneity justifies a random effects model I calculate Cochran Q statistic (Cochran 1937) and use the pertinent chi-square distribution (under the null hypothesis of adequacy of a fixed model): a Q value of 61.49 with 6 degrees of freedom (see Table 5.1) does lend some support to the idea that the studies are heterogeneous and that a random effects model is more appropriate. I then proceed to calculate a weighted estimated effect under a random effects model. The corresponding statistic is d_2 in Table 5.1 which is calculated using w^*_i as weights instead of w_i. The new weights include a component of between-study variance, θ, which is estimated using an estimator suggested by DerSimonian and Laird (1986). The value of d_2 is close to d_1 and so is its variance, σ_2. As a result the z-value associated with the random effects statistic is fairly similar to that obtained with the fixed effects model.

In summary, the results do suggest that the sample of studies contains some bias against the null hypothesis of no relation between deforestation and population pressure.

Are there any patterns in the estimated levels of significance and size of effects? Can one say that some primary hypotheses tests are more likely than others to show up significant and sizable effects? The first conjecture is that studies where the control variables include mediating institutions or factors that are likely to create spurious relations or intermediate mechanisms or contingencies should reveal effects of population growth that are closer to 0. Thus, for example, in studies where there are controls for demand for fuelwood and land clearing due to shifting cultivation (or for suitable proxies), the residual variation in deforestation attributable to population growth *after controlling* for these two mediating factors should shrink significantly. The second conjecture

FIGURE 5.2   Funnel Plot of Effect Size (D).

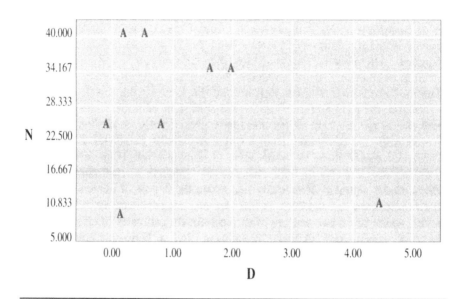

is that when population growth is used as an indicator of population pressure the results should be different from those obtained when using population density.

To explore these conjectures, I proceed in two different stages. In the first stage I calculate statistics that reveal whether or not the variability of results in the various studies is significantly high. Some of this was already accomplished with the estimation of Cochran's Q. The first panel of Table 5.2 displays the results of two additional statistics to assess between-study heterogeneity. Consistent with the results obtained before, the chi-squares values are high and significant indicating the presence of non-trivial heterogeneity in the results. In the second stage I break down the analysis by strata. I use two different stratification principles are used to define the strata. The first refers to the inclusion in the models of controls for potential confounders. The strata are defined as follows: tests with appropriate controls (strata I) and tests without appropriate controls (strata II). The second stratification principle is the type of indicator for population pressure. The strata are as follows: tests using population density (strata III) and tests using population growth (strata IV). Although it would have been desirable to discriminate also among studies according to their treatment of outliers, this was not possible since in none of them was there an explicit treatment of the problem.

The results displayed in the second panel of Table 5.2 show that the expectations are supported by the information. In fact, although the Z statistic is lower in strata II than in strata I, the D statistic is, as expected, much higher. In both cases the Z statistic is significant at less than .001. The results obtained with strata III and IV also behave in accordance with expectations: in strata III the Z value is low and insignificant, whereas in strata IV the Z value is statistically significant. Similarly, the statistic D is about 30 percent higher in strata IV than in strata III. It should be noted that the statistic d_1 instead of D, leads to the same conclusions (results not shown).

TABLE 5.2 Specialized Results of Meta-Analysis

Panel 1: Diffuse Comparisons of Significance and Size of Effects.

a) Chi-Square for Significance Levels	= 20.74	p= .004
b) Chi-Square for Size Effects	= 28.92	p= .000

Panel 2: Focused Comparison Using C and P as Stratifying Criteria.

a) Strata I: with $C = 1$	Strata II: with $C = 0$
Z= 4.66	2.56
p= .000	.001
Z_{Fisher}= .38	.76
r^2= .13	.41
D = .78	1.8

b) Strata III: with $P = 1$	Strata IV: with $P = 0$
Z= 2.05	5.46
p= .019	.000
Z_{Fisher}= .40	.54
r^2 = .15	.25
D = .83	1.15

Panel 3: Imposing a Linear Structure on the Data.

a) Use of C as a predictor:
   Best line for D is : 2.19-1.7378*C and correlation is -.43.
   Z for Size of Effects is .10 and p=.45

b) Use of P as a predictor:
   Best line for D is : 1.18-.05*P and correlation is -.09.
   Z for Size of Effects is .85 and p=.19

The relation between statistics from primary hypothesis tests and presence (absence) of appropriate controls and utilization of growth or density is examined further in the third panel of Table 5.2. There I have calculated the parameters of a linear structure imposed to describe the relation between D values and membership in the strata. As can be seen, the p-values are relatively high, suggesting that the likelihood that the outcomes of the 8 studies varied linearly as a function of the predictors (either presence/absence of suitable control or use of population growth or population density) is quite low.

To summarize: these simple tests show that the primary studies were performed on samples where population pressure exerts a gross positive effect on deforestation. But this finding should be qualified even within the limitations of the procedures used. First, the magnitude of the effects of population pressure is at best modest. Second, although the tests performed were not completely decisive on this point, there is evidence suggesting that the sample of primary hypotheses tests is biased against the null hypothesis. Third, the meta-analysis indicates that, although no linear structure can be invoked, the results change according to the nature of the indicator used for the independent variable and depending on the presence (absence) of suitable controls. This is important since, as is shown later, case studies do confirm that population growth may be related to deforestation but only if some other conditions prevail. In this sense the results of the meta-analysis are consistent with the results to be presented later.

How accurate are these results? There are two considerations of some importance. The first is that the procedures used to test for heterogeneity and to group the studies although useful to illustrate the main ideas are not the most sophisticated that one could employ. Indeed, methods for detecting nonexchange-ability of studies and (when adequate) for pooling them, have moved beyond the frequentist approaches described before. Advances in Bayes and empirical Bayes procedures as well as new techniques within classical approaches to multiple comparisons have been formulated and could be used (National Research Council 1992).

The second consideration of importance is that as I was not able to test the importance of the role played by outliers in primary studies, I am left only with the possibility of tentatively assessing its potential impact. Other studies (see Chapter 6) have suggested that the relation of population and deforestation is unduly influenced by outliers. In particular, it has been claimed that the estimated partial effect of population growth is much lower when conspicuous outliers are removed. If this is uniformly so, two complementary conclusions follow. The first is that the principal results of the meta-analysis should stand *even when the role of outliers is accounted for*. That is, the *magnitude* of the effects of population pressure on deforestation is by all accounts small, and if anything, the analysis should provide a conservative, upper-bound estimate of its effects.

In the absence of heavy measurement errors, the occurrence of outliers can only be explained by misspecification of the main model. Thus if, as I suspect, outliers are indeed present in at least some of the 8 studies, it follows that a focused meta-analysis will reveal that part of the strong heterogeneity found in the primary results is fully attributable to exclusion (inclusion) of outliers. This reasoning leads to a second conclusion, namely, that a meta-analysis carried out with full information on outliers would make more precise the nature of the factors to be included in the model but will not be inconsistent with the finding regarding the presence of underlying contingencies or mediations.

A Qualitative Approach

A quantitative approach that seeks to estimate the magnitude of association between variables is limited by the fact that it cannot shed much light on complex causal connections where the outcomes are functions of multiple joint contingencies and mediations. Recall that I argued that simple linear functional representations are unlikely to suffice as representational devices of causal mechanisms. This is not just an issue that can be solved with better measurement, by adding more variables to an equation, or by invoking nonlinear functions, no matter how complex the equation ends up being. Rather it is a problem rooted in the type of causal complexity that the approach can (or cannot) handle. Derived from families of production functions, the postulated equations are not compelling enough as devices to capture the environment of institutions that play decisive roles in the ultimate effects of population on deforestation (and vice versa).

An alternative approach is to examine the findings of case studies designed to account more fully for the characteristics of actors, or institutions that accompany the occurrence/non-occurrence of an event of interest. The results of several case studies can then be examined to retrieve patterns or sequences of conditions that seem to occur (or be absent) together with an outcome. A tool of analysis is Boolean algebra which has been proposed and systematized by Ragin (1987). In what follows I apply this methodology to between 50 and 55 case studies to disentangle the causal mechanisms that are likely to be involved in processes of deforestation. As will be shown later, findings indicate that although population growth is clearly in the background, it is far from being a very powerful force at all. Instead, the case studies reveal the recurrent presence of a handful of mediations and contingencies that sometimes in isolation from population growth and sometimes jointly with it precipitate instances of accelerated deforestation. The meta-analysis revealed that there are only weak grounds to attribute much causal relevance to population pressure and that there are other factors that could be playing a significant role. What quantitative studies did not reveal (but rather obscured) is the type of mediations and contingencies with the potential to swamp the effects of population growth.

A Brief Introduction to Qualitative Algebra

Most of the empirical research on the relation between population pressure and deforestation is based on case studies: self-contained descriptions of technology and culture and of the social, political, economic, and demographic characteristics that are associated with activities leading to clearing of forests. The studies differ in terms of the units of analysis, the geographic and temporal location, the historical detail and the depth of description and use of quantitative materials. They have in common, hover, a holistic approach to identify the configuration of causes that trigger the outcome of interest. Another shared feature is that, by and large, they rely heavily on measurement strategies on discrete rather than continuous spaces so that constructs are measured as presence (absence) of some characteristics or outcomes.

Assume that one is able to identify 15 case studies of deforestation. Deforestation, D, is operationalized as a dichotomy: it can be rapid, in which case $D = 1$, or slow or nonexistent, in which case $D = 0$. Case studies have uncovered or sought to identify the role of the following three dichotomic factors:

1. Land distribution (L) can be unequal with a plentiful supply of landless peasants, in which case $L = 1$. Alternatively, land might be more or less equally distributed, though pockets of "hacienda" or "fundo" or "latifundio" property and of landless peasants may persist. In this case $L = 0$.
2. Titling policies (T) can be well defined and based on long-term agreements more or less independently of year-to-year yields, sustained on an ample credit supply. In this case $T = 0$. Or, alternatively, it can be ill-defined, based on current exploitation and yields and eroded by a shortage of credit and imperfect credit and capital markets. In this case $T = 1$.
3. Population growth (R) can be high, with high rates of new household formation and high rates of entrance into the labor force ($R = 1$), or it can be mild with a slow rate of family formation and of entrance into the labor force ($R = 0$).

A study of 15 hypothetical case studies, each associated with one of the 8, (2^3), possible causal configurations, reveals the following patterns:

L	T	R	D	Frequency (No. of Studies)
1	0	1	1	3
0	1	0	1	2
1	1	0	1	1
1	1	1	1	2
1	0	0	0	1
0	0	1	0	1
0	1	1	0	3
0	0	0	0	2

According to the first row, in three studies deforestation occurred in the presence of land inequality, well-defined titling and high population growth; according to the second row, in two case studies deforestation occurred in the absence of land inequality and population growth but when titling was ill-defined, and so on.

A few remarks about the hypothetical "data base" are in order. First, note that although the case studies illustrate all possible combinations of conditions, this is not necessary and, moreover, it will seldom occur. By and large, the diversity of empirical realizations is less than the diversity generated by all possible combinations of configurations. Second, the hypothetical case studies include some that reveal deforestation (D =1) and also others where the outcome does not occur (D = 0). Further, without exception, each configuration is unequivocally associated with one and only one outcome whereas, in practice, each configuration could yield a distribution of cases, some with D=0 and others with D=1. As I show later, this is an important factor that influences the strength of causal inferences. Third, all outcomes are clear-cut: there is no case where the outcome of interest (D) is questionable. This is a radical simplification for, as should be evident from the measurement complications discussed before, it will not be uncommon to find empirical instances where the true outcome is difficult to assess.

To avoid excessive cluttering I adopt the following notational convention: if any condition, say condition L, attains value 1, I use the capital letter that identifies the condition, L. On the other hand, if L=0 then I symbolize this event by using the lower-case letter l. In addition is the symbol "^" for the Boolean operator for intersection of sets ("and") and " + " to symbolize the Boolean operator for union of sets ("or"). Thus, for example, the Boolean expression $\{<L^R> + <T^r>\}$ indicates the occurrence of two sets of conditions: land inequality (L) and high population pressure (R) *or* ill-defined property titling (T) in the absence of population pressure (r). Finally, the Boolean expression $\{D=<L^R>+<T^r>\}$ indicates that deforestation occurs, (D), either in the presence of land inequality *and* high population pressure *or* when no population pressure is accompanied by ill-defined titling of property.

Very simple principles of reduction applied to the table of data can be used to determine prime implicants and then to obtain a parsimonious expression for the outcome of interest, namely D. These principles *do not take into account* the frequency with which each combination of conditions occurs, only their empirical occurrence; one occurrence is as potent to establish a combination of conditions as five may be. In the hypothetical example described above the most parsimonious description of the conditions leading to D is as follows:

$$\{D = L^{\wedge}R + T^{\wedge}r\} \qquad (3)$$

According to (3), population pressure is neither a necessary nor a sufficient condition for deforestation to occur: population growth only matters *if it occurs in conjunction with* land inequality. Instead, distorted titling legal codes and policies lead to deforestation even in the absence of population pressures of any sort.

Analysis of Case Studies of Deforestation

Sampling Principles. The search for case studies to perform qualitative analysis followed the same principles stated before for the meta-analysis. First I located review studies and then applied the ascendant and descendant methods. In a second stage I used citations that appeared in the more prominent journals in the area. In all I was able to gather 55 studies. Then three exclusionary principles were systematically applied:

1. First, were excluded studies that were purely descriptive, too general, or only cursory summaries of quantitative information,
2. Also excluded were all studies that were multiple reportings of the same main study but appeared in different publications. In these cases an attempt was made to maximize the amount of information supplied in the various versions of the study,
3. Finally, I excluded studies where the extent of deforestation or population growth was assessed only informally, using reports from others or what appear to be uncorroborated guesses of the investigators.

The 38 studies that remained in the sample were all in-depth analyses of one or a few countries, areas, or villages/districts. There is a fair representation of all the developing regions where deforestation seemingly is or could become a problem. In all these studies attempts were made to assess the extent of deforestation using either official statistics based on topographical maps or special surveys, aerial photography, or satellite imagery. The magnitude of population pressures (growth and density) was measured from official statistics.

Most of the studies identify other causal factors that appear to trigger rapid deforestation. However, in only a handful of studies is there a precise assessment of the nature of *all* these factors. In most studies there is a blend of factors that are brought to bear, some of which are merely identified as being present (or absent) from examination of secondary sources, interviews, or direct observation by the researcher whereas others are more precisely characterized.

The Pitfalls of the Sample of Case Studies. The final sample of case studies is prone to some of the same errors that were described earlier and to some new ones as well. In particular, unidentified studies could disproportionately represent instances of only one type of relation between the various causal factors and deforestation. Unlike the situation faced in meta-analysis, however, there is no well-defined procedure to gauge the extent of the problem. As indicated below, it is likely that the sample is somewhat biased but in ways that are impossible to correct with the tools available. The only solution to this problem is to formulate a study from scratch where deforestation and nondeforestation are possible outcomes of a set of causal conditions of interest.

Unlike case studies dealing with revolutions or strikes, where the occurrence of the event is never in doubt, case studies of deforestation begin on an ambiguous note: the degree or even the existence of deforestation is not always well established. An important consequence of this ambiguity is that one may obscure outcome differences that could be revealing of the types of human activities and social institutions that are involved. An analogous but distinct ambiguity is related to what is referred to as the intensity of deforestation. As discussed before, different *types* of deforestation cause different degrees of damage to the ecosystem: the wholesale destruction of a tropical forest to gather a few cubic meters of precious roundwood cannot be assigned the same status as equivalent areas (volume) deforested through the collection of fuelwood for the simple reason that their implications for sustainability and regeneration are totally different. In virtually no study was there a direct attempt to qualify the intensity of deforestation. To mitigate somewhat the effects of this uncertainty I classified the studies according to the *dominant* type of activity implicated in the clearing of forests. Although admittedly the correlation between long-term environmental implications and type of activities leading to deforestation is far from tight, explicit control for the latter should reduce potential biases. Table 5.3 contains information on the immediate causes of deforestation and, in addition, on the geographic area of the study and the source used for the assessment of deforestation.

An additional difficulty presented by case studies is the recurrent presence of ambiguities in what can be referred to as time and space lags. Although the timing of events may be unquestionable, the precise lags are rarely well defined. For example, if titling policies are convincingly implicated in the acceleration of the rate of deforestation one will not always know from the case studies themselves what was the time elapsed between implementation of the policies

TABLE 5.3 Summary of Selected Characteristics of Case Studies[a]

Characteristics	Frequency (Number of Studies)
a. Region/Area:	
Himalayas (Nepal)	6
Central America	2
Brazil (Amazons)	5
Sub-Saharan Africa	8
Andean Regions	2
Philippines, Indonesia and Thailand	9
Others (Thailand, China, India, Pakistan and Haiti)	6
b. Assessment of deforestation	
Aerial Surveys	2
Topographical Maps-Surveys	17
Landsat	8
Not mentioned	11
c. Types of deforestation	
D(1)	35
D(2)	5
D(3)	18
D(4)	11

[a]The sources for the case studies appear in the list of references under the following: Anderson 1987; Bautista 1990; Bee 1987; Collins 1986; Collins & Pointer 1986; Denevan 1982; Donner 1987; Eckholm 1976; Eder 1990; FAO 1986a; Fleming 1983; Ghildyal 1979; Hall 1989; Hallsworth 1982; Hosier 1988; International Institute for Environment and Development & World Resources Institute 1986; Ives and Messerli 1989; Kleinert 1987; Lele and Stone 1989; Lundahl 1979; Lundgren and Lundgren 1982; Martens 1983; McNeil 1988; McNicoll 1990; Mullen 1989; Postel 1984; Smil 1983; Southgate, Sieva and Brown 1983; Thomson 1988; Tucker 1988; Westoby 1989; Whitney 1987; Williams 1989.

and the initiation of the acceleration. "Space lags" are even more difficult to pin down. They occur when a local outcome, such as clearing of dense tropical forests, follows as a consequence of events that take place elsewhere, such as excess growth of the labor force in some urban areas. Naturally, these two ambiguities not only surround in-depth case studies but are also responsible for weaknesses in more conventional cross-sectional analyses.

Perhaps the most troublesome feature of the case studies is that all of them identify an instance of deforestation (or D=1). I did not include case studies examining a case of no deforestation (D=0). To be sure, there was some variability in the "gravity" of forest clearing but, for the most part, the distinction was qualitative and informal and made no attempt to use it. The result is a set of observations that have been seemingly sampled on the dependent variable so that only those with D=1 are included. I simply do not know the extent to which the same combination of conditions that yields a "positive" (D=1 or deforestation) outcome also leads, in other contexts, with other units of observation, to a "negative" (D=0 or no deforestation) outcome. This limits the usefulness of the inferences since I will not be in a position to provide solid tests of internal validity. Strictly speaking, Boolean expressions for "positive" outcomes could be manipulated and translated into Boolean expressions for the "negative" outcomes. And unless one is able to observe the realization of both positive and negative outcomes the exact boundaries within which the inferences apply cannot be assessed. The results will hold true only if the (unaccounted) cases of "negative" outcomes represent a lower proportion than those of "positive" outcomes for *each* combination of conditions that contributes to the formulation of causal statements. To limit the damage that this problem creates, I use a series of experiments to establish inferences with minimum associated errors.

Data Reduction. Before attempting to draw inferences from the case studies in the sample, I identify a set of conditions to which in alternative combinations, are attributed important influences on the likelihood of deforestation. What follows is a brief description of these conditions. In the quest toward simplification I have reduced the set so that each of the elements contained in it can attain only two values depending on whether it is present or absent. Although this may do injustice to the richness of the case studies, it is an almost unavoidable simplification to enhance parsimony while simultaneously illustrating the application of the procedure.

Population pressure (R). Attains a value 1 if there is evidence of rapid population growth or high population density in urban or rural areas or both and 0 otherwise.

Unequal distribution of land (L). L attains a value of 1 if there is evidence of a highly skewed distribution of land for cultivation and 0 otherwise. Excessive land fragmentation and the resulting decreasing yields and sheer landlessness are at the root of shifting cultivation that progresses at the expense of forests or that literally follows the tracks left by the logging industry into low fertility soils. Landlessness and fragmentation are partly the outcome of differential population growth but they are also the direct result of preservation of land tenure regimes biased toward inefficient latifundia and large cattle ranches.

Unequal access to credit, technology and markets (C). C is present in social contexts where small proprietors have little or no access to credit for investment in factors of production to increase yields (fertilizers, pesticides), or when they have difficult or no access to technological innovations such as irrigation and mechanization of operations, or, finally, when commercialization of the products is restricted by lack of control over transportation, storage, and pricing mechanisms. If at least one of the three conditions is satisfied then C = 1. Otherwise C attains a value of 0.

Titling policies (T). The variable T attains the value 1 when there is evidence of allocation of property to small cultivators under rules or codes that prevent direct appropriation of the land and its full use in commercial transactions. Under these conditions title over property may be gained only after direct and extensive use (clearing), but it may not include provisions to enable the cultivator to offer it as collateral. Thus, while the legal proprietor may have rights of use, he has no claim to rights of exchange. Faulty titling policies have been blamed for eroding incentives to apply proper conservation management and for actually promoting (making rational) the indiscriminate exploitation of the natural resource base.

Biased government policies (G). This is a complex dimension under which are included several conditions that, in very thorough studies, are kept strictly separate (e.g., see Repetto 1988; Repetto and Gillis 1988). G attains a value of 1 if one of the following conditions exist:

1. central government policies of rent recovery, royalties, and tax incentives that grossly undervalue the current and future discounted value of forest assets;

2. central government policies granting subsidies and tax exemptions for the exploitation of forests for production geared to export markets. These subsidies and tax exemptions make profitable for individual proprietors operations that are socially unprofitable;

3. inappropriate investments or incentives for the design and implementation of proper management for conservation;

4. misguided settlement policies that reshuffle the labor force with no or only scarce provision for the socially rational exploitation of forest lands.

Admittedly, this is a highly stylized "reduction" of conditions that are invariably more complex and difficult to disentangle. Take as a first example the case of government policies. The dimensions included here do not distinguish between two very different though connected types of actions.

Actions in the first type operate through juridical interference of central governments that transform a common property resource into a public resource and in so doing inevitably undermine traditional rights over forest resources, consequently enhancing problems of open access to common property. The second type of action is flawed valuation of resources that precedes or accompanies the implementation of investment incentives, credit concessions, and tax and royalty provisions designed to generate immediate material gains in the form of tangible returns or by alleviating pressure over land.

A second example is the case of unequal distribution of land. Cases of inequality may differ widely. Even if one were able to standardize for the degree of land inequality, there remains the possibility that fertility of the soil, predominant type of production, optimal factor inputs combination, and division of labor will accentuate or attenuate the effects of skewed property distribution. As suggested before, community and household-level mechanisms, possibly including the role of women and children, may serve to temporarily alleviate the consequences of population pressure or, equivalently, of land fragmentation in highly unfair and inflexible property regimes.

Furthermore, the logical status of these conditions is not identical. Some are mediating factors (for example L) whereas others operate as contingencies (for example T). Similarly, the verification of the presence (absence) of any one of them from examination of case studies reports is seldom straightforward. Two factors that may appear to be identical in several studies may be, in reality, sufficiently different to require different categories. But although the conceptualization may not be as rigorous as one would like, it will suffice if the aim is only to highlight the role that these factors play in the production of the outcome of interest (deforestation) rather than producing a correct explanation for it.

Finally, an additional data reduction issue needs to be discussed and clarified. To be consistent with the previous discussion I should introduce a distinction between types of deforestation. Indeed, the most immediate activities that lead to the clearing of forest not only have different impact on the ecosystem but are themselves the results of heterogeneous processes. In keeping with this conjecture, to which I assigned a prominent role in our previous discussion, I classify the activities into four types: clearing by landless peasants and small landholders for subsistence production or for monocultivation (cash crops for export markets); clearing for the operation of large cattle ranches or for extensive cultivation of one or two products; clearing for fuelwood and charcoal manufacturing; and clearing for logging. These will be referred to as D(1), D(2), D(3) and D(4) respectively. The descriptions contained in each case study contain remarkable imprecision on this issue. With a few exceptions, deforestation is always attributed to a combination of two or more deforestation

TABLE 5.4 Distribution of Conditions for Sample of Case studies

Cases	*Conditions*					*Outcomes and Frequencies*			
	R	L	C	G	T	D(1)	D(2)	D(3)	D(4)
I	1	0	0	0	0	1(14)	1(0)	1(11)	1(0)
II	1	1	0	0	0	1(6)	1(3)	1(2)	1(1)
III	1	0	0	1	0	1(6)	1(0)	1(2)	1(3)
IV	1	0	0	1	1	1(1)	1(0)	1(1)	1(0)
V	1	1	1	0	0	1(1)	1(0)	1(1)	1(1)
VI	1	1	1	0	1	1(2)	1(0)	1(2)	1(0)
VII	1	1	0	1	0	1(2)	1(0)	1(0)	1(2)
VIII	1	1	1	1	0	1(1)	1(0)	1(1)	1(1)
IX	0	0	0	0	1	1(1)	1(0)	1(1)	1(1)

Note: The numbers in parentheses are the frequencies of cases for the corresponding outcome. See text for a definition of symbols.

activities, but without ever assigning prominence to any one of them. If my conjecture is correct, this ambiguity in the description of the outcome can only obscure the causal analysis.

Deforestation as the Outcome of Complex Sets of Causal Factors. Table 5.4 displays the results of reducing the data from the case studies into a form suitable for the application of qualitative algebra.

Applications of simple rules of Boolean algebra lead to the following causal statements describing each of the outcomes:

$$\{D(1) = <G^{\wedge}(l^{\wedge}c^{\wedge}t^{\wedge}r)> + <R^{\wedge}(c^{\wedge}g^{\wedge}t)> + <(R^{\wedge}G)^{\wedge}(l^{\wedge}c)> +$$
$$<(R^{\wedge}L^{\wedge}C)^{\wedge}(g)> + <(R^{\wedge}L^{\wedge}G)^{\wedge}(t)>\} \quad (4a)$$
$$\{D(2) = <(R^{\wedge}L)^{\wedge}(c^{\wedge}g^{\wedge}t)> + <G^{\wedge}(l^{\wedge}c^{\wedge}t^{\wedge}r)>\} \quad (4b)$$
$$\{D(3) = <R^{\wedge}(c^{\wedge}g^{\wedge}t)> + <(R^{\wedge}G)^{\wedge}(lc)> + <(R^{\wedge}L^{\wedge}C)^{\wedge}(g)>\} \quad (4c)$$
$$\{D(4) = <(R^{\wedge}L)^{\wedge}(c^{\wedge}t)> + <(R^{\wedge}G)^{\wedge}(l^{\wedge}c^{\wedge}t)> + <G^{\wedge}(r^{\wedge}l^{\wedge}c)>\} \quad (4d)$$

The first feature of the expressions that deserves comment is that D(1), D(2), and D(4) are outcomes that can obtain *in the absence* of population growth. In the other case, D(3), deforestation only occurs if there is detectable population pressure. This should not be surprising since, as remarked before, the relation between fuelwood consumption and population growth is perhaps more straightforward and direct than all others.

The second feature is that population pressure alone (in the absence of the other factors) can by itself produce only outcomes D(1) and D(3). In all the other cases population pressure must act in conjunction with either L, T, or G to generate deforestation. The third feature is that, as hypothesized, the conjunction of causal mechanisms leading to deforestation is not identical across outcomes and that a single causal representation is insufficient to capture all the processes. Note for example that government policies can by themselves generate D(4), even in the absence of any of the other conditions, but that this is insufficient to generate other outcomes.

Finally, in addition to R the two pivotal conditions across outcomes are government policies (G) and land inequality (L). Titling policies do not appear to be necessary when other conditions are present and access to credit seems to play a minor role throughout.

Flaws of the Analysis: Multiple Outcomes and the Occurrence of Nondeforestation. A potentially serious problem is that the case studies did not establish a one-to-one correspondence between types of deforestation and sets of causal conditions. Thus, in those cases where more than one type of outcome was detected, the explanatory argument was focused on the dominant type of outcome only, not on the secondary ones. Instead, Table 5.4 was constructed including *all* outcomes detected in the study and associating them with the causal configuration invoked to explain the dominant outcome, if no other was given. The consequence of this is that Table 5.4 may *overrepresent* the causal configuration for a particular outcome. Thus, for example, the configuration of conditions for D(4) may be overdetermined in the sense that some of them are included merely because D(4) occurred simultaneously with, say, D(1), not because they are necessary for D(4) to occur. To check on the effects of this reconstruction, I prepared a different table of truth in which only dominant outcomes were represented. The results of the analysis of that table are only trivially different from the ones already presented and are not examined further.

What about the impact of cases of non-deforestation? Discussion of this issue begins with a clarification. The problem is generated not by excluding cases where the outcome is reforestation of degraded territories. Indeed, reforestation belongs to a different class of phenomena that requires an idiosyncratic explanation; it is simply not an outcome *alternative* to deforestation. The difficulty is a subtler one since it is rooted in a possibly nonempty set of cases where a particular causal configuration in Table 5.4 *does not* lead to deforestation. Suppose, for example, that through an experimental design or through detailed historical investigation one ascertains that the configuration represented by Case I in Table 5.4 is associated not only with 14 cases of D(1)=1 but with 14 cases of D(1)=0. Leaving aside for the moment the fact that the endeavor that would lead to such finding may be impossible to undertake, one may well ask what would be the outcome associated with the causal configuration? Perhaps, the simplest solution is to neglect it, since its probable outcome is questionable. Less problematic would be the case in which the same difficult endeavor unearthed not 14 but only 2 cases of nondeforestation. On purely probabilistic grounds, one would be on safe ground associating to Case I the response D(1) =1.

It follows that Table 5.4 leads to the selection of explanations that may be highly sensitive to the occurrence of nondeforestation. This is because one implicitly assumes that there is no causal configuration where the frequency of "negative" outcome (nondeforestation) is larger than or equal to the frequency of cases of deforestation. To assess the nature of the potential bias I repeated the analysis using only those causal configurations in which there were more than 10 combined cases of deforestation and excluded those that were represented by a very small number of cases. To make the exercise feasible, I disregard the distinction between the various outcomes and consider only a generic one, namely D. Only the first three causal configurations in Table 5.4 are coded $D=1$; all the others are coded $D=0$. The explanation inferred from the new coding scheme lends more emphasis to the role of population growth in that it supports the idea that deforestation ($D=1$) does not occur in its absence, although its presence does not always lead to the outcome. The explanation, however, is much less supportive of or illuminating about the role of other factors.

Are There Alternative Strategies to Draw Inferences from Qualitative Case Studies? Each case study could be conceived as an elaboration of an "expert opinion" on the relations between population growth and deforestation. If so, the outcomes associated with each case study can, in principle at least, be studied with a conventional approach, such as the Delphi method, or with less conventional approaches involving Bayesian and non-Bayesian methods (National Research Council 1992). But in addition to questions arising about the actual applicability of these procedures to the case studies collected, it is unclear how they could be applied with ease to studies ("or expert opinions") that are essentially nonquantitative. One of the advantages of the Boolean algebra approach is that it is naturally adaptable to inherently qualitative evaluations. It remains to be explored under what conditions more quantitative approaches involving Bayesian procedures could be applicable to qualitative case studies.

Summary and Conclusion

The application of two relatively new procedures to the analysis of the relation between deforestation and population pressure reveals that while population pressure is an important force leading to deforestation, it rarely acts alone to produce the outcome. Other determinants appear to be necessary as mediation and contingencies for population growth (or density) to have a discernible impact.

The quantitative analysis suggested that even if the effects of population growth are statistically significant, their magnitude is quite modest. And this is a conservative conclusion for the sample of studies examined is likely to be biased against the null hypothesis of no relation. In addition, none of the studies included in the sample incorporated factors that the qualitative analysis showed to be strategic, and none of them supplied adequate information about outliers. The qualitative analysis made possible the identification of important social institutions that create an environment where population pressure may or may not affect deforestation or, alternatively, through which population pressure could affect the rate of deforestation. The very tentative analysis I performed suggests the usefulness of separating types of deforestation and points to the importance of government policies, land distribution and access to credit and technology. But this type of analysis has important limitations, particularly with regard to the inability to observe "negative" outcomes and their associated causal configurations.

Meta-analysis is limited by the power of the quantitative studies that yield the hypothesis tests: if simple quantitative models are not fit to represent the relations, then meta-analysis will produce little of value. Similarly, the application of qualitative algebra is only as powerful as the case studies themselves: imprecise characterization of outcomes, sloppy conceptualization and measurement of conditions, and poor identification of factors will, in one way or another, affect the set of propositions that can be derived. This suggests that better-designed quantitative and qualitative studies incorporating some of the factors identified here are necessary before meta-analysis and qualitative algebra could yield optimal results.

References

Allen, J.C. and Barnes, D. (1985). The Causes of Deforestation in Developing Countries. *Annals of the Association of American Geographers* 75(2):163-184.

Anderson, D. (1987). *The Economics of Afforestation. A Case Study in Africa.* Baltimore: The Johns Hopkins University Press.

Arcia, G., Bustamante, G. and Paguay, J. (1991). URBIQUITO: Un analysis de la relacion entre poblacion y medio ambiente. Quito: Direccion de Planificacion, I. Municipio de Quito.

Barbier, E.B., Burguess, J.C. and Markandya, A. (1990). The Economics of Tropical Deforestation, *Ambio* 20(2):55-58.

Barrows, C.J. (1991). Land Degradation. *Development and Breakdown of Terrestrial Environments.* Cambridge: Cambridge University Press.

Bautista, G. M. (1990). The Forestry Crisis in the Philippines: Nature, Causes, and Issues. *The Developing Economies* 28(1):67-94.

Bee, O.J. (1987). Depletion of the Forest Resources in Philippines. Field Report Series No. 18. Manila: Institute of Southeast Asian Studies.

Begg, C.B. and Berlin, J.A. 1988. "Publication Bias: A Problem in Interpreting Medical Data (with discussion)," *Journal of the Royal Statistical Society*, Series A, 151:419-463.

Biswas, A.K. (1986). Land Use in Africa. *Land Use Policy* 3(4):247-259.

Blaikie, P. and Brookfield, H. (1987a). Questions from History in the Mediterranean and Western Europe. In *Land Degradation and Society*, eds. P. Blaikie and H. Brookfield, London and New York: Methuen.

Blaikie, P. and Brookfield, H. (1987b). Approaches to the Study of Land Degradation. In *Land Degradation and Society*, eds. P.Blaikie and H. Brookfield. London and New York: Methuen.

Bowonder, B. (1985). Deforestation in Developing Countries, *Journal of Environmental Systems* 15 (2):171-191.

Bremer, J. (1984). Fragile Lands: A Theme Paper on Problems, Issues, and Approaches for Development of Humid Tropical Lowlands and Steep Slopes in the Latin America Region. Washington D.C.: Development Alternatives, Inc.

Budowski, G. (1982). Socio-Economic Effects of Forest Management on the Lives of People Living in the Area: The Case of Central America and Some Caribbean Countries. In *Socio-Economic Effects and Constraints in Tropical Forest Management*, ed. E.G. Hallsworth. New York: John Wiley and Sons, Ltd.

Chattopadhyay, S. (1985). Deforestation in Parts of Eastern Ghats Region (Kerala), India, *Journal of Environmental Management* 20:219-230.

Cochran, W.G. (1936). "Problems Arising in the Analysis of a Series of Similar Experiments," *Journal of the Royal Statistical Society*, Supplement 4:102-118.

Collins, J.L. (1986). Smallholder Settlement of Tropical South America: The Social Causes of Ecological Destruction, *Human Organization* 45(1):1-10.

Collins, J.L. and Painter, M. (1986). Settlement and Deforestation in Central America: A Discussion of Development Issues. Working Paper 31. Washington D.C.: Cooperative Agreement on Human Settlement and Natural Resources System Analysis.

Denevan, W.M. (1982). Causes of Deforestation and Forest and Woodland Degradation in Tropical Latin America. Washington D.C.: U.S. Office of Technology Assessment.

DerSimonian, R. and Laird, N.M. (1986). "Meta-analysis in Clinical Trials," *Controlled Clinical Trials*, Vol 7:177-188.

Diaz, C.P. (1982). Socio-Economic Thrusts in an Integrated Forest Management System: The Philippine Case. In *Socio-Economic Effects and Constraints in Tropical Forest Management*, ed. E.G. Hallsworth. New York: John Wiley and Sons, Ltd.

Donner, W. (1987). *Land Use and Environment in Indonesia*. Honolulu: University of Hawaii Press.

Eckholm, E. (1976). *Losing Ground. Environmental Stress and World Food Prospects*. New York: W.W. Norton and Company, Inc.

Eckholm, E. (1979). Planting for the Future: Forestry and Human Needs. Worldwatch Paper 26. Washington D.C.: Worldwatch Institute.

Eder, J.F. (1990). Deforestation and Detribalization in the Philippines: The Palawan Case. *Population and Environment* 12(2):99-115.

FAO (1986a). Natural Resources and the Human Environment for Food and Agriculture in Africa. Environment and Energy Paper 6. Rome: FAO.

FAO (1986b). *African Agriculture: The Next 25 Years*. Annex II. The Land Resource Base. Rome: FAO.

Feeny, D. (1988). *Agricultural Expansion and Forest Depletion in Thailand, 1900-1975.* In World Deforestation in the 20th Century, eds. J.F. Richards and R.P. Tucker. Durham, NC: Duke University Press.

Fleming, W.M. (1983). Phewa Tal Catchment Management Program: Benefits and Costs of Forestry and Soil Conservation in Nepal. In *Forest and Watershed Development and Conservation in Asia and the Pacific,* ed. L.S. Hamilton. Boulder, CO: Westview Press.

Ghildyal, U.C. (1979). The People and the Forests. In *Man and the Forests in India,* eds. K. M. Gupta and D. Bandhu. New Delhi: Today and Tomorrow Printers and Publishers.

Ghosh, R.C. (1982). Socio-Economic Effects and Constraints in Forest Management. In *Socio-Economic Effects and Constraints in Tropical Forest Management,* ed. E.G. Hallsworth. New York: John Wiley and Sons, Ltd.

Goucher, C.L. (1988). The Impact of German Colonial Rule on the Forests of Togo. In *World Deforestation in the 20th Century,* eds. J.F. Richards and R.P. Tucker. Durham: Duke University Press.

Hall, A.L. (1989). *Developing Amazonia. Deforestation and Social Conflict in Brazil's Carajas Programme.* Manchester: Manchester University Press.

Hallsworth, E.G. (1982). The Human Ecology of Tropical Forest. In *Socio-Economic Effects and Constraints in Tropical Forest Management,* ed. E.G. Hallsworth. New York: John Wiley and Sons, Ltd.

Hosier, R.H. (1988). The Economics of Deforestation in Eastern Africa, *Economic Geography* 64(1):121-136.

International Institute for Environment and Development and World Resources Institute (1986). *World Resources 1986.* New York: Basic Books.

Ives, J.D. (1987). The Theory of Himalayan Environmental Degradation: Its Validity and Application Challenged by Recent Research, *Mountain Research and Development* 7:189-199.

Ives, J.D. (1988). Development in the Face of Uncertainty. In *Deforestation: Social Dynamic in Watershed and Mountain Ecosystem,* eds. J. Ives and D.C. Pitt. London: Routledge.

Ives, J. and Messerli, B. (1989). *The Himalayan Dilemma. Reconciling Development and Conservation.* London: Routledge.

Kaul, S. K. (1979). Human Aspects of Forest Development. In *Man and the Forests in India,* eds. K. M. Gupta and D. Bandhu. New Delhi: Today and Tomorrow Printers and Publishers.

Khattak, G.M. (1983). The Watershed Management Program in Mansehra, Pakistan. In *Forest and Watershed Development and Conservation in Asia and the Pacific,* ed. L.S. Hamilton. Boulder: Westview Press.

Kleinert, C. (1987). Settlement Pressure and the Destruction of the Forest in Rwanda (Eastern Central Africa), *Applied Geography and Development* 29:93-105.

Lele, U. and Stone, S.W. (1989). Population Pressure, the Environment and Agricultural Intensification. Variations on the Boserup Hypothesis. MADIA Paper 4. Washington D.C.: The World Bank.

Lundahl, M. (1979). *Peasants and Poverty. A Study of Haiti.* (Chapter 5: Erosion.) New York: St. Martin's Press.

Lundgren, B. and Lundgren, L. (1982). Socio-Economic Effects and Constraints in Forest Management: Tanzania. In *Socio-Economic Effects and Constraints in Tropical Forest Management*, ed. E.G. Hallsworth. New York: John Wiley and Sons,Ltd.

Martens, J. (1983). Forests and Their Destruction in the Himalayas of Nepal. Miscellaneous Paper 35. Kathmandu: Nepal Research Centre.

McNeil, J.R. (1988). Deforestation in the Araucaria Zone of Southern Brazil, 1900-1983. In *World Deforestation in the 20th Century*, eds. J.F. Richards and R.P. Tucker. Durham, NC: Duke University Press.

McNicoll, G. (1990). Social Organization and Ecological Stability under Demographic Stress. Working Paper, Center for Policy Studies, The Population Council.

Morgan, W. B. and Moss, R.P. (1972). Savanna and Forest in Eastern Nigeria. In *People and Land in Africa South of the Sahara*, ed. M. Prothero. New York: Oxford University Press.

Mullen, B. (1989). *Advanced BASIC Meta-Analysis*. Hove and London: Lawrence Erlbaum Associates.

Myers, N. (1980). Deforestation in the Tropics: Who Gains, Who Loses. In *Where Have All the Flowers Gone? Deforestation in the Third World*, ed. Department of Anthropology. Williamsburg: College of William and Mary.

National Research Council (1992). *Combining Information: Statistical Issues and Opportunities for Research*. Wahsington, DC: National Academy Press.

Osenobo, G.J. (1988). The Human Causes of Forest Depletion in Nigeria, *Environmental Conservation*, Vol 15 (1):17-28.

Postel, S. (1984). Protecting Forests. In *State of the World 1984*, ed. Worldwatch Institute. New York: W.W. Norton and Company.

Ragin, Ch. (1987). *The Comparative Method: Moving Beyond Qualitative and Quantitative Strategies*. Berkeley: University of California Press.

Regional Office for Asia and the Pacific (RAPA) and FAO (1989). *Environmental Problems Affecting Agriculture in Asia and Pacific Region (World Food Symposium)*. Rome: FAO.

Repetto, R. and T. Holmes (1983). The Role of Population in Resource Depletion, *Population and Development Review* 9(4):609-632.

Repetto, R. and M. Gillis (1988). *Public Policies and the Misuse of Forest Resources*. Cambridge: Cambridge University Press.

Repetto, R. (1988). *The Forest for the Trees? Government Policies and the Misuse of Forest Resources*. Washington, D.C.: World Resources Institute.

Rosenthal, R. (1979). The 'File Drawer Problem' and Tolerance for Null Results, *Psychological Bulletin* 86:638-641.

Rudel, T.K. (1989). Population, Development, and Tropical Deforestation: A Cross-National Study, *Rural Sociology* 54(3):327-338.

Schram, G. (1987). Managing Urban/Industrial Woodfuel Supply and Demand in Africa. *The Annals of Regional Science*, Vol 21(3):60-79.

Singh, J.S., V. Pandey and A.K. Tiwary (1984). Man and Forests: A Central Himalayan Case Study, *Ambio*, Vol 13(2):80-87.

Smil, V. (1983). Deforestation in China, *Ambio xii* (5):226-231.

Southgate, D., Sierra, R. and Brown, L. (1989). *The Causes of Tropical Deforestation in Ecuador: A Statistical Analysis.* Ohio State University.

Stonich, S. (1989). The Dynamics of Social Processes and Environmental Destruction: A Central American Case Study, *Population and Development Review* 15 (2):269-296.

Sutlive, V. H., Jr. (1980). Introduction. In *Where Have All the Flowers Gone? Deforestation in the Third World*, ed. Department of Anthropology. Williamsburg: College of William and Mary.

Thiesenhusen, W.C. (1991). Implications of the Rural Land Tenure System for the Environmental Debate: Three Scenarios, *Journal of Developing Areas* 26(Oct):1-24.

Thompson, M. and Warburton, M. (1988). Uncertainty on the Himalayan Scale. In *Deforestation: Social Dynamic in Watershed and Mountain Ecosystem*, eds. J. Ives and D.C. Pitt. London: Routledge.

Thomson, J.T. (1988). Deforestation and Desertification in Twentieth-Century Arid Sahalian Africa. In *World Deforestation in the 20th Century*, eds. J.F. Richards and R.P. Tucker. Durham: Duke University Press.

Tucker, R.P. (1988). The British Empire and India's Forest Resources: The Timberland of Kumoan and Assam, 1914-1950. In *World Deforestation in the 20th Century*, eds. J.F. Richards and R.P. Tucker. Durham: Duke University Press.

Westoby, J. (1989). *Introduction to World Forestry.* London: Basil Blackwell.

Whitney, J.B.R. (1987). Impact of Fuelwood Use on Environmental Degradation in the Sudan. In *Lands at Risk in the Third World: Local-Level Perspectives*, eds. P.D. Little and M.M. Horowitz, with A.E. Nyerges. Boulder and London: Westview Press.

Whitlow, J.R. (1980). Deforestation in Zimbabwe: Problems and Prospects. Supplement to *Zambezia*. Salisbury: University of Zimbabwe.

Williams, M. (1989). Deforestation: Past and Present, *Progress in Human Geography: Global and Regional Changes in the Biosphere over the Past 300 Years* 13(2):176-208.

Williams, M. (1990). Forests. In *The Earth as Transformed by Human Action*, eds. B.L. Turner, W. C. Clark, R.W. Kates, et al., Cambridge and New York: Cambridge University Press with Clark University.

World Resources Institute, International Institute for Environment and UNEP (1991). *World Resources 1990-1991: Forest and Rangelands.* New York: Basic Books.

World Resources Institute, World Bank and UNDP (1985). *Tropical Forests: A Call for Action. (Part I: The Plan).* Washington D.C.: World Resources Institute.

Population and Environment:
Reviews and Case Studies

Introduction to Part Two

The overviews and methodologies in Part One only begin to grapple with the case study materials available on this topic and their inadequacies. The following chapters consider case study materials and do so in a way that draws out the complexities in linkages between population and the environment discussed earlier.

Many of these chapters rely on the seminal work of Blaikie and Brookfield (1987). Those authors were among the first to argue that environmental deterioration is a social and political-economic issue rather than merely an environmental process. Their work was principally about soil erosion but has applications to other forms of environmental degradation.

Each of these chapters focuses on a specific constellation of environmental concerns--agricultural intensification, land degradation, deforestation, extraction of resources, and urbanization--but the chapters, as a group, do not attempt to review all possible environmental problems. There are other problems of the greatest importance, for example, air pollution and waste disposal, that should be studied further in a similar way.

All of the chapters point to topics for future research, and emphasize the importance of multi-disciplinary research. As a group, the chapters highlight the findings of Palloni's study in Part One; despite the abundance in the literature of microlevel case studies, the kinds of comparable data across time and space from which robust generalizations can be made are largely lacking. These issues are considered further in the concluding chapter.

Reference

Blaikie, Piers M., and H. C. Brookfield, *Land Degradation and Society* (London: Methuen, 1987).

6

Population Change and Agricultural Intensification in Developing Countries

Richard E. Bilsborrow and Martha Geores

In recent years there has been an explosion of concern about environmental degradation in the world. Accounts of environmental problems include those of the annual reports of the United Nations Environmental Program (beginning with WCED 1987), the World Resources Institute, the World Bank (1992), the World Conservation Union (IUCN et al.--see *Caring for the Earth* 1991), the Worldwatch Institute, and many environmental organizations. Moreover, it appears that environmental problems continue to get worse. For example, estimates of the net annual rate of deforestation in the world (almost all in the developing countries) were reported to be some 11 million hectares per year around 1980 but are now estimated to be 17 million (FAO 1991) to 17-20 million hectares per year (World Bank 1991). On the surface, the broad litany of environmental problems appears necessarily related to the size and characteristics (including location and concentration) of the human population as well as its practices (technology of resource extraction, resource use, and disposal). Yet existing knowledge of these relationships is almost entirely of a descriptive, ad hoc nature. This chapter attempts to advance this limited knowledge base by first reviewing the ways in which population change can influence land use (and thereby the environment), then reviewing the literature on recent changes in land use, and, lastly, sketching what currently available cross-country data indicate about those relationships.

The issue is important for several reasons: rates of population growth continue to be high in many developing countries (UNFPA *Annual Report* 1992); the rapid pace of human settlement of so-called frontier areas, especially tropical moist forests, threatens to change the world's climate and reduce biodiversity and the gene pool; the prevalence of poverty and hunger continues to be high, especially in rural areas (e.g., World Bank 1990; Chen et al. 1990); strong linkages between high population growth and the social status of women, their economic roles and their involvement in the use of natural resources; and increasing reservations about the sustainability of existing agricultural practices based on high inputs of chemicals, water (especially through irrigation), and capital-intensive machinery.

The second section of this paper discusses conceptual issues regarding the effects of demographic processes on patterns of land use during the course of socioeconomic development. The complexity of the processes involved is illustrated, as well as the points where policy and contextual factors condition the nature of the response. The third section reviews recent literature on agricultural intensification, mainly in Latin America and Africa. The fourth uses cross-national data to explore the extent to which population processes and agricultural intensification may be related. The conclusion indicates some promising directions for future research.

Conceptual Linkages Between Population, Land Use, and the Environment

Conceptually, the interrelations between demographic processes, land use practices, and the environment in rural areas of developing countries are very complex. Since many aspects of those interrelations have been discussed previously (Pingali and Binswanger 1987; Bilsborrow 1987; Bilsborrow and Geores 1990; Cain and McNicoll 1990), only a short summary is provided here. The focus is on the interrelations between population change and land use, and only incidentally on the implications for the environment.

The relationships between population growth and agriculture have long attracted the attention of scholars, going back at least as far as the Zoroastrians around 325 B.C., the Indian sage Kautilya around 300 B.C., the Bible, and Aristotle in his *Politics* (Parsons 1992:3-6). But perhaps the first comprehensive theoretical approach was that of Thomas Malthus, who in 1798 postulated that whereas human populations have a tendency to grow geometrically (whenever economic conditions improve temporarily), the "means of subsistence" grows only arithmetically, the former thus tending to outstrip the latter over time. It is important to note, however, what Malthus and the classical economists meant by growth in the means of subsistence. Their argument was as follows: as population grows, demand for food grows, which can be met by either bringing new land into cultivation or cultivating existing farmland more intensively through the application of more labor to each unit of land. When the first strategy is followed, since land varies in soil quality and topography, the most fertile land is assumed by Malthus to be cultivated first. Therefore, as extensification proceeds, the average quality of land and therefore production per unit of land area tends to fall; moreover, as intensification proceeds through adding labor to land, output per worker must fall because of the principle of diminishing returns, viz., if one factor of production is fixed (land), increasing applications of the other variable factor (labor) eventually result in smaller increments of output per additional unit of labor. In either case, agricultural production per worker falls over time, leading, according to Malthus, to an increase in mortality (a "positive" check) and thereby reducing population growth back toward its initial level of zero.

It is curious that subsequent discussions of Malthus have focused upon the effects of intensification on land use via the principle of diminishing returns, though its implications were not clearly stated by Malthus (only by David Ricardo and other classical economists). On the other hand, Malthus did clearly indicate his belief in expansion occurring at the so-called "extensive margin" (increase in the land area under cultivation) through both internal and international migration from densely populated areas to "new" areas (Malthus, *Second Essay*, 1960:151-157; see also Bilsborrow and Okoth-Ogendo 1992, fn. 8). However, there is no basis for accepting Malthus's explicit assumption that land at the extensive margin is necessarily of lower fertility than land already in use (at the "intensive margin"): the rent or market value of land is determined by *both* its inherent fertility and its location relative to markets (see also Ghatak and Ingersent 1984:253). Many factors, including location vis-à-vis a body of water/road/railroad, colonial or local political decisions, and topography and climate, also influence where human population concentrations and hence markets are located.

History has proved Malthus and the classical economists wrong. The "Malthusian crisis" has not occurred, due largely to the effects of technological advances. These have shifted up production functions relating outputs to units of labor in such a way that the static principle of diminishing returns to labor has been overwhelmed in the long run, and agricultural output per unit of labor has continued to rise for two centuries, mostly during the past 50 years.

While Malthus viewed both the increase in the land area cultivated and increasing intensity of cultivation as induced by population growth, the argument was not well developed for either. Recently, scholars have postulated more clearly specific types of responses to population growth in rural areas, focusing on the responses to increased population *density*. Such increases are viewed as creating pressures upon living standards, which, following the historian Arnold Toynbee, tend to stimulate a response. Kingsley Davis developed the "theory of the multiphasic response," postulating that families respond by altering their demographic behavior: postponing marriage; reducing fertility within marriage by whatever means available; or outmigrating. Davis viewed the responses as "multiphasic" in the sense that several could occur simultaneously, but he did not consider the possibility of nondemographic responses, nor the full implications of outmigration (1963). Ester Boserup filled in the first gap, noting that as population grows relative to land, there is a tendency for farm households to use the land more intensively by reducing the time during which land is held idle or fallow and by changing technology in ways that facilitate increasing labor per unit of land (1965). The term "intensification of agriculture" has come to be associated with these processes as well as with her name. The literature spawned by the "Boserup thesis" includes critiques, defenses, and mathematical representations of her model, and cannot be further elaborated here (see, e.g., Lee 1986; Robinson and Schatjer 1984; Gleave 1992; and references cited therein).

But as Malthus pointed out, there is a third possibility--the "extensification of agriculture," or the extension of the land area under cultivation, often associated with population mobility and the appropriation of new lands. Indeed, this has been, since the development of sedentary agriculture some 10,000 years ago, the major means by which agricultural production has increased (Wolman and Fournier 1987). Internal migration, the neglected demographic variable in recent decades, is the key link between population pressures in one area and agricultural expansion in another.

However, because the responses to a growing population are multiphasic, the extent to which any *one* of the three categories of responses occurs--demographic, land intensification, or land extensification--depends on the extent of the other two responses, and hence also on the many factors influencing those responses. This greatly complicates attempts to analytically relate a change in any one of the possible "explanatory factors" (of which population growth or density is only one) to any specific response of interest, such as induced land intensification. Induced technical change also occurs through the operation of market economic forces, as shown by Ruttan and Hayami (1989) in examining the effects of changes in relative prices of inputs on factor proportions in agriculture (i.e., relative intensity of use of labor and land) over the past century in both the U.S. and Japan. Regarding the theory of induced innovation, see also Binswanger and Ruttan (1978).

The fact that there are a number of alternative forms of response cited in the text may be viewed as consistent with the "peasant survival strategy" literature that evolved in the 1980s in Latin America (see Arguello 1981). A useful comparative discussion of alternative theoretical approaches is found in Walker (1992).

The extent to which the processes described above lead to environmental degradation depends on the preexisting density of human habitation, the quality or characteristics of the land (its so-called "carrying capacity": c.f. Higgins et al. 1982), and land use practices. This is evidently true of areas of origin of potential migrants as well as areas of destination, though the recent literature has focused on the latter because areas of destination are, increasingly, tropical rainforests characterized by a rich biodiversity (Myers 1984, 1986), and other fragile lands on steep hillsides and semi-arid areas. The resulting processes of deforestation, soil degradation, and dessication are described widely in the literature and need not be repeated here (see references in initial paragraph of this paper).

Focusing instead on areas of origin, farmers with larger surviving families are likely to first respond by reducing fallow times, and clearing (deforesting) more of their own land, thus reducing the vegetative cover that retains moisture and protects the soil. Eventually, increased soil erosion and decreases in soil fertility tend to occur. In modern times, additional forms of agricultural intensification are also possible responses in areas of origin, at least in those developing countries that Higgins et al. describe as being at an "intermediate" level of technology; these responses are in the form of increases in the use of fertilizer or irrigation. However, to the extent that fertilizer is applied excessively, as it sometimes is by poorly educated farmers, there is chemical runoff into nearby bodies of surface water, seepage into groundwater supplies, or chemical poisoning of the soil. Also, as the irrigated area expands more than the available supply of water allows, salts build up in the soils (salinization). Such forms of land degradation are described widely in the literature, though their full extent is not known (e.g., see annual FAO and Worldwatch reports).

Figure 6.1 illustrates the linkages described above, beginning with population growth in rural areas in the left central area. In combination with the local availability of land and the prevailing system of land tenure and land distribution, population growth determines rural population density. Over time, population growth increases density to the point at which pressures to adapt arise. We postulate that the first and most immediate response, since it involves less stress ("cognitive dissonance") and less disruptive changes in behavior, is to either extend the area cultivated *in situ*, if the family has additional lands of its own or other nearby lands that can be appropriated, or reduce fallow time. Extending the area cultivated requires clearing additional land, if the land still has trees, and thereby increases local deforestation; the implied loss of vegetative cover increases the risk of soil erosion. Similarly, depending on soil resilience and specific cropping practices, reducing fallow time may lead to a decline in soil fertility. Under the usual conditions prevailing in developing countries, compensating changes in agricultural technology and land management do not occur autonomously, so that a time comes when accumulating demographic pressures stimulate additional responses. In the absence of technological change induced or stimulated by public policy, families have had to abandon their plots and migrate elsewhere. To the extent that they migrate to other rural areas and establish new farm plots by clearing land, land extensification occurs, often involving the appropriation of marginal lands--in lowland rainforests, steep slopes, or semi-

176

FIGURE 6.1 Illustration of Possible Land Use Responses to Population Growth in Rural Areas of Developing Countries

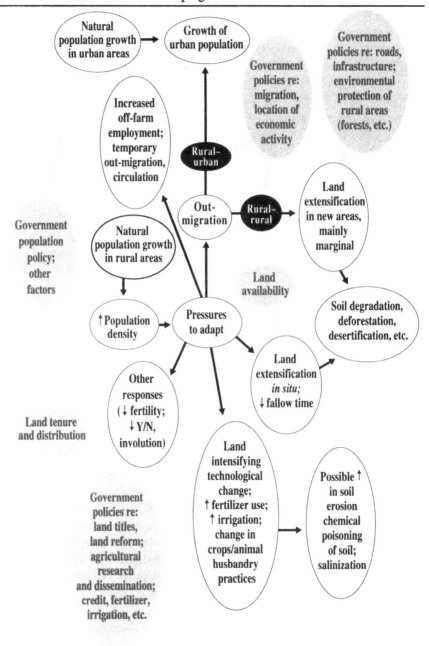

arid lands. While the environmental effects vary in each case with the density of human habitation, land practices, and soil properties, varying degrees of deforestation and soil dessication and erosion may occur.

Yet, as noted earlier, other forms of induced or endogenous response to demographic pressure are also possible, as indicated by the two lower solid arrows in Figure 6.1. Moreover, because of technological advances and the increased role of the state, both of these other forms are more likely in developing countries now than a few decades ago. First, responses in the form of postponing marriage or reducing fertility within marriage may occur, usually over the course of a generation, or about 30 years. Second, land-intensifying technological change--increased applications of labor per unit of land, perhaps accompanied by changes in crops grown or increased use of hybrid seeds, fertilizer, and irrigation--may occur. This may be facilitated by agricultural research and government extension policies. Lele and Stone have recently recommended that governments in sub-Saharan Africa take an active responsibility in stimulating land-intensifying technological change because of the absence of sufficient induced (endogenous) change associated with the high rates of rural population growth prevailing in the region (1989). A critical issue, however, is the extent to which the poor, *small* farmers will have access to the new technology and the additional complementary inputs it usually requires. The dotted lines in Figure 6.1 thus indicate the role of public policies, which also include influencing fertility decline, facilitated by the advent of modern fertility-regulation technology. To the extent that exogenous policies contribute to either a land-intensifying increase in labor absorption and agricultural output or to a reduction in human fertility, the inherent pressures for the alternative response, outmigration, are mollified. The converse is also true, following the multiphasic argument.

A further elaboration of the factors influencing the adoption of technologies involving land intensification (see the causal or conditioning arrows indicated by gray lines at the lower left of Figure 6.1) would include the following (not listed in order of priority, and borrowing from Bilsborrow 1987:188-190):

(a) The availability and accessibility of untapped, potentially cultivable land. Communications and transportation systems and thereby government road-building programs, such as those of Brazil and other countries in the Amazon, play a major role in determining the impact of this factor on migration flows and land extensification, and hence on easing the pressures to intensify *in situ*.

(b) Land tenure, the existing size of landholdings, and their distribution. When land distribution is highly skewed, the potential for bringing more land into cultivation via land reform is greater in the area. Studies of the relationship between farm size and productivity have found that smaller farms generally have much less land idle and also have higher productivity per acre in use than larger farms. In addition, the more unequal the land distribution, the slower the adoption and diffusion of new technology.

(c) Government policies that affect migration directly or indirectly help determine the extent to which people migrate in response to population pressures rather than intensify agricultural land use. Examples of such policies include the widespread colonization programs in countries sharing the Amazon basin, the transmigration program in Indonesia, and plantation settlement projects in Nigeria and Sri Lanka. On the other hand, policies to discourage migration to "frontier" areas (or to cities) put more pressures on farmers to adopt technological changes. In the absence of appropriate agricultural policies (see [f] below), farmers' efforts to increase the intensity of use of existing land lead to increased soil "mining" and immiserization (similar to Geertz's "involution" [1963]).

(d) The availability of off-farm, local employment opportunities. This factor was not considered in any of the theoretical discussions above. Small-scale artisan/industrial activities can provide important employment opportunities in rural areas, which may allow families with increasingly fragmented farms resulting from cumulative population growth and partible inheritance to maintain their standard of living without recourse to *any* of the multiphasic responses listed above. Temporary or seasonal outmigration or circulation can also provide off-farm sources of income that help sustain the household without recourse to permanent outmigration, as noted by Rudel (1990), and by Gould (1992) in regard to western Kenya.

(e) The availability of urban employment opportunities and the capacity of urban areas to receive migrants. This determines the extent to which rural-urban migration functions as an "escape valve" from rural population pressure. However, such migration does nothing to stimulate the increased rural food production needed to feed the growing urban population.

(f) Particular factors affecting the prospects for labor-intensive, land-saving technological change. These include soil conditions, household circumstances (including the availability of untapped labor), and the availability, accessibility, and cost of adopting new forms of land-intensifying agricultural technology. The latter, in turn, depend on many factors: the indigenous agricultural research in the country (and efforts to adapt the findings of research conducted through the 17-center international Consultative Group on International Agricultural Research [CGIAR] funded by the World Bank and various governments); the number, training, geographic distribution, and specific activities of agricultural extension agents; the availability and price of fertilizers, pesticides, etc., through either domestic production or imports, or the (cultural) practice of using animal manure and its supply; the availability of credit, especially for small farmers, to assist in the adoption and implementation of new technology; the existence of large-scale government irrigation schemes or support for small ones to facilitate multiple cropping; market prices of agricultural commodities relative to the costs of inputs and prices of nonagricultural products (the internal terms of trade, regularly distorted against rural producers by government price-fixing of food and the "urban bias" in development policies in general, as discussed by Lipton [1977] and others); and import and export, tax and subsidy policies regarding agriculture, which also influence relative prices and the profitability of agricultural operations. Various studies have also found that the size of holding is directly related to the adoption of new technology (e.g., Binswanger and Ruttan 1978; Lee and Stewart 1983), in part because of the usual allocation of credit in favor of large farm-owners.

(g) The strength of factors sustaining high rural fertility. In countries where desired family sizes in rural areas are lower than actual family sizes and where cultural factors favoring high fertility are not very strong, the (subsidized) provision of modern contraceptives and sterilization in rural areas is likely to contribute to a fertility reduction, as has happened during the past two or three decades in much of Latin America and Asia. However, in most of Africa and the Middle East, desired family sizes continue to be high and family planning programs have not yet led to sizable fertility declines. Aside from such programs, other factors contributing to a fall in fertility include improvements in women's education, employment opportunities, and status, and measures to reduce infant mortality.

(h) The broader institutional structure. This may have important effects on the type of response prompted by increased population pressure, since it influences farmers' access to not only land but also credit, agricultural inputs (including labor), and product markets. This category also encompasses a host of community-level or contextual factors affecting productivity and welfare, such as the local availability and cost of schools, health care, transportation, social programs, cooperatives, religious and political organizations, etc. A wide range of cultural factors, including traditional versus "modern" world views and collective (village or tribe) land and natural resource management systems, are also important in influencing technology adoption and land use (Blaikie and Brookfield 1987; Bernard 1988; Cain and McNicoll 1990).

Given the complexity of the process in Figure 6.1, in which each possible response has multiple causes or explanatory factors--with population pressure only one--and given that the multiple possible responses are linked to each other, it is very difficult to empirically demonstrate a convincing linkage between growing population pressure and changes in land use in developing countries. Thus, in analyzing the factors responsible for any *one* possible response to population pressure, such as land intensification, it is necessary to consider all the *other* possible responses *and* the roles of the various factors potentially influencing them. That is, a far more holistic approach is needed than that evinced by the literature to date. Moreover, the relative importance of the various possible explanatory factors on any particular response varies with the natural environment and many contextual and institutional factors. Some of these factors are quite difficult to measure. A related problem is the lack of adequate and linkable population-land use-environmental datasets, at either the cross-country or sub-country level. This stems from the lack of recent agricultural censuses, which provide data on land use (satellite imagery is very useful for measurement but does not provide any information on factors influencing land use decisions), as well as the lack of good data on land degradation, or even adequate measures, such as of soil degradation (Moran 1987; Blaikie and Brookfield 1987; Bilsborrow and Geores 1990; Clarke and Rhind 1992).

Population Effects on the Intensification of Agriculture and Land Quality

This section reviews evidence relating to Boserup's hypothesis regarding the intensification of agriculture (1965). Boserup postulates that as the rural population grows and population density in rural areas increases, farm households adopt a variety of land-intensifying innovations, which raise production per unit of land area by increasing labor and other inputs and changing methods of production. In both pre-modern and contemporary societies, such technological adaptations have taken many forms, including reductions in fallow periods and greater applications of labor to the weeding and care of plants, intercropping and multicropping, increasing small-scale irrigation, terracing steep slopes, using crop varieties requiring or responding to higher labor inputs, and applying more natural fertilizer. In the contemporary era of chemical technology, increasing the use of chemical fertilizers, pesticides and herbicides, new hybrid seeds, and large-scale irrigation constitute other forms of land-intensifying technological change. Since a positive relation between population density (the preferred measure of population pressure on land) and labor inputs per hectare is typical (e.g., it is observed for all six African countries in the recent study of Lele and Stone 1989:34) and does not in itself necessarily imply technological change, the discussion below concentrates on the other forms.

The most measurable forms of Boserup's land intensification are a reduction in the fallow period and a shift from annual to multiple cropping. Therefore, the review that follows first focuses on these changes in developing countries, citing country studies that shed light on demographic aspects. A systematic assessment of the environmental implications of the observed changes in land use is beyond the scope of this paper, although some environmental effects are indicated below. Muscat (1985) and many others have noted, for example, the pervasiveness of land degradation associated with intensive agricultural practices over long periods of time in China, northern India, Java, and parts of South and East Africa and the Andean highlands. Leonard (1987) describes the pervasiveness of deforestation and soil erosion throughout Central America, Joly (1989) in the lowlands of Panama, Ibrahim (1984) in the Western Sudan, Repetto and Holmes (1983) in Java, and Lele and Stone (1989) in six African countries. Other examples are cited in Bilsborrow and Geores (1990).

Shortened Fallow Periods and Increased Cropping Intensities

The traditional form of agriculture in much of the tropics is swidden agriculture, which is well suited to the tropical environment under conditions of low population density. Under swidden systems, soil fertility is maintained by interspersing short periods of cultivation with long periods of fallow. In recent years, population growth and the consequent increases in population density have often rendered swidden systems inadequate to meet food demands, requiring that more intensified agricultural techniques be used or that the fallow period be reduced. A shorter fallow period means that cultivation is re-initiated before nutrient-supplying vegetation has a chance to regenerate. The results are lower crop yields, weed invasions, and erosion.

In terms of recent empirical evidence, only smatterings of data are available, some cross-sectional and some over time. For example, because of population growth, fallow periods were reduced from 12-15 years to 4 years in Sierra Leone, from 10-30 years to 5 years in northeast India (Russell 1988), and from 18 years to 5-15 years in Belize (Arnason et al. 1982), resulting in each case in lower land productivity. In Mexico, 50 percent of the arable land was fallow in the 1930s, but only 35 percent by 1960 (Yates 1981) and far less now.

In sectorial farming systems, land in highland areas that is not allowed to regenerate its soil fertility by lying fallow (common in the Andes) gradually loses the nutrients in the topsoil and subsoil by erosion, so that increased gullying results (for Bolivia, see Godoy 1984). Inequitable land tenure also contributes to the need to shorten fallow periods; with most farmers lacking sufficient land, growing crops on land that should remain fallow has increasingly become a necessity in the short term and a disaster in the long term. Soil erosion has been linked to increases in population in the context of inequitable land tenure systems, whereby plot sizes per household are too small to support the families living on them. Smallholders and tenant farmers cannot practice soil conservation methods because they need to cultivate all the land available to them, involving deforestation, loss of groundcover, and soil erosion. Studies have shown a connection between inequitable land tenure and erosion in Bolivia, El Salvador, Guatemala, Pakistan, the Philippines, and Taiwan (Crosson 1983; Leonard 1987; Preston 1969). Foy and Daly (1989) link land degradation to inequitable land tenure in the fertile regions of Costa Rica and El Salvador, because of excessively intensive cultivation of marginal lands by the poor. They observe, however, that in Haiti, the effects of maldistribution (although substantial) are clearly overshadowed by those of population pressure. Erosion is a major problem in many areas where marginal land is intensively cultivated. In Latin America, Stonich (1989) attributes a 58 percent decline in fallow time in southern Honduras between 1952 and 1974 to population growth, and reports fallow periods less than half as long on small as on large farms (data are similar for Guatemala, see Bilsborrow and Stupp 1988).

In Africa, Adepoju states that in many regions of Nigeria, growing populations are leading to fragmentation of landholdings, shorter fallow periods, and increasing landlessness (1978:11-12). Tiffin (1975, quoted in Gleave 1992:5) observed a shift from leaving land fallow to permanent cultivation in northern Nigeria over the period 1916-1963 as population quadrupled. Bernard (1988) reports that in Meru district, Kenya, population growth in the highlands led to decreases in the fallow periods of adjacent lowlands traditionally used for grazing, resulting in soil degradation and desertification in those areas, where soil fertility was low to start with. Ibrahim (1984) and Horowitz and Salem-Murdoch (1987) discuss the effects of population growth on declining fallow periods in Darfur province and other areas of the Sudan. Such declines have led to reductions in soil fertility and millet yields (Bilsborrow and DeLargy 1991). In an earlier cross-sectional study of a number of areas in eastern Nigeria, Morgan (1963, cited in Gleave 1992:3) observed average fallow periods of seven years in areas with a population density of 72 persons per sq. km, five years where density was 102-189, and only three years where it was 306. In a recent study of six countries in sub-Saharan Africa, Lele and Stone (1989) observed the expected relationships between population density and fallow times in Cameroon and Malawi but not in Senegal or Tanzania (the lack of a relationship in the latter was ascribed to state interventionist policies).

In Asia, declines in fallow lands have occurred over the long run in China, India, Indonesia, and Sri Lanka as populations have grown (Perkins 1969; Kumar 1973). Kumar (1973) presents estimates (albeit now dated) on the extent of fallow land in developing countries, both for continents and for eight countries in Africa: the latter data show a strong correlation across countries between rural population density and the percentage of fallow land. The African countries with the highest density (Kenya and Lesotho) had by far the lowest percentages of fallow land in 1960 (6 and 9 percent, respectively, versus 57 to 99 in the other six countries in western and central Africa). Kumar's estimates at the continental level also show that the percentage of fallow land is far lower in Asia than in Africa or the Middle East, and that fallow land declined with population growth in all regions between 1950 and 1960 (1973:166-167).

Conceptually similar to a reduction in fallow periods is an increase in crop intensity (the proportion of time a given area of land is in use). Thus, there has occurred substantial switching from annual to double-cropping (two crop cycles on the same area of land per year) and triple-cropping (or, in general, "multiple cropping"). Such a switch has been widespread in the cultivation of rice in Asia over recent decades. Crop intensity may be defined as the ratio of total cropped area divided by net cropped area times a hundred, a measure of the time(s) the land is used per year. Ratios computed from Kumar (1973) and Grigg (1985) for seven densely-populated Asian countries plus Egypt show high levels of multiple cropping in all countries, with the highest ratios (145 to 180) in areas with the highest rural population density--Taiwan, Egypt, Bangladesh, and China (Bilsborrow 1987:195). Moreover, crop intensity rose in all countries over the periods considered. In their detailed study of Nepal, Ives and Messerli (1989:194) note that crop intensity increased as a direct consequence of population growth. Such an increase was quite dramatic: from a mean crop intensity of 110 in 1965-1980 to 166 in 1985/1986. However, the best-known study of intensification in Asia is probably that of Geertz (1963), who observed intensification of labor and double- and triple-cropping of rice crops as responses to increased population density in Indonesia. However, these responses did not compensate for population growth, with the result that standards of living still declined, an outcome described by Geertz as a condition of shared poverty, or "involution."

While many of the studies cited here note that the increased intensity of land use is associated with land degradation, the relationship between land degradation and population pressures has not been specifically investigated. While this relationship needs investigation, it is hampered by the lack of adequate measures of land degradation and conceptual problems associated with, for instance, the fact that population pressure and land degradation (e.g., soil erosion) in one site may result in environmental damage in another (as when deforestation in the upper slopes of a watershed causes increased flooding downstream) makes the causal linkages difficult to detect. Soil erosion is usually measured based on the extent of sedimentation or dissolved solids found in the water downstream from the area of interest. But such a measure is an *areal* measure, not usually feasible for micro- or farm-level measures. What has been recommended is the use of yield or productivity data to reflect presumed changes in soil fertility due to changes in soil erosion or nutrient loss (see e.g., Blaikie and Brookfield 1987). But this, of course, makes it impossible to relate soil conditions (or anything else) to changes in yields over time. Another indirect indicator of changes in yields is the proportion of the farmer's output sold in the market (Okafor 1987), which presumes the farmer produces his own subsistence crops and meets those needs first, only selling in the market whatever surplus is left. This is not true of most Third World farmers. In addition, land use practices themselves are crucial determinants of land degradation, and vary widely across population groups having similar densities. Furthermore, natural forces also cause soil degradation, as they have been doing for hundreds of millions of years. The loss of soil fertility sometimes leads to switches in the crop mix by, for instance, replacing certain crops with ones less demanding in terms of nutrients, as has happened widely in West Africa with the replacement of yams by cassava (Gleave and White 1969).

A study by Turner et al. (1977) reflects one of the few efforts to formulate a quantitatively testable model to examine the effects of population pressure on agricultural intensity, taking into account other factors. It therefore deserves a more detailed assessment. The authors collected data from a number of studies by geographers and anthropologists who investigated a number of population groups living as tropical subsistence agriculturalists (27 groups in 29 areas of 10 countries, on all four developing continents plus two Pacific islands). Population density (persons per sq. km of area actually used for crops) varied from 1 in eastern Peru and central Brazil to 200 or more on the Tonga Islands (Pacific) and the famous (made so by Boserup) Uhara Island in Lake Victoria. Agricultural intensity was computed as the proportion of time that each unit of land area was in the cropping (vs. fallow) phase. Thus, an area with nine years fallow for every year cropped had an intensity of 0.1, an area cropped one year and fallow the next had a value of 0.5, etc. When the observed values were multiplied by 100 to convert to percentages, they ranged from 5 to 150.

Agricultural intensity was then plotted against population density, suggesting a strong, positive relationship. A multiple regression analysis was performed, yielding:

$$\log A = .79 + .109 \; D + .125 \; S + .0042 \; P + .003 \; LP - .0007 \; DP \quad (R^2 = .87)$$
$$\qquad\quad (.024) \quad (.068) \qquad (.0008) \quad (.001) \quad (.0003)$$

where A represents agricultural intensity, P population density, D length of dry season, S soil quality, L whether the population raised livestock, 0.0042 is the regression coefficient for P, and 0.0008 its standard error. The coefficient of P is statistically significant at the 0.01 level, showing that agricultural intensity was higher in areas with higher population density. Other variables examined included rainfall and use of root crops. The fact that the positive relationship between intensity of land use and population density exists even with the inclusion of the other explanatory variables suggests the relationship is robust. While there are problems regarding the peculiar sample composition and the necessity of inferring time-series relationships from cross-country evidence (Bilsborrow and Geores 1990), the results are intriguing and indicate a promising direction for more intensive country-level research, in particular, to examine changes *over time.*

Increased Use of Fertilizer, Irrigation, and Other Modern Inputs

A number of agricultural inputs can be applied to land to increase its productivity (apart from more labor, which is not discussed here, but is usually necessary to apply the other inputs). These include chemical and organic fertilizers, pesticides, herbicides, and fungicides; irrigation also facilitates a more intensive use of land. While dramatic changes in agricultural methods are occurring in many areas of low-income countries, their possible relationship to population factors has been examined hardly at all. In fact, there has been little research on the causes of these changes at a level above that of the individual farm, and investigations at the latter level have not included assessments of demographic factors (e.g., differences, or changes over time, in population per hectare).

Technological change may occur in the form of adoption of new crop varieties and new seeds. The introduction of cash crops and the commercialization of agriculture also contribute to intensification and agricultural change, as seen in many areas of Africa (Gleave 1992:3-4, and references cited) and elsewhere. If these changes also require increases in the labor/land ratio in their application (e.g., hybrid rice that requires a regular supply of irrigation water, fertilizer, or pesticides and herbicides), their introduction in areas of high population density could provide evidence of population-induced technical change. The evidence on Latin America appears mixed: hybrid seeds of corn/maize were first brought to the region, to Mexico, in the 1950s; by the 1970s, one-quarter of the corn planted in Mexico came from new hybrid seeds, as did 30 percent in El Salvador (Grigg 1985:195). Maize yields in Mexico were estimated by FAO to have increased by one-third between 1960 and 1980, but it is not clear what role the new seeds played. Regarding other crops, semidwarf varieties (e.g., of wheat), which require regular supplies of water and fertilizer, have been adopted much less in Latin America (only two-fifths of the total area) than in more densely populated Asia. Their adoption in Latin America has been mostly in Argentina and Mexico, where in both countries yields quadrupled between 1950 and 1975 (Grigg 1985:195). Given Mexico's growing scarcity of land (even more true of El Salvador, see Higgins et al. 1982), such an adoption of new, land-intensifying seeds may provide some evidence of population-induced technological change, though exogenous factors were evidently also involved. In most of the rest of Latin America, continuing land extensification and other factors appeared to be reducing pressures toward intensification.

Other forms of adaptation have often accompanied or preceded intensification and, as stated above, intensification is not an inevitable consequence of growing population density. Rushahigi (1985) observes that density has been high and growing for some time in Rwanda, with little adaptation in technology. He attributes this to the fact that the development of more advanced forms of technology depends on the *economic* capacity to acquire and implement the innovation. This capacity is very limited in poor and isolated countries such as Rwanda, unless there is significant outside assistance. Chipande (1988) and Mortimore (1989) noted similar reasons for the negligible adoption of modern technological inputs in Malawi and Nigeria, respectively.

Surprisingly, there has been practically no research on the relationship between demographic processes and the use of chemical fertilizers or other modern chemical inputs, even though the latter are intended to extract a larger value of output per unit of land. In fact, there is little research on the determinants of changes over time in fertilizer use in general. Lele and Stone (1989:34ff), however, observed a positive association between population density and use of fertilizer in Kenya and Cameroon, but none in four other countries, because of either government interventions (to increase fertilizer use in low-density areas) or the lack of ability of small farmers to pay for fertilizer (viz., Malawi).

Another form of intensification that may be induced in part by population growth is irrigation: areas of high population density are more likely to develop irrigation systems to extend the length of the growing season and allow multiple cropping. Since the latter has been discussed above, we only note here the lack of research specifically aimed at understanding the linkages between irrigation and population pressure. Although a high correlation exists at the country level between population density and percentage of land irrigated, whether recent increases in density are associated with *increases* in irrigation within countries is not known.

Lastly, some limited evidence exists that areas with high density are more likely to be planted with high-yielding or high-value crops, as observed by Lele and Stone for Cameroon and Kenya and by Mortimore for the Kano area of Nigeria. It is well known that agricultural areas around cities throughout the world are utilized intensively to produce high-value crops such as vegetables.

The relationship between population growth and increased inputs depends on economic conditions and capacity prevailing in rural areas. Where economic resources are sufficient, increasing population density can help stimulate positive technological changes of the land-intensifying sort. But where rural poverty and small landholdings dominate, growing density seems more likely to force survival strategies involving more temporary outmigration, off-farm employment, or permanent outmigration, depending on the various contextual factors listed above.

While there is some hope for developing more sustainable agricultural procedures for low-fertility, thin, tropical soils (Sánchez and Benites 1987; Morán 1987; Browder 1989), and while methods have long been known for achieving more sustainable uses of highland slopes (viz., rice terraces in Asia), these methods often require substantial infrastructure or labor investments (cf. Collins 1988, on the effect of outmigration on depleting the labor required for maintaining terraces in Peru). But the increased infrastructure is beyond the scope of government budgets in most low-income countries, which may require, as in China, the mobilization of human labor beyond the capability of democratic governments. Thus, linkages between population growth and soil degradation are real and not always subject to the simple "technological fix" espoused by many agricultural economists and agronomists--at least, not in the short run and not without vast increases in economic and technical aid to low-income countries.

A Look at Some Cross-Country Data and Relationships

The focus here will be on the relationship between demographic data and data on land use and agricultural technology. Land use data, such as land tenure and the size distribution of holdings, the proportion of a country's land area in agricultural use (following FAO, land which is Arable or in Permanent Crops, or A&P land), the use of fertilizer, and forest cover, are usually available from agricultural censuses. Unfortunately, these censuses were carried out less frequently in the 1980s than earlier, and generally are conducted less frequently than population censuses: few countries in Latin America, for example, held an agricultural census in the 1980s. Estimates of land use are therefore prepared by countries or by the FAO on the basis of limited data from the occasional agricultural survey or even only production data. Consequently, the data are of varying quality across countries and are rarely available in a reliable form for recent periods.

Tables 6.1 and 6.2 present the available data. The growth of the rural population is lower than that of the overall population in all countries, being highest in sub-Saharan Africa. Using medians as the preferred measure of central tendency (because of the skewed nature of the data), the following average annual rural population growth rates for the different regions are obtained: Asia, 1.9; Middle East, 1.4; Latin America, 1.6; and sub-Saharan Africa, 2.2. Countries experiencing negative rural population growth were the Republic of Korea in Asia, Argentina, Brazil, Chile, Cuba, and Uruguay in Latin America, and Gabon and Liberia in Africa.

The column Change in A & P Land reflects land extensification, or increases in the percentage of the total land area of the country devoted to agricultural uses between 1965 and 1987. The final column shows land per agricultural worker in 1985, which serves as a measure of rural population density. Median changes over time in the percentage of land in agricultural use were as follows for the different regions: Asia, 0.7; Middle East, 0.4; Latin America, 1.6; and sub-Saharan Africa, 0.6. The much larger change in Latin America is noteworthy, reflecting a much greater process of extensification of agriculture. Land per agricultural worker (only the 1985 levels are shown here, see Bilsborrow and Geores 1990) generally declined because of positive rural population growth, particularly in Asia and sub-Saharan Africa--from a median of 0.38 ha/person in 1965 to 0.30 ha/person in 1985 in Asia, and from 1.44 to 1.35 ha/person in sub-Saharan Africa. Yet, because of an increase in agricultural land in some countries and net rural outmigration in others, in both the Middle East and Latin America the medians actually rose: from 1.17 to 1.39 ha/person in the Middle East, and from 2.3 to 2.7 ha/person in Latin America.

Table 6.2 provides data on two measures of intensification of particular interest for this paper, fertilizer use and irrigation. We first examine the change in average kilograms of fertilizer used per hectare of cropland between 1975/1977 and 1985/1987. Again, the regional data are highly skewed, reflecting differing technological and economic abilities to produce or import fertilizer. The highest median increase was 39 kg/ha in Asia, followed by 18 kg/ha in the Middle East, 9 kg/ha in Latin America and only 1 kg/ha in sub-Saharan Africa. The greater increase in Asia is probably related to both greater population pressures and faster economic growth in the 1980s, which facilitated greater purchases of fertilizer.

The column Change in % A & P Irrigated provides data on trends in the percentage of agricultural (A&P) land irrigated between the 1970s and 1980s. The median changes were 5.0 in Asia, 1.5 in the Middle East, 2.3 in Latin

America, and 0 in sub-Saharan Africa. In each region, countries experiencing dramatic increases appear to be all countries of high density, with little space for land extensification: Bangladesh, Nepal, the Republic of Korea, and Vietnam in Asia; Costa Rica, Cuba, El Salvador, Mexico, and Trinidad in Latin America; and Madagascar in Africa.

The last column provides a constructed land productivity index, which shows the relative change in land productivity over the period 1965-1985--the ratio of the country's land productivity in 1985 to that of 1965 (see table note). In general, Asia experienced the highest median increase across countries in the relative land productivity index of 1.71, followed by the Middle East at 1.56, sub-Saharan Africa with 1.32, and Latin America with 1.22. The larger change in Asia reflects a combination of greater adoption of Green Revolution technology and increases in fertilizer application. While Latin America did trail in both, the major factor explaining the difference across the two continents was the policy of land extensification in Latin America, where lands with substantially lower fertility were brought under cultivation, particularly in the lowland rainforests. This policy alleviated pressures toward the intensification of agriculture.

The relationships between demographic and land use variables can be conceptualized either statically or dynamically. The static perspective is of less interest because of our interest in the relationships over time, but it is still a useful starting point. Thus we expect a positive relation between, for instance, population density and the use of fertilizer and irrigation, even though many other factors also influence the latter two. It is evident from Table 1 that high correlations across countries do indeed exist between population density and levels of intensification (both being higher in Asia). Although existing data do not permit a truly dynamic analysis, we can examine the relationships between *changes* over time in demographic variables and *changes* in land use and the environment by means of a comparative statics analysis. A series of graphs, presenting the two-way relationships (particularly useful for identifying "outliers") were prepared for various alternative measures of the "independent" (population) and "dependent" variables: this exploratory analysis was necessary given the uncertainties regarding the reliability of the measures used and the differences in the sample composition of countries available for each measure. The definitions of variables ultimately used are found in the notes to the tables.

Figure 6.2 indicates that land intensification tends to be positively related to population pressures. In the absence of data on changes over time in the average fallow period, changes in fertilizer use (FERX) and changes in the percentage units of agricultural land irrigated are used as measures of land intensification (so that a change from 0 to 5 is the same as one from 50 to 55).

TABLE 6.1 Cross-Country Data on Population and Agricultural Extensification

Country	Rural Population Growth 1960s-1980s[a]	Change in A&P Land 1965-87[b]	Land per Agricultural Worker, 1985[c]
Asia			
Cambodia	.[d]	0.7	0.60
China	1.6	-0.6	0.13
Indonesia	1.9	2.0	0.25
Korea, DPR	.	3.4	0.30
Korea, Rep	-1.7	-1.2	0.18
Laos	.	0.3	0.29
Malaysia	2.3	2.0	0.80
Mynamar	.	-0.2	0.54
Philippines	1.9	3.0	0.29
Thailand	2.3	14.2	0.60
Vietnam	.	2.5	0.17
Afghanistan	2.4	0.2	0.84
Bangladesh	.	0.7	0.12
India	1.9	3.2	0.34
Iran	1.8	-0.3	1.09
Nepal	2.2	3.6	0.15
Pakistan	.	1.7	0.37
Sri Lanka	1.8	0.0	0.22
Middle East			
Iraq	0.9	1.3	1.39
Jordan	.	0.7	1.51
Lebanon	.	0.4	0.96
Syrian Arab Rep	2.6	-5.3	1.94
Turkey	1.3	1.8	1.13
Algeria	0.8	0.0	.
Egypt	2.0	-0.2	.
Libya	2.9	0.1	.
Morocco	1.4	2.7	.
Tunisia	1.1	3.8	.
Latin America			
Costa Rica	2.1	0.7	2.1
El Salvador	.	3.5	1.2

(*continues*)

TABLE 6.1 (*continued*)

Country	Rural Population Growth 1960s-1980s	Change in A&P Land 1965-87	Land per Agricultural Worker, 1985
Latin America			
Guatemala	2.1	3.0	1.5
Honduras	2.8	2.5	2.2
Mexico	1.5	0.7	2.7
Nicaragua	2.0	0.7	2.9
Panama	2.2	0.3	2.7
Cuba	-0.6	13.6	3.9
Dominican Rep	1.6	8.8	1.8
Haiti	0.9	6.3	0.5
Jamaica	.	-2.1	0.8
Trinidad	.	4.5	3.1
Bolivia	.	1.6	3.8
Brazil	-0.2	3.1	5.7
Colombia	.	0.3	1.9
Ecuador	2.3	0.5	2.7
Guayana	0.3	0.7	5.9
Paraguay	2.0	3.2	3.5
Peru	2.0	0.8	1.6
Venezuela	0.8	0.5	5.0
Argentina	-0.7	2.4	28.9
Chile	-0.5	1.6	9.4
Uruguay	-0.4	0.2	8.5
Sub-Saharan Africa			
Botswana	2.4	0.4	4.52
Burundi	1.5	6.7	0.78
Ethiopia	3.1	0.7	1.35
Kenya	2.9	0.5	0.47
Lesotho	.	-2.4	0.30
Madagascar	1.9	1.2	0.84
Malawi	.	3.0	1.01
Mozambique	.	0.2	1.21
Rwanda	2.6	11.5	0.44
Somalia	.	0.1	0.75
Sudan	1.1	0.3	2.84

(*continues*)

192

TABLE 6.1 (continued)

Country	Rural Population Growth 1960s-1980s	Change in A&P Land 1965-87	Land per Agricultural Worker, 1985
Africa			
Tanzania	2.4	0.5	0.87
Uganda	.	3.7	1.30
Zambia	1.4	0.3	3.66
Zimbabwe	.	0.6	1.76
Angola	.	0.0	3.25
Cameroon	2.4	2.2	2.19
Central African Rep	2.2	0.2	1.80
Rep of Congo	.	0.2	3.72
Gabon	-1.2	0.8	2.27
Zaire	1.2	0.2	0.71
Burkina Faso	.	1.3	0.86
Chad	2.7	0.2	2.21
Mali	3.0	0.3	0.63
Mauritania	0.4	-0.1	0.47
Niger	.	0.6	2.31
Benin	0.4	2.2	2.41
Ghana	2.5	0.2	1.26
Guinea	.	0.0	0.09
Liberia	-1.3	0.0	0.74
Nigeria	.	0.9	1.96
Senegal	.	2.9	3.00
Sierra Leone	.	4.1	2.09
Togo	2.1	0.9	1.94

Sources: column (1): UN, 1988; columns (2) and (3): FAO, 1984, 1988.

[a]Column 1, Rural Population Growth Rates, 1960s-80s. Average annual rural population growth rate based on available census data from each country for a census in the 1960s and the 1980s. The definition of "rural" depended on each country's own definition, which varied greatly; also, not all countries reported separate rural and urban populations.

[b]Column 2, Change in A&P Land, 1965-87. The change in the percentage of total land in the country which was in cropland and pasture, between 1965 and 1987.

[c]Column 3, Land per Agricultural Worker, 1985. Hectares of A&P land per economically-active person engaged in agriculture, 1985.

[d]"." means data not available.

TABLE   6.2   Cross-Country   Data   on   Agricultural   Intensification

Country	Per Capita GNP 1987	Change in Fertilizer Use 1976-86	Change in % A&P Irrigated 976-86	Land Prod. Index 1985/1965
Asia				
Cambodia	.	0	0.0	0.89
China	294	121	3.0	2.28
Indonesia	444	73	8.0	1.92
Korea, DPR	910	84	4.0	1.94
Korea, Rep	2689	61	10.0	1.59
Laos	166	2	7.0	2.10
Malaysia	1820	86	1.0	2.13
Mynamar	212	13	1.0	2.21
Philippines	589	16	4.0	1.63
Thailand	850	12	5.0	1.26
Vietnam	200	3	10.0	1.71
Afghanista	220	3	2.0	.
Bangladesh	164	39	8.0	1.36
India	311	30	5.0	1.63
Iran	1756	40	3.0	2.14
Nepal	161	14	16.0	1.07
Pakistan	353	49	7.0	2.12
Sri Lanka	406	57	5.0	1.60
Middle East				
Iraq	2400	28	2.0	1.24
Jordan	1560	18	2.0	0.80
Lebanon	.	7	3.0	.
Syrian Arab	1645	27	2.0	2.98
Turkey	1213	19	1.0	1.56
Algeria	2629	18	2.0	1.96
Egypt	678	159	0.0	1.89
Libya	5453	3	1.0	2.56
Morocco	615	13	1.0	1.19
Tunisia	1182	11	2.0	1.81
Latin America				
Costa Rica	1608	20	16.5	1.61
El Salvador	842	22	13.0	1.30

(continues)

TABLE 6.2 (*continued*)

Country	Per Capita GNP 1987	Change in Fertilizer Use 1976-86	Change in % A&P Irrigated 1976-86	Land Prod. Index 1985/1965
Latin America				
Guatemala	947	18	1.3	1.44
Honduras	808	1	0.5	1.40
Mexico	1825	24	7.0	1.45
Nicaragua	829	34	5.2	1.68
Panama	2239	- 2	2.0	1.72
Cuba	.	73	7.7	0.81
Dominican Rep	734	17	2.9	1.07
Haiti	362	2	2.2	1.10
Jamaica	940	-26	4.4	1.30
Trinidad	4149	- 1	7.0	0.51
Bolivia	496	1	0.1	0.86
Brazil	2021	10	2.0	0.36
Colombia	1238	27	4.6	1.67
Ecuador	1044	15	2.3	1.51
Guayana	389	- 9	-4.0	0.87
Paraguay	995	4	-0.4	0.92
Peru	1467	-13	-6.3	0.91
Venezuela	3226	88	2.2	1.59
Argentina	2394	2	1.0	1.12
Chile	1358	20	-2.3	1.11
Uruguay	2198	9	4.2	1.23
Sub-Saharan Africa				
Botswana	1059	- 2	0	0.93
Burundi	241	1	1	1.35
Ethiopia	126	2	0	1.23
Kenya	330	24	0	1.46
Lesotho	355	7	.	1.26
Madagascar	207	1	10	1.18
Malawi	164	7	0	1.42
Mozambique	146	- 2	2	1.06
Rwanda	301	1	0	1.37
Somalia	290	- 1	1	1.42
Sudan	331	- 2	1	1.56

(*continues*)

TABLE 6.2 (continued)

Country	Per Capita GNP 1987	Change in Fertilizer Use 1976-86	Change in % A&P Irrigated 1976-86	Land Prod. Index 1985/1965
Africa				
Tanzania	180	3	2	1.38
Uganda	260	0	0	0.97
Zambia	248	4	0	1.33
Zimbabwe	585	8	4	.
Angola	.	0	.	0.82
Cameroon	966	4	0	1.27
Central Afr.	334	0	.	1.34
Rep of Congo	873	1	1	1.31
Gabon	2733	3	.	0.78
Zaire	153	- 1	0	1.43
Burkina Faso	191	2	0	1.38
Chad	139	0	0	1.12
Mali	200	10	3	1.23
Mauritania	439	0	0	1.52
Niger	258	0	0	0.85
Benin	305	5	0	1.80
Ghana	393	- 6	0	1.38
Guinea	316	- 1	0	1.20
Liberia	451	-10	0	1.70
Nigeria	368	8	0	1.71
Senegal	510	- 5	0	1.36
Sierra Leone	249	1	1	1.13
Togo	286	5	0	0.98

Sources: columns (1-3): WRI: 1990; column (4): FAO: 1977, 1988.

ᵃColumn 1, Per Capita GNP, 1987. Per capita gross national product in 1987 US dollars.

ᵇColumn 2, Change in Fertilizer Use, 1976-86. Change in average kilograms of fertilizer used per hectare of cropland, based on three-year average figures for 1975-77 and 1985-87.

ᶜColumn 3, Change in % Land Irrigated, 1976-86. Change in percent of A&P land irrigated from 1975-77 to 1985-87.

ᵈColumn 4, Land Productivity Index, 1985/1965. Relative index of land productivity in 1985 divided by land productivity in 1965, where land productivity in, e.g., 1965, is taken as the index of FAO agriculture production for 1966/1967 and divided by the amount of A&P land in 1965 (similarly for the 1985 index).

While a positive relation is indeed observed between FERX and rural population growth, it is weak (see Figure 6.2). For sub-Saharan Africa, however, it is strongly significant, perhaps reflecting the greater relative strength of the relation at lower levels of development, where farmers are closer to subsistence levels and extraneous market forces are less likely to intervene. Changes in fertilizer use are also strongly related to concurrent growth in GNP in the overall sample, but within regions the relation is strong only for Asia (results not shown). However, those changes were not found to be related to either population density or inequality in land distribution.

Perhaps more interesting is whether there exists a trade-off between intensification and extensification, as implied by the expanded multiphasic theory proposed earlier. Thus we would expect an *inverse* relationship between changes in fertilizer use and increases in A&P land: countries where more land is being brought into cultivation would be less likely to increase fertilizer use, while countries where additional land is not available would have to increase inputs. Figure 6.3 suggests that such a trade-off does exist (contrast Brazil with Egypt and China), although it is weakened by Jamaica (though no more land is available to add to cultivation, a foreign exchange crisis may have impeded increases in fertilizer use) and by the many countries in Africa reporting zero changes in both A&P land and fertilizer use (perhaps because the estimates available were not updated from one period to the next).

The other cross-country measure of intensification available, change in percentage of land irrigated, is not related to population growth (Figure 6.4) and only slightly to GNP growth and population density. It is interesting to examine the extent to which increases in fertilizer use and increases in irrigation are alternative (substitute) responses to population growth and to what extent they are complementary or occur together. In fact, we observe a strong positive correlation (coefficient of 0.24, significant at the 3 percent level), suggesting they are usually complementary. Thus increases in fertilizer use and in area irrigated have tended to occur together in developing countries in recent decades.

It is extremely difficult to use cross-country data to investigate the relationships between population pressure and land intensification or extensification (see also Bilsborrow 1992), partly because of serious deficiencies in the data. A few outliers tend to determine the nature of the relationships, making it problematic to draw general inferences. Perhaps as better data become available in the future, those problems may be reduced. But in any case, cross-country--and for that matter, cross-region or cross-household--relationships only reflect static relations. What is desired is evidence on changes over time and the role of population and other factors in such changes.

FIGURE 6.2   Relationship between Change in Fertilizer Use and Rural
Population Growth

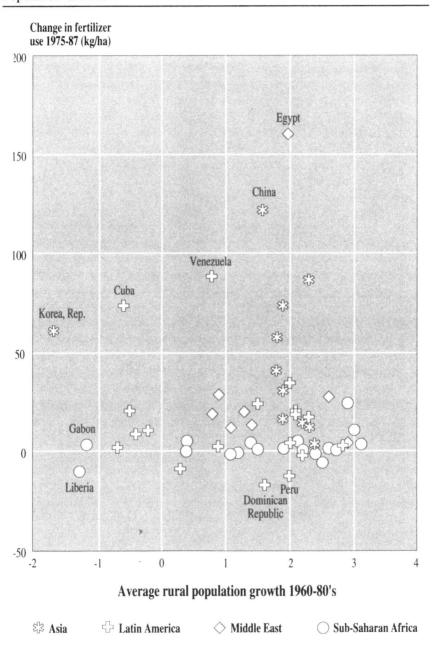

Average rural population growth 1960-80's

198

FIGURE 6.3 Relationship between Change in Fertilizer Use and Change in A&P Land

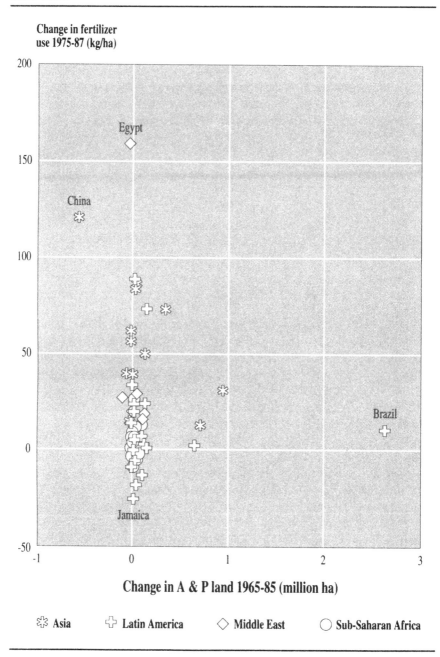

Change in fertilizer use 1975-87 (kg/ha)

Change in A & P land 1965-85 (million ha)

✳ Asia   ✚ Latin America   ◇ Middle East   ○ Sub-Saharan Africa

FIGURE 6.4 Relationship between Change in Irrigated A&P Land and Rural Population Growth

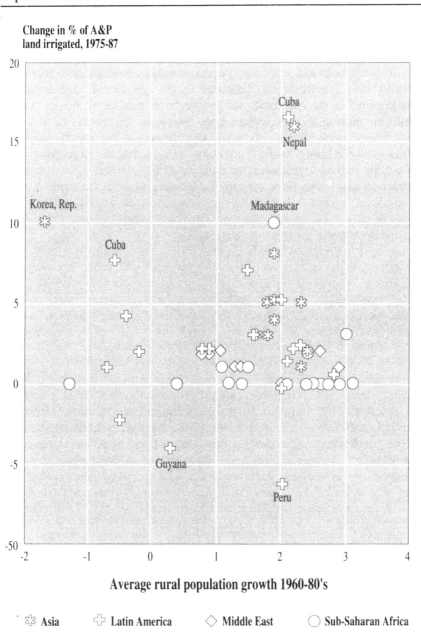

Change in % of A&P
land irrigated, 1975-87

Average rural population growth 1960-80's

⁂ Asia   ✚ Latin America   ◇ Middle East   ◯ Sub-Saharan Africa

Conclusion

The discussion above indicates that some positive relationships exist between population growth (especially rural growth) and land intensification (in the form of increases in fertilizer use), both across and within countries. However, these relationships are complicated by the *interrelationships* between land intensification and other concomitant responses to population pressures, including land extensification (increases in the agricultural land area) and outmigration, on the one hand, and by the large number of factors besides population pressure that also influence the responses. Moreover, the observed cross-country relationships are weak and dependent on the inclusion or exclusion of particular "outlier" countries. Given that the underlying data, particularly on land degradation, are unreliable, the weakness of cross-country studies provides *prima facie* support for focusing research on more detailed country-level studies. But up to now country-level studies have usually examined only a few, partial aspects of the relationships and have, moreover, rarely demonstrated *quantitative* relationships.

The review presented here helps identify the relationships that need to be investigated at the country or sub-country level, as well as the advantages of such efforts. Cross-country data are too aggregated for a proper investigation of the relationships: overall population density (at a country level or even population relative to arable land) is less important than *where* the population is concentrated, whether density thresholds (plots too small to be economically viable or human habitation too dense for the sustainability of the ecosystem) are being surpassed in *particular* areas as population continues to grow. What are the consequences in those local areas? In order to delineate the linkages between demographic responses and land intensification, it is crucial to investigate the full range of responses (in addition to intensification, as indicated in Figure 6.1) as well as the effects of other potential explanatory factors (besides population change) that influence any of the processes. It is also necessary to observe how these responses and explanatory factors have changed over time and to conduct analyses to elucidate these relationships over time. Examples of truly multiphasic responses or processes occurring over time in the developing countries abound in the literature. Two examples include western Kenya, as described by Gould (1992), and Tanzania, as described by Maro (1975, cited in Gleave 1992:5). In both cases, extensification, intensification, and increasing resort to off-farm employment and temporary outmigration and circulation all occurred. However, since the nature of the relationships depends on a number of *contextual* factors that differ from one country to another, deriving generalizations will again be difficult--at least until a number of detailed country studies have been carried out. The importance of looking at individual countries separately derives from the fundamental relevance of the context, in all its dimensions, in conditioning the nature of the linkages. Indeed, an increasing number of Latin American countries are, or soon will be, experiencing net *negative* rural population growth, as outmigration continues and rural fertility rates decline further. The nature of the population-agricultural technology linkage is certain to differ in such contexts from the usual one focused upon in the text, and may or may not involve reverse processes, viz., outmigration inducing the use of more extensive agricultural methods (including lower agricultural worker-land ratios and more use of machinery). Shortages of agricultural labor are being experienced in some areas already, and can also contribute to environmental decline where environmental management is labor-intensive. An interesting example of the latter, for Peru, is found in Collins (1988).

Sub-national data for provinces, or preferably even smaller geographic units such as districts or communities, are better suited for such an endeavor than country-level data. The use of local data facilitates so-called "ecological" or cross-area analyses, which can potentially measure--and therefore permit controlling for--a number of additional economic, institutional, and contextual factors, for which data are available only within countries, not across countries. Intra-country analysis also facilitates the inclusion of variables that reflect relevant local environmental characteristics (e.g., soil quality, topography, forest cover remaining) and a wide range of measures reflecting potential policy variables, including agricultural policies, road building and other government infrastructure, local population policy variables (family planning, zoning restrictions, limitations on inmigration), and attitudinal variables reflecting local norms (e.g., attitudes towards the environment, desired family sizes, attitudes toward inmigrants or "outsiders," etc.). Most of the latter variables can be collected only at the most micro or community level, where data can also include information from interviews with community leaders and informants on long-term historical processes in the community, which should be taken into consideration in analyzing contemporary processes of change. If information is obtained for a sufficiently large number of communities (say, a minimum of 50)--which is recommended given the low cost involved--empirical tests of hypotheses can be carried out (see Bilsborrow, Standing, and Oberai 1984). Although community-level information is likely to involve qualitative as well as quantitative variables, which complicates estimating relative effects, such analyses can better take into account the broader historical and cultural factors than either the more macro- or more micro-level analyses.

If adequate data can be obtained for a large number of small geographic areas or communities, multivariate analyses may be useful to assess the *relative* importance of the factors hypothesized as influencing land intensification and alternative related processes. It is important to not only have data over time but also to take into account lags in the relationships. Since the type of analysis suggested has not yet been carried out anywhere and is likely to be quite complicated, it will require both intensive data collection and a multidisciplinary team of collaborators. Still, such an approach is needed to ensure that the holistic aspects of the process are properly taken into account. Short of this, cross-area analyses of sub-national units, based upon existing data sets, are already feasible in some countries. Such analyses would allow a more detailed investigation of the relevant relationships than can be done using cross-country data, and may also provide results useful for policy formulation provided variables reflecting policy decisions are included.

Nevertheless, there are some important advantages in investigating the interrelations between demographic factors, land use practices, and environmental degradation at an even more disaggregated level, that is, at the household or farm level. The household is the locus for both demographic and resource-use decisions, so households are the immediate actors whose behavior must be better understood to address the rural land use and environmental problems of developing countries. That is also the level at which the most useful theorizing exists (though further work is needed) and hypotheses can be elaborated and formally tested. It is the level at which the role of women in environmental management and its linkages to her economic activities (farm work and agricultural technology, off-farm work, collection of fuelwood and water) and demographic behavior (fertility, migration, and consequent household size and characteristics) can best be examined. Last but not least, it is at the household level that full multidisciplinary collaboration in both the data collection phase and the analysis phase--involving ecologists, agriculturalists, and social scientists--can be most useful and is most needed.

References

Adepoju, A. (1978). "New conceptual approaches to migration in the context of urbanization: case of Africa south of the Sahara," in *Seminar on new conceptual approaches to migration in the context of urbanization*, Bellagio: IUSSP Committee on Urbanization and Population Redistribution, July.

Arguello, O. (1981).   "Estrategias de supervivencia: un concepto en busca de su ontenido." *Demografía y Economía* 15(2).

Arnason, T., J. D. H. Lambert, J. Gale, J. Cal and H. Vernon, (1982). "Decline of soil fertility due to intensification of land use by shifting agriculturists in Belize, Central America." *Agro-Ecosystems* 8:27-37.

Bedoya Garland, E., (1987).   "Intensification and degradation in the agricultural systems of the Peruvian Upper Jungle: the Upper Huallaga case." In *Lands at Risk in the Third World*, edited by P. D. Little and M. M. Horowitz, pp. 290-315. Boulder, Colo.: Westview Press.

Bernard, F. E., (1988). "Population growth and agricultural change in Meru district, Kenya." Paper prepared for Population Growth and Agricultural Change in Africa Conference, Florida, May 1-3.

Bilsborrow, R. E., (1987). "Population pressures and agricultural development in developing countries: a conceptual framework and recent evidence." *World Development* 15(2):183-203.

————, (1992). "Poulation growth, internal migration, and environmental degradation in rural areas of developing countries. *European Journal of Population*, vol. 8, no. 2, pp. 125-148.

____, G. Standing and A. Oberai, (1984). *Migration Surveys in Low-Income Countries: Guidelines for Survey and Questionnaire Design*. London and Sydney: Croom Helm, for the International Labour Office.

____ and P. W. Stupp, (1988). "The effects of population growth on agriculture in Guatemala." Carolina Population Center Papers, No. 88-24, Chapel Hill, North Carolina: University of North Carolina at Chapel Hill, Carolina Population Center (also presented at PAA, 1989).

____ and P. F. DeLargy, (1991). "Land use, migration and natural resource deterioration in the Third World: the cases of Guatemala and Sudan." *Population and Development Review*, supplement to vol. 16 on *Resources, Environment and Population*, edited by K. Davis and M. Bernstam, pp. 125-147.

____ and M. E. Geores, (1990). *Population, Environment and Sustainable Agricultural Development*. Monograph prepared for the U.N. Food and Agricultural Organization, Rome (unpublished).

____ and M. Geores, (1991) . *Population, Land Use and the Environment in Developing Countries: What Can We Learn from Cross-National Data?* Paper prepared for NAS Workshop on Population and Land Use, Washington, D.C., December 4-5.

—— and H.W.O. Okoth-Ogendo, (1992). "Population-driven Changes in Land Use," *Ambio*, vol. 21, no. 1, pp. 37-45.

Binswanger, H. P. and V. W. Ruttan, (1978). *Induced Innovation: Technology, Institutions and Development*. Baltimore, Md.: Johns Hopkins University Press.

Blaikie, Piers and H. Brookfield, eds. (1987). *Land Degradation and Society*, London: Methuen.

Boserup, E., (1965). *The Conditions of Agricultural Growth: The Economics of Agrarian Change under Population Pressure*. London: Allen and Unwin.

Browder, J. O., (1989). "Development alternatives for tropical rain forests." In *Environment and the Poor: Development Strategies for a Common Agenda*, edited by H. J. Leonard, pp. 111-133. New Brunswick, N.J.: Transaction Books.

Cain, M. and G. McNicoll, editors, (1990). *Population and Development Review*. Supplement to vol. 15 on *Rural Development and Population: Institutions and Policy*.

Chen, Robert S., W.H. Bender, Robert W. Kates, E. Messer and Sally R. Millman (1990), *The Hunger Report: 1990*. Providence, RI: World Hunger Program, Brown University.

Chipande, G., (1988). "The impact of demographic changes on rural development in Malawi." In *Population, Food and Rural Development*, edited by R. D. Lee, W. B. Arthur, A. C. Kelley, G. Rodgers and T. N. Srinivasan, pp. 162-174. Oxford: Clarendon Press.

Clarke, John and David Rhind (1992). *Population Data and Global Environmental Change*. ISSC/UNESCO Series 5. International Social Science Council.

Collins, Jane (1988). *Unseasonal Migrations: The Effects of Rural Labor Scarcity in Peru*. Princeton, NJ: Princeton University Press.

Crosson, P., (1983). *Soil Erosion in Developing Countries: Amounts, Consequences and Policies*. Madison, Wisc.: University of Wisconsin, Center for Resource Policy Studies.

Davis, K., (1963). "The theory of change and response in modern demographic history." *Population Index* 29(4):345-366.

Food and Agricultural Organization (FAO), United Nations (1977, 1984, 1988). *FAO Production Yearbook*. Rome.

Food and Agricultural Organization, United Nations (1991). "Second interim report on the state of tropical forests," Presented at 10th World Forestry Congress, Paris, France, September. Rome: FAO Forest Resources Assessment 1990 Project.

Foy, G. and H. Daly, (1989). *Allocation, Distribution and Scale as Determinants of Environmental Degradation: Case Studies of Haiti, El Salvador and Costa Rica*. Environment Department Working Paper No. 19, The World Bank.

Geertz, C., (1963). *Agricultural Involution: The Processes of Ecological Change in Indonesia*. Association of Asian Studies monographs and papers, no. 11, Berkeley, Calif.: University of California Press.

Ghatak, Subrata and Ken Ingersent (1984). *Agriculture and Economic Development*. Baltimore, MD: The Johns Hopkins University Press.

Gleave, Barry M. (1992). "Population density, population change, agriculture and the environment in the Third World," Presented at British Society for Population Studies Annual Conference 1992 on Population and Environment. Oxford, England, 9-11 September.

--------- and H.P. White (1969). ."Population density and agricultural systems in West Africa," in *Environment and Land Use in Africa*, M.F. Thomas and G.W. Whittington, eds. London: Methuen.

Godoy, R., (1984). "Ecological degradation and agricultural intensification in the Andean highlands." *Human Ecology* 12(4):359-383.

Gould, Bill, (1992). "Population growth, environmental stability and migration in Western Province, Kenya," Paper presented at British Society for Population Studies Annual Conference 1992. Oxford, England, 9-11 September.

Grigg, David, (1985). *The World Food Problem 1950-1980*. Oxford: Basil Blackwell.

Higgins, G. M., A. H. Kassam, L. Naiken, G. Fischer and M. Shah, (1982). *Potential Population Supporting Capacities of Lands in the Developing World*. Rome: Food and Agriculture Organization.

Horowitz, M. M. and M. Salem-Murdock, (1987). "Impact of fuelwood use on environmental degradation in the Sudan." In *Lands at Risk in the Third World*, edited by P. D. Little, M. M. Horowitz and A. E. Nyerges, pp. 95-114. Boulder, Colo.: Westview Press.

Hubbell, Stephen P., and R. Foster (1992). "Short-term dynamics of a neotropical forest: why ecological research matters to tropical conservation and management," *Oikos*, vol. 63, no. 1, pp. 48-61.

Ibrahim, F., (1984). *Ecological Imbalance in the Republic of the Sudan - with Reference to Desertification in Darfur.* Bayreuther Geowissenschaftliche Arbeiten, Vol. 6, Bayreuth, Federal Republic of Germany.

IUCN/UNEP/WWF (1991). *Caring for the Earth. A Strategy for Substainable Living.* Gland, Switzerland: The World Conservation Union (International Union for the Conservation of Nature).

Ives, J. and B. Messerli (1989). *The Himalayan Dilemma.* London: Routledge.

Joly, L. G., (1989). "The conversion of rain forests to pastures in Panama." In *The Human Ecology of Tropical Land Settlement in Latin America,* edited by D. A. Schumann and W. L. Partridge, pp. 86-130. Boulder, Colo.: Westview Press.

Kumar, J., (1973). *Population and Land in World Agriculture: Recent Trends and Relationships.* Population Monograph Series No. 12, Berkeley, Calif.: Institute of International Studies, University of California.

Lee, L. and W. Stewart (1983). "Landownership and the adoption of minimum tillage," *American Journal of Agricultural Economics,* vol. 65, pp. 1070-1075.

Lee, Ronald (1986). "Malthus and Boserup: a dynamic assessment," in *The State of Population Theory: Forward from Malthus,* pp. 96-130, D. Coleman and R. Schofield, eds., Basil Blackwell.

Lele, Uma and Stephen W. Stone (1989). "Population pressure, the environment and agricultural intensification: variations on the Boserup hypothesis," *Managing Agricultural Development in Africa/MADIA.* Discussion Paper No. 4. Washington, DC: The World Bank.

Leonard, H. J., (1987). *Natural Resources and Economic Development in Central America.* Washington, D.C.: International Institute for Environment and Development.

Lipton, Michael J. (1977). *Why Poor People Stay Poor: A Study of Urban Bias in World Development,* London: Temple Smith.

Little, P. D. , M. M. Horowitz, and A.E. Nyerges editors, (1987). *Lands at Risk in the Third World: Local-Level Perspectives.* Boulder, Colo.: Westview Press,Inc.

Malthus, T. R., (1960). *On Population* (First Essay on Population, 1798, and Second Essay on Population, 1803). New York: Modern Library, for Random House.

Maro, P.S. (1975). "Population growth and agricultural change in Kilimanjaro, 1920-1970," *BRAALUP Research Paper 40,* Dar es Salaam: University of Dar es Salaam, Tanzania.

Morán, E. F., (1987). "Monitoring fertility degradation of agricultural lands in the lowland tropics." In *Lands at Risk in the Third World,* edited by P.D. Little, M.M. Horowitz and A.E. Nyerges, pp. 69-91. Boulder, Colo.: Westview Press.

Morgan, W.T.W. (1963). "The 'White Highlands' of Kenya," *Geographical Journal,* no. 129, pp. 140-155.

Mortimore, M., (1989). *The Causes, Nature and Rate of Soil Degradation in the Northernmost States of Nigeria and an Assessment of the Role of Fertilizer in Counteracting the Processes of Degradation.* Environment Department Working Paper No. 17, Washington, D.C.: The World Bank.

Myers, N., (1984). *The Primary Source: Tropical Forests and Our Future.* New York: W. W. Norton.

206 Richard E. Bilsborrow and Martha Geores

--------, (1986). "Tropical deforestation and a mega-extinction spasm." In *Conservation Biology: The Science of Scarcity and Diversity*. Edited by M. E. Soule, pp. 394-409. Sunderland, Mass.: Sinauer Assoc., Inc.

Okafor, Francis, (1987). "Population pressure and land resource depletion in Southeastern Nigeria," *Applied Geography*, no. 7, pp. 243-256.

Parsons, Jack, (1992). "Population growth as a factor leading to conflict over land and other natural resources," Presented at British Society for Population Studies Annual Conference on Population and Environment, Oxford, England, 9-11 September.

Perkins, D. H., (1969). *Agricultural Development in China, 1368-1968*. Chicago: Aldine Publishing Co.

Pingali, P. L. and H. P. Binswanger, (1987). "Population density and agricultural intensification: a study of the evolution of technologies in tropical agriculture." *Population Growth and Economic Development: Issues and Evidence*. Edited by D. G. Johnson and R. D. Lee, pp. 27-56. Madison, Wisc.: University of Wisconsin Press.

Preston, D. A., (1969). "The revolutionary landscape of highland Bolivia." *The Geographical Journal* 135(1):1-16.

Repetto, Robert and Thomas Holmes (1983). "The role of Population in resource depletion in developing countries," *Population and Development Review*, vol. 9, no. 4, pp. 609-632.

Robinson, Warren, and Wayne Schutjer (1984). "Agricultural development and demographic change: a generalization of the Boserup model," *Economic Development and Cultural Change*, vol. 32, no. 2, pp. 355-366.

Rudel, Thomas K. (1990). "Rapid Population growth and environmental degradation in rural areas of developing countries: evidence and research needs," Paper prepared for the United Nations Population Division, Department of Economic and Social Development.

Rushahigi, C., (1985). "Les obstacles lies au programme de population et de production au Rwanda." *Famille, Sante, Developpement* (3):44-49.

Russell, W. M. S., (1988). "Population, swidden farming and the tropical environment." *Population and Environment* 10(2):77-94.

Ruttan, V. W. and Y. Hayami, (1989). "Rapid population growth and technical and institutional change," in U.N. Proceedings of U.N. Expert Group Meeting, New York, August 23-26, 1988, pp. 393-428.

Sánchez, P. A. and J. R. Benites, (1987). "Low-input cropping for acid soils of the humid tropics." *Science* 238:1521-1527.

Stonich, Susan (1989). "The dynamics of social processes and environmental destruction: a central American case study," *Population and Development Review*, no. 15, pp. 269-195.

Tiffen, M. (1975). "Population movements in the twentieth century: a Nigerian case study," in *The Population Factor in African Studies*, R.P. Moss and R.J.A.R. Rathbone, eds. London: University of London Press.

Turner II, Billie Lee, R.Q. Hanham and A.V. Portararo (1977). "Population pressure and agricultural intensity," *Annals of the Association of American Geographers*, vol. 67, no. 3, 384-396.

United Nations (1988). *World Population Prospects as Assessed in 1988.* New York: United Nations.

Uyanga, J., (1985). "The demographic impacts of the Cross River Plantation Projects in Nigeria." In *Impact of Rural Development Projects on Demographic Behavior,* edited by R. Bilsborrow and P. DeLargy. New York: UNFPA, pp. 99-132.

Walker, Robert (1992). "Population and environmental degradation," Institute of Tropical Forestry, Rio Piedras, Puerto Rico (unpublished).

Wolman, M. G. and F. Fournier, (1987). *Land Transformation in Agriculture.* John Wiley and Sons.

World Bank, (1990). *World Development Report - Focus on Poverty.* Washington, D.C.: The World Bank.

--------, (1991). *Forest Sector Review.* Washington, DC: The World Bank.

--------, (1992). *World Development Report - Development and the Environment,* Oxford: Oxford University Press.

World Commission on Environment and Development (1987). *Our Common Future.* Oxford, England: Oxford University Press.

World Resources Institute (1990). *World Resources 1990-1991.* New York: Oxford University Press.

Yates, P. L., (1981). *Mexico's Agricultural Dilemma.* Tucson, Ariz.: University of Arizona Press.

7

The Social Context of Land Degradation ("Desertification") in Dry Regions

Peter D. Little

Most reports about desertification base their arguments on a litany of statistics, themselves derived from conflicting definitions. In a highly variable environment, how could one measure the advance of the desert edge, when moreover that edge itself is ill-defined (to say nothing about the poverty of the data)? (Warren and Agnew 1988:3).

This chapter attempts to assess current understandings of land degradation in dry regions, with an emphasis on social science contributions. As used here, a "dry" region is an area where annual rainfall is less than 800 mm. This interpretation approximates the climatic definition of arid and semiarid lands. While the focus of this volume is on "population and the environment," it is argued here that factors other than those based on demography are considerably more important in accounting for land degradation in dry regions. The unwarranted emphasis on population growth as the primary cause of desertification (Steeds 1985) diverts attention from more exigent social and

I wish to acknowledge the assistance of Matthew Richard, who helped me gather many of the bibliographic sources for this chapter, and Vivian Carlip, who edited an earlier version of this chapter. In addition, very helpful comments on an earlier draft were provided by David Brokensha, Michael Horowitz, Michael Painter, Marianne Schmink, and Michael Watts.

210   *Peter D. Little*

political processes. I suggest that tautological arguments, including those based
on the "population hypothesis," have dominated the discourse over degradation
in dry areas and that an unhealthy reliance on aggregate data and macro theories
has distorted what is actually happening in local communities deemed to be
threatened by desertification (for example, those of regions covered by the
desertification maps, see UNCOD 1977a).

Part of the problem has been that dry regions have been defined as
environmentally problematic ("fragile") areas, and this has influenced the types
of research questions that have been addressed. This chapter stresses that the
negative connotations associated with land use in dry regions have focused too
much attention on environmental issues and not enough attention on the social
and economic relations that influence resource use. This is particularly tragic
and misdirected since, perhaps, the most noteworthy characteristic of dry
regions is their correlation with poverty and underdevelopment. As Kates and
Haarmann demonstrate so well, "The incidence of poverty and dryness correlate
closely in developing countries . . . Overall the [world's] poorest countries are
primarily dry" (1992:7).

It is suggested here that a focus on *population distribution* rather than
population growth and fertility rates is more appropriate for examining
relationships between environmental change and demography. A review of case
studies supports Mortimore's finding, based on more than 20 years of fieldwork
in northern Nigeria, that regarding the desertification issue the "population
pressure hypothesis is . . . a gross oversimplification" (1989:211). In fact, in
a country like Kenya, where drylands account for more than 70 percent of the
territory, population densities and growth rates are considerably lower in the dry
regions than in other parts of the country. While Kenya's annual population
growth rate 1979-1989 is estimated at 3.34 percent, the five largest arid districts
have growth rates during this period--Turkana (2.25 percent), Garissa (-0.40
percent), Mandera (2.64 percent), Wajir (-1.06 percent), and Marsabit (2.64
percent)--that are well below the national average (Kenya 1991:32). Those
increases that have taken place in these districts are in the main towns. For
example, in the two districts, Wajir and Garissa, that experienced negative rates
between 1979 and 1989, the population of district capitals (towns) increased by
more than 90 percent and, in the case of Wajir town, the growth was in excess
of 200 percent (Kenya 1991:35). Such uneven rates of population growth, even
at the level of a district, highlight the importance of disaggregating demographic
data to determine patterns of distribution across different locations.

Land degradation, then, may be associated with how populations are
distributed within dry regions and between dry and other areas. Social,
political, and economic forces that restrict population mobility, concentrate
populations around certain locations while leaving vast expanses of territory
underpopulated, and result in movements of people from higher-rainfall,
"overcrowded" parts of a country into dry areas underlie many of the

environmental problems in dry zones. As will be shown later, residents of dry regions--particularly pastoral herders--generally have lower fertility rates than other populations (Hill 1985; P. Little 1989). Of course, exceptions to this pattern can be found. Leslie et al., for example, suggest that pastoral Turkana of northern Kenya have higher fertility rates than those practicing a sedentary life-style (1988).

The paper is divided into four parts. The first section briefly reviews the origins of the debate and the contradictory concepts and assumptions that surround it, and suggests that the lack of consensus on definitions and concepts has hindered research on land degradation in dry regions by failing to distinguish between manmade and climate-induced transformations and by confusing irreversible with reversible ecological change. The second part offers a conceptual framework that highlights the important social and institutional factors underlying land degradation in dry regions. It suggests that a coherent framework is required to filter the prolific variety of human experience and institutions and to focus on the most relevant variables.

Part three presents three case studies of desertification drawn from Africa and Asia that demonstrate the salience of social and economic variables, and that synthesize biophysical and socioeconomic data. The case studies attempt to contextualize locally--in space and time--a debate that is often conducted at a level of abstraction and aggregation that has little meaning for local process and field research. Areas were selected that appeared by global assessments to have "high" to "very high" rates of desertification (UNCOD 1977b; Odingo 1990a). Because of the large number of African-based studies and because this continent is depicted as the most affected by desertification (according to some estimates 55 percent of the continent is threatened by desertification [Rosanov 1990:59; also see Mabbutt 1989:103]), two of the examples are drawn from this region. Asian countries with large expanses of arid and semi-arid territories, such as China and Afghanistan, have received some attention in the literature (Cincotta, Yanqing, and Xingmin 1992; Dregne 1983; Odingo 1990a; UNEP 1984), but very little field research has been carried out recently in these countries and virtually none of it by social scientists. The lack of data on the social dimensions of degradation in these countries makes it difficult to include them as case studies. Yet even on the best-documented continent, Africa, the lack of longitudinal data on land use and environmental change inhibits definitive statements about degradation.

In the final section, I present some of the policy implications derived from the paper's findings and discuss how governments are attempting to deal with problems of land degradation. The extent to which government and development agencies have responded (or not responded) to new research findings about degradation is also addressed. The discussion concludes by outlining an agenda for future research on desertification that incorporates both policy concerns and the interests of the larger social science community.

Two important disclaimers are necessary. First is that this paper, written by a social scientist, cannot do justice to the biophysical research on degradation and the subtle biological arguments that are essential for assessing the extent to which, borrowing from Binns (1990), desertification is "fact" or "myth." While there are a few truly interdisciplinary attempts that draw on both social and biological perspectives (Abel and Blaikie 1989; Scoones 1990; and the research in India sponsored by International Crops Research Institute for the Semi-Arid Tropics (ICRISAT) and the Central Arid Zone Research Institute--see Jodha 1980; Spooner and Mann 1982), most studies emphasize either the socioeconomic or biological dimensions of the phenomenon. Even in so-called "interdisciplinary" projects on desertification, such as the Integrated Project in Arid Lands (IPAL) program in Marsabit, Kenya, a synthesis of the social and biological is difficult to achieve, in this case resulting in conflicting positions by members of the same research team (see O'Leary 1984, 1985; IPAL 1984; and Lamprey and Yussuf 1981). Since the biophysical evidence is often invoked to demonstrate the inadequacies of local social practice and institutions, I do make an attempt to assess the adequacy of the biological data and the logic of ecological arguments. This means making the physical scientist accountable to the same standards of data analysis and quality and logic of argument by which any scholar (social scientist or other) should be scrutinized. It does not imply that a social scientist can substitute, for example, for a range scientist or agronomist in determining the biophysical processes underlying land degradation.

The second disclaimer deals with the complexity of the issue at hand and the difficulty of addressing it within the confines of a relatively short essay. A bibliography on the topic alone would run to several thousand entries (see Glantz and Orlovsky 1986; Mortimore 1989). In order to make the paper manageable, I have narrowed the topic by limiting the discussion to dry regions of Africa and Asia and by emphasizing certain economic activities (e.g., pastoralism) over others (irrigated agriculture). While desertification is now seen as a concern in the sub-humid and humid tropics (see Goodland and Irwin 1975; Mabbutt 1989; Cloudsley-Thompson 1988), these higher rainfall areas are not included, in part because their economic activities and productivity differ considerably from those of the drier areas. The paper briefly touches on the relationship between urbanization (settlement growth) and desertification, but cannot give it sufficient attention in a relatively short essay. Recent research on peri-urban land use in India and Nigeria (Mortimore 1989) demonstrates the marked effects--often negative--that urban growth can have on dryland ecologies. In some cases, I may have been guilty of overgeneralization and insensitivity to local variation, the same accusations that I attribute to the literature on desertification (see the discussion later in the paper).

The Debate

Few intellectual debates about the environment are marked with more ambiguity, scientific "self-righteousness," and ideology than the debate about land degradation in dry regions or what is more commonly called "desertification." First noted as a problem in Africa in the 1930s (see Stebbing 1935, 1938), desertification emerged as a major environmental issue in the 1970s with the publicity surrounding the prolonged Sahelian drought and the speculation regarding its causes and ecological effects (Glantz 1977; Eckholm and Brown 1977; Grove 1977; MAB 1975). Concerns about desertification were voiced in the Middle East as early as the nineteenth century. Blame for land degradation in the region was placed unambiguously on the Bedouin nomads: "To call him a 'son of the desert' is a misnomer; half the desert owes its existence to him" (Palmer 1977:297, cited in Horowitz and Little 1987:65). The convening of a major UN conference on desertification in the mid-1970s (UNCOD 1977a), the establishment during the same period of a research and policy unit--the "Desertification Branch"--at the United Nations Environment Programme (UNDP), and the increased availability of funding for research and projects on desertification heightened awareness of it among the general and scientific communities and propelled it into the center of the environmental dialogue.

Scholars and practitioners of several disciplines, from history and political science to range management and geography, engaged in the debate, while the "field" gained further status by the introduction of journals devoted to the topic (e.g., *Desertification Control Bulletin*) and its treatment by such prestigious academic organizations as the National Academy of Sciences (1975) and the American Association for the Advancement of Science (AAAS) (see Reining 1978). An entire paper easily could be devoted to how the desertification debate was constructed, how political and institutional factors contributed to its production and reproduction, and how "science" was invoked to justify the excessive funds and projects allocated to such an elusive issue. Projects were hastily designed and implemented, with little clear understanding of the nature of the problem or agreement over what constituted desertification; many of them received strong political and financial support from multilateral (e.g., UN bodies and the World Bank) and bilateral foreign aid programs (e.g., US, French, and Nordic aid agencies). Henry Kissinger, the former US Secretary of State, spoke in the 1970s of "rolling back the desert" in Africa (cited in Watts 1987:172), while as recently as 1986 a member of the European Parliament "declared that aid must go to the Sahel, because the desert was advancing at 8 km a year" (Warren and Agnew 1988:2). Since assumptions about the causes of desertification often pointed to humans and their institutions as culprits, social scientists were increasingly drawn into the dialogue (Grove 1977; Horowitz 1976; Sandford 1982).

Several publications examine the origins of the debate about desertification, the different arguments that have been posited, and the ways that governments, aid agencies, and scientists have responded to the perceived "crisis" (Glantz 1977; Spooner and Mann 1982). The topic has received so much attention during the past 25 years that several recent works are devoted solely to reviewing the history and development of this "school" (Odingo 1990b; Spooner 1989; Glantz and Orlovky 1986; Helldén 1990). While not trying to rehearse old stories, it is nonetheless worthwhile to summarize some of the major points and events that have influenced the debate about land degradation in dry regions. How have understandings of degradation changed in the last 50 years? What new concepts and theories, including those derived from population studies, have been advocated in recent years? And what roles have ideology and politics played in shaping the debate about desertification?

History

Concerns about land degradation in dry regions were voiced in many colonial states, fueled in part by the publicity surrounding the "dust bowl" crisis of the 1930s in the U.S. As noted earlier, Stebbing was perhaps the first to publicize the threat of desertification, referring to the "creeping desert" of northern Nigeria in the 1930s, but concerns about dryland degradation were voiced elsewhere at about the same time. Writing on the Baringo area of northern Kenya, for example, Maher warns about the ecological effects of overgrazing, which had turned the district into a dust bowl, as well as the "agricultural slums of Kenya" (1937). The soil erosion and conservation "campaigns" of the 1930s in Africa, a favorite theme of Africanist historians (Beinart 1984; Anderson and Grove 1987; McCracken 1982), dealt in part with land degradation in dry regions. One of the motivating factors in East Africa for an increased attention to conservation at the time was the heightened concern that "the region was becoming progressively more arid" (Anderson 1984:323). In Kenya particular attention was focused on the semi-arid areas and on the need to halt the "overstocking" of pastures by herders. As in West Africa and elsewhere, the environmental degradation of dry regions that was observed was attributed to human agency. Local husbandry methods, therefore, had to be altered, which usually implied forced destocking, compulsory pasture rotation, and/or the construction of soil conservation works. The first so-called land rehabilitation projects in dry regions of Africa can be traced to this period.

The "creeping desert" scenario of Stebbing and other advocates (Lamprey 1975; Rapp, Le Houérou, and Lundholm 1976) was replaced in the 1980s by a more subtle and complex portrait of the phenomenon that depicted desertification as "pulsating deteriorations . . . radiating out from centers of excessive population pressure" (Nelson 1988:1). Rather than "waves" of sand (e.g., Lamprey's southward movement of the Sahara [1975]), degradation in drylands was seen in terms of "pockets" of deterioration, often surrounding major water points or human settlements (Bernus 1981; IPAL 1984; Coughenour et al. 1985). An important exception to this seems to be UNEP and related groups, which still adhere to alarmist notions of an advancing desert (UNEP 1991). In many cases, however, the culprit remained human agency, and whether perceived in terms of waves or pockets of degradation, the blame for desertification was attributed to the activities and institutions of herders and farmers.

Considerable doubts about whether or not desertification actually existed (or still exists) began to be voiced strongly by the mid-1970s (Warren and Maizels 1977; Sandford 1976), and picked up considerable momentum in the 1980s, when the results of several long-term studies became available (see Ellis and Swift 1988; Homewood and Rodgers 1987; IPAL 1984; Abel and Blaikie 1990). Natural scientists began to question the biological evidence proffered, the parameters that were being measured, and the simplicity with which arguments about desertification were being stated. Even among the most strident believers, a consensus seemed to be reached that the extent and seriousness of land degradation in dry regions, as reflected in the papers of the UN Conference on Desertification in 1977 and in other publications of the 1970s and 1980s, were grossly overstated (Nelson 1988). Figures on the southward expansion of the Sahara, for example, had been greatly exaggerated because of the harsh climatic conditions of the 1970s and early 1980s (Nicholson 1982; Tucker, Dregne, and Newcomb 1991), and doomsday specters of environmental and economic collapse in many dry regions were misdirected. The Saharan advance that had been attributed to processes of desertification was later found to be more a result of a prolonged period of low rainfall than of mismanagement of land (Ahlcrona 1988; Tucker, Dregne, and Newcomb 1991; Warren and Agnew 1988). The failure to distinguish drought-induced changes, reversed when precipitation reappears, from secular changes in the environment has muddled many arguments about desertification, particularly those presented in the 1970s when the Sahelian drought occurred (Glantz 1977; National Academy of Sciences 1975; UNCOD 1977b). Indeed, it is more than a coincidence that the environmental alarm over degradation in dry regions of Africa emerged during two particularly dry periods of this century, the late 1920s to 1930s and the late 1960s to 1970s. As will be shown later, this correspondence reflects fundamental misunderstandings about arid land ecology, climate, and economy that have plagued scientific arguments during the past several decades.

Social scientists, too, have been skeptical, questioning many of the behavioral assumptions regarding the causes of desertification; they have also voiced concerns about some of the environmental arguments (Horowitz 1979; Horowitz and Little 1987; Sandford 1982; Watts 1987). Perhaps the most publicized of the social science debates concerns land tenure and the extent to which common property ownership, a prevalent form of tenure in dry regions, fosters resource abuse. The term "common property" masks the considerable complexity of these tenure systems, which can range from loosely regulated ("open access") to strictly regimented institutions where rights to land can approach the definition of private use. For a useful assessment of current understandings of common property systems, see National Research Council (1986). Characterizing certain situations as the "tragedy of the commons," Hardin (1968) and others (Hare 1977; Hopcraft 1981; Cook 1983) speculate that much of the degradation in dry regions results from the contradictions inherent when animals are owned privately but land and other resources are held in common. The maximizing individual, therefore, can optimize output from his/her private herd without assuming any of the costs (externalities) of resource degradation that stem from overuse. According to this logic, degradation in dry regions is caused by inadequate property systems and incentives, which need to be altered if conditions are to improve.

In opposition to this view, a considerable amount of social science research demonstrates that tenure systems in dry regions: regulate access to users and do sanction abusers; have mechanisms in place to conserve resources at certain times of the year and to guard against mismanagement; and are usually economically, socially, and environmentally more effective than more exclusive (private) forms of ownership (see Sandford 1983; McCabe 1990; Little and Brokensha 1987; Scoones 1989a; and deRidder and Wagenaar 1984). While the evidence against a "tragedy of the commons" in dry regions appears overwhelming--especially to those who have actually lived and carried out fieldwork in low-rainfall areas--it still has support among many scholars and policymakers. For example, in a recent publication on the Sahel, Le Houérou argues: "Water and pastures are actually common public resources, whereas animals are individually and privately owned; so it is in every user's interest to draw a maximum and immediate profit from communal resources without considering what may happen in the long run . . . this is the 'tragedy of the commons' or the 'looting strategy'" (1989:239). This position also appears to have received recent support from the World Bank (see Gorse and Steeds 1987). The "tragedy of the commons" and variations on this theme remain salient in debates about the causes of desertification.

Other assumptions about the causes of land degradation have attracted considerable attention from social scientists. One of these is the relationship between overpopulation and desertification, which has been debated in the literature since at least the 1930s. Excessive population was frequently cited as a cause of land degradation in the British colonial states, and it was the impetus for some of the earliest resettlement schemes in Africa and Asia. Population continues to figure prominently in current debates about the causes of desertification (Jodha 1991; Mabbutt 1989; Cloudsley-Thompson 1988; Steeds 1985), although the Malthusian warnings of catastrophic famines and die-offs caused by desertification have been tempered. As Caldwell (1984) points out, instead of a Malthusian scenario many of the countries said to be most seriously threatened by desertification (e.g., India and the Sahelian states) experienced population growth rates during the 1970s that differed little from other regions. As in most debates about the relationship between overpopulation and resource management, questions of access and distribution are rarely acknowledged (an important exception is Blaikie and Brookfield 1987). The so-called degradation caused by overuse and reduced land/people ratios (population growth) may be a function of inequities in land distribution or a historical process that alienated large parcels of land for alternative uses and crowded the "natives" into less fertile reserves (in Zimbabwe and Kenya, for example). By focusing on a variable like the land/people ratio, the primary social and historical causes underlying the figure are mistakenly avoided.

Toward a Definition of Degradation

Discussions about desertification have been hampered by the lack of a clear definition of the phenomenon. The term has been defined in so many different ways that a literature devoted to reviewing and posing different definitions of desertification exists (Ahlcrona 1988; Odingo 1990b; Spooner 1989). In a retrospective article, Glantz and Orlovsky, for example claim that academics and policymakers use more than 100 definitions of desertification (1986). While they point out that definitions are frequently contradictory, there is a consensus that desertification is a negative process--a change from "a favored or preferred state (with respect to quality, societal value, or ecological stability) to a less preferred one" (1986:216-217). The negative descriptions range from the deterioration of ecosystems (Reining 1978), to a spread of desert-like conditions (Rapp, Le Houérou, and Lundholm 1976), to a reduction of resource potential on arid and semiarid lands (Helldén 1990). Because some definitions focus on changes in the composition of vegetation, the term "green desertification" has come to mean a change in vegetation quality, even when green biomass may actually increase (Warren and Agnew 1988:6). Earlier definitions of desertification allowed for natural or climatically induced degradation, but recent interpretations limit the process to irreversible degradation that has been caused principally by human agency (see Helldén 1990; UNEP 1991). Even so, most analyses of land degradation in dry regions still confuse reversible change, often instigated by climate and human activities, with desertification.

This paper follows the lead of Warren and Agnew (1988) and opts to use the less value-laden term "degradation" over desertification. Degradation can encompass desertification, but "is undoubtedly more widespread and, in aggregate, much more the serious menace to sustainability" (Warren and Agnew 1988:1). The creation of "desert-like" conditions, implied in the term desertification, is not very prevalent and has little applicability to the types of ecological changes that are actually taking place in dry regions.

The definition of degradation that I use here and find useful for describing processes in arid and semi-arid areas is that of Abel and Blaikie (cited in Behnke and Scoones 1991:16):

> Range degradation is an effectively permanent decline in the rate at which land yields livestock products under a given system of management. "Effectively" means that natural processes will not rehabilitate the land within a timescale relevant to humans, and that capital or labour invested in rehabilitation are not justified . . . This definition excludes *reversible* vegetation changes even if these lead to temporary declines in secondary productivity. It includes effectively irreversible changes in both soils and vegetation.

Although this definition pertains to rangelands and livestock production, it is generally applicable to dryland ecologies and to other land use activities. Three important points related to desertification can be made about Abel and Blaikie's interpretation. First is the recognition, in contrast to other definitions, that both natural and human factors can contribute to degradation. By not excluding natural processes from their statement, Abel and Blaikie's (1989) description more closely approximates the realities of dry regions. Land use decisions in these zones are dramatically influenced by physical processes, such as rain-induced erosion, and by the unpredictability of nature, and these variables should be included in explanations of degradation (this is discussed in detail later in this section).

A second important element of Abel and Blaikie's definition is that it excludes reversible environmental change. In environments characterized by highly erratic rainfall and uncertainty, much of the ecological change is climatically induced, reversible, and, therefore, not really degradation. However, degradation does include changes that require either unrealistic levels of capital and labor or an unrealistic time frame for rehabilitation. Degradation in dry regions, then, cannot be determined from one or two years of observation, and it is questionable whether even a decade of data collection is sufficient.

The third, and perhaps most important, part of Abel and Blaikie's definition is that it introduces an economic and social element by using the phrase "under a given system of management." As Behnke and Scoones (1991) correctly point out, degradation assessment, as defined here, does not try to determine which land use system is "best" for a rangeland. Rather it "attempts to assess the capacity of a given management system to maintain those features of the natural environment which are essential for its continued well-being" (1991:16). What is critical, therefore, is identification of the goals and objectives of the particular management system. Is it to maximize the number of people that can be maintained on a certain land area (a concern of many indigenous pastoral systems) or to achieve maximum profits from yields of beef (a concern of most commercial ranching enterprises)?

Abel and Blaikie leave it to the policymakers to determine a "socially acceptable trade-off between the interests of present and future generations" (1990:20) and to producers and communities to determine what are acceptable levels of resource utilization and conservation consistent with expected levels of social welfare. Their interpretation recognizes that degradation is a continuous, dynamic process that cannot be captured by simplistic concepts like "carrying capacity" or stocking rates. Like the explanations of Biot, Sessay, and Stocking (1989) and Sandford (1982), this definition must be meaningful to local users and to decision makers who formulate policies for these regions. What is environmentally optimal in terms of slowing the process of degradation may be economically and socially so unrealistic that it warrants little serious attention.

New Understandings of Dryland Ecosystems

The definition discussed above reflects recent insights into rangeland (arid and semi-arid) ecologies that are significant departures from the past. In addition to Abel and Blaikie's (1990) study, Westoby, Walker, and Noy-Meir (1989), Biot, Sessay, and Stocking (1989), Ellis and Swift (1988), Behnke and Scoones (1991), and Scoones (1990) challenge the conventional wisdom that has been used to assess ecological change in dry regions. It is important to summarize the findings of this recent work before turning to what are more strictly socioeconomic and institutional concerns.

The first, and perhaps most important, outcome of recent ecological research concerns the stability of arid and semi-arid environments. Much of the literature on desertification assumes that a stable environment, where animal stocking and human population rates can conform to a specified threshold or carrying capacity, is highly desirable. Exceeding this threshold (the stable level) can result in degradation and instability for the system. By contrast, data collected over the last decade reveal that most dryland zones are inherently unstable and, therefore, attempts to adjust conditions to some notion of "stability" violate the natural order and in themselves are destabilizing. These findings particularly confront the "carrying capacity" concept for its failure to recognize the variability and patchiness of dryland ecologies (Scoones 1989b) and for its requirement of quantifying a process as dynamic as the feeding habits of different livestock species (Bartels, Perrier, and Norton 1991; de Leeuw and Tothill 1990). Moreover, contrary to conventional wisdom, dry regions "are intrinsically 'resilient' compared with more 'stable' ecosystems" (Abel and Blaikie 1990:20), because they must deal with such climatic perturbations and instability. They are not the "fragile" ecologies that much of the literature portrays (Hare 1977; UNEP 1991).

Second, the inherent instability of these systems raises questions about the "proper" use of arid vegetation and lands. Westoby, Walker, and Noy-Meier (1989) have argued that because standard vegetation models view equilibrium and advanced/"climax" vegetation as desirable, they are inappropriate. Instead, we should question if climax stages of vegetation are good and treat other accepted truths as problematic. Other research, for example, has shown that commonly used indicators of land degradation, such as the replacement of perennial grass species by annuals or the encroachment on grasslands by bush species, are questionable (Penning de Vries et al. 1982; Scoones 1989a). Annual grasses can contribute more protein to an animal's diet than perennials, while bush ("woody") plants provide important browse resources during the dry season and can help stabilize and provide nitrogen to soils (Scoones 1989a:11). These findings indicate that we know far less about what a degraded rangeland looks like--in terms of plant cover--than earlier work had led us to believe. In short, a degraded or desertified parcel of land may have little to do with many of the standard vegetation indicators that had been previously used.

Finally, innovative research on soil erosion, a common symptom of degradation, also raises doubts about earlier understandings of dryland ecologies. Biot (1990), for example, demonstrates with data from Botswana that some degree of soil erosion is necessary to sustain productivity in arid regions. Rather than focus on site-specific rates of erosion, larger units, such as a catchment, must be observed, as well as the redistribution of soil from one location to another. "Thus, soil is not so much lost from, but redistributed within, catchments" (Biot 1990:4), and transported to areas where vegetation has a better chance of growing, either due to geologic or topographic factors. From this perspective, a dry ecosystem is likely to be more sustainable if there is a varied landscape, with some areas being *net losers of soil* ("runoff" areas: see Penning de Vries and Djiteye 1982; Pickup 1985) and others *net gainers of soil* ("runon" areas: see Biot 1990). In this sense, desertified locations contribute to the *diversity* of the landscape and may be necessary to sustain the system as a whole. Because rainfall is low and erratic in these areas, vegetation growth is better promoted by redistributing ("eroding") soil to areas that are most suitable for plant growth.

I can hardly do justice here to the complexity of the ecological arguments summarized above. Yet, what seems to be implied by recent research "is the dumping of 50 years of conventional wisdom" (Drinkwater 1990:10) about rangeland ecologies. This is "wisdom" that has figured prominently in the construction of the desertification debate. While considerable uncertainty remains about arid and semi-arid environments--in many instances more than exists about the sociology and political ecology of these areas--the new information helps to illuminate many of the behavioral and institutional processes observed in these areas. The chapter now turns to those social and cultural variables that mediate natural resource use, and that are important for understanding the causes of degradation in dry regions.

Social and Institutional Factors

In analyzing the social dimensions of land degradation in dry regions, it is tempting to start with the assumptions and premises that have defined the desertification debate. Several excellent works have been devoted to doing precisely this, that is, critically examining the assumptions about population growth, land tenure systems, and other sociological factors that have been invoked in the desertification dialogue (see Horowitz 1979; Homewood and Rogers 1987; Watts 1987). This body of work, not surprisingly, reveals that many of the social premises are pathetically flawed, on both empirical and theoretical grounds. Rather than repeat work that has already been skillfully covered, it is critical here to pose new sets of questions, and to establish new premises and an analytical framework for examining land degradation. Since an important goal of this paper is to examine the micro dimensions of the desertification debate, variables that have been examined in the field and for which historical (longitudinal) data exist are emphasized. The goal of this section, therefore, is to pose a set of questions and a list of variables that can be used to assess the case studies reviewed later in the paper.

An Analytical Framework

Notwithstanding the complexity of the literature that now exists on the sociology and political economy of desertification, the issue of land degradation in dry regions seems to center on relationships surrounding three basic factors: land, labor, and income (or material welfare). With few exceptions (Blaikie and Brookfield 1987; Watts 1987), theoretical discussions of land degradation are generally lacking and often ill-conceived. The "political ecology" literature, for example, has helped to focus attention on important historical and structural factors but it is increasingly ill-defined in terms of theory, incorporating ideas and concepts from neoclassical economics, human ecology, and marxism. Relationships involving land (including the resources on it, such as water) are enmeshed in the political question of access, while the distribution of labor and welfare influence how resources are actually allocated. Using these three variables (land, labor, and income), a framework for examining land degradation can be constructed that incorporates both structural and behavioral variables. Recent work on the political ecology of land degradation is instructive here, since it highlights the structural and political conditions that facilitate resource abuse (Schmink and Wood 1987; Blaikie and Brookfield 1987). However, this perspective needs to be complemented by careful analyses of local resource use and of the behavioral factors that affect degradation. It is not enough to understand the structural conditions underlying environmental degradation--the "political ecology"; we also need to examine how these are played out locally in terms of institutions and land and labor use. To my knowledge, this type of integrated analysis of land degradation has only rarely been carried out.

For illustrative purposes, I discuss this model in the context of Baringo, Kenya, a region often depicted in the literature as a "man-made desert" (Maher 1937). The structural and historical parameters affecting land, labor, and income relations include such factors as state appropriation of land, unfavorable access to markets, and policies that encourage concentrated settlements--often dominated by outsiders--in dry areas. In Baringo, while the causes of low, unequally distributed income tend to be structural and historical, analysis at the macro level alone does not reveal the behavioral processes that influence the direction of change in Baringo. It is necessary to trace the ways in which low incomes and labor shortages affect homestead decisions about production, marketing, and land management, as well as the ability of homesteads to earn a living without degrading the resource base.

The model starts with low-income homesteads that confront problems of producing sufficient food and income to meet consumption and expenditure needs (for a more detailed discussion, see Little 1992). To make ends meet, they divert labor from herding to wage employment and other nonpastoral activities. They also begin to mortgage their futures by selling off productive animals (assets) from their herds and by entering into client-like relationships with wealthy absentee herd owners who overuse local pastures. The latter group does little to enhance the resource base, but instead pursues short-term goals of accumulating profit that reinforce land management problems. Diversification into wage employment and other activities constrains the amount of labor available for livestock and agricultural production in Baringo (for a Latin American example, see Collins 1987).

Labor shortages, in turn, result in land and herd management practices that may be harmful to the local ecology, ill-adapted to drought conditions, and economically (if not ecologically) unsustainable over time. They provoke the concentration of human and livestock populations around settlements because of constraints to moving herds onto distant, underutilized rangelands. Wealthy herders and powerful outsiders (such as merchants and government officials) benefit from the local impoverishment by purchasing animals cheaply at stress (drought) sales, hiring local pastoralists to manage their herds, and alienating common pasture for private use. The population problem in the dry zones of Baringo is a consequence of this phenomenon, as well as of excessive immigration from neighboring, overcrowded areas. The migratory process heightens competition over land and other resources, resulting in sharp, sometimes violent, conflict between the resident population and migrant and outsider groups. As impoverishment and differentiation increase, indigenous tenure institutions are weakened and the customary "division of labor" is altered, so that females are increasingly obligated to herd livestock while men are drawn off to wage employment outside of the region. Household structures are fragmented and women are burdened with more and more work. The net results for the majority are problems that become evident physically in localized environmental problems and socially in increased impoverishment and conflict.

While specific to one particular dry region and vastly oversimplified, the framework discussed above does point to certain social variables that should be examined. A more careful assessment of these factors is worthwhile, as their local expression may vary depending on specific cultural, historical, and environmental conditions. As indicated earlier, the noteworthy issues center around land, labor, and income relationships, which themselves are being rapidly transformed by regional and global market forces.

Land Relationships

Considerable research recently has been conducted on land tenure systems in dry regions. Much of it has been carried out by anthropologists and geographers (Downs and Reyna 1988; Peters 1988; Bassett 1988), but economists also have shown an interest in the topic (Runge 1986). The recent concern with common property systems has resulted in considerable work on arid land economies and ecologies, since dry rangelands are an ideal case of a public (common) resource (see National Research Council 1986). While the "tragedy of the commons" argument remains prominent in policy circles, current evidence suggests that a tragedy of the "tragedy of the commons" may now be taking place. The tenure reforms--such as private enclosure of communal lands--that are being implemented as a solution to the "commons" enigma actually facilitate the kinds of tenure ambiguities and environmental pressures that were originally associated with common property systems. The onslaught that is taking place in many dry regions stems partially from actions taken by individuals and governments to confront a perceived "tragedy of the commons."

Current emphasis should be placed on identifying trends that narrow land and water rights and provide for more exclusive use of land in dry regions, either through spontaneous individuation, outright land reform (the individual land titling of rangelands in Kenya, Botswana, and India), or other restrictions on use. Population mobility is an essential strategy for confronting the spatial and seasonal heterogeneity of dry regions, and therefore a narrowing of use rights that restricts mobility and removes critical resources from collective use creates land management problems (perhaps even desertification).

The role that the state (political economy) plays in facilitating changes in land relationships also needs careful assessment. The study of local community/state relations takes on special significance in dry regions because their inhabitants play such a peripheral role in most states (important exceptions to this, of course, are found in the Middle East and North Africa). For example, sub-Saharan African herders rarely have had the political power that other groups have experienced (Horowitz and Little 1987). As Bishop points out in the case of Mali, "Pastoralists . . . frequently suffer loss of grazing rights, when disputes with farmers lead to administrative intervention" (1988:7). With few exceptions, the policies of African states (colonial and independent) regarding land can be construed as antipastoral, especially vis-à-vis policies for settled agricultural and urban populations (Galaty et al. 1981). In many dry regions obvious examples of state intervention are found, including the loss of rangelands due to state development schemes, wildlife parks, and planned settlements that promote irrigation or dryland agriculture (Anderson and Grove 1987; Bassett 1988; Hitchcock 1980). Like the processes discussed earlier, these actions give rise to land management problems because they encourage unfavorable land/people ratios and excessive use of remaining lands.

To what extent are local regulations on resource use affected by state interventions? Have local institutions in dry regions been able to withstand interventions by the state and outside groups? Often local regulations on resource use have proved ineffective in halting encroachment by outside groups and organizations, including those promoted by the state. Governments have sometimes undermined the power and autonomy of local organizations vested with resource management responsibilities, leaving a vacuum that outsiders use to increase their control of resources. Some informative data are now available on how the state has heightened uncertainties over land rights and removed control from local organizations. In the Sudan, for example, the state took control of range regulation from local authorities in the late 1960s (Haaland 1980), while in Botswana indigenous institutions have been supplanted by District Land Boards that currently regulate access to land and water (Gulbrandsen 1980). In both cases outsiders have been able to benefit from the changes. Civil servants, traders, and urban-based herd owners take advantage of the uncertainties surrounding land rights by staking claims to water points and land and overusing them (cf. Peters 1988; White 1987; Behnke 1985; Hogg 1987). The usurpation of power by African states usually has been incomplete, creating for farmers and herders in dry regions what Runge (1981) calls a problem of "assurance." In such situations producers lack confidence in the capacity of either state or local institutions to regulate access to resources.

Labor Relationships

Recent transitions in labor relationships have significantly influenced land use and management in dry regions. Before examining these changes, however, I will discuss a related topic--demography--that is fundamental to understanding the dynamics of labor supply.

The Demographic Question. As mentioned earlier, much of the debate about the causes of desertification centers on population growth, although there is little definitive data on the relationship between such growth and degradation in dry regions. What data do exist seem to show that fertility and growth rates among residents in dry regions, especially pastoralists, are generally lower than among settled communities in higher-rainfall areas (Hill 1985; Caldwell 1984). The information does not suggest that desertification has caused lower population growth rates, nor does it imply that overpopulation has extended degradation. Additional population growth in dry regions, as compared to other environments, is difficult to absorb by intensifying either local agricultural or pastoral production systems (India and other areas of Asia are important exceptions; see later discussion). While pastoral livestock systems can achieve higher productivity per land unit by emphasizing goats and sheep--rather than cattle and camels (Little 1992)--dryland economies are limited in the extent to which they can intensify in the absence of costly technologies.

Irrigation is, of course, one way that additional population can be absorbed--indeed, irrigation is responsible for relatively high population densities in dry parts of North Africa, the Middle East, and South Asia--but in many areas it is either technically or financially unrealistic. Ironically, the promotion of irrigation to combat desertification in what were formerly pastoral areas has actually increased resource problems by encouraging population concentration, cutting off important water points and grazing areas for livestock, and limiting mobility. A sizable literature suggests that land degradation is widespread around irrigated settlements in dry regions (Hogg 1986, 1987; Merry 1987; Spooner and Mann 1982).

What is the normal response to increased population and/or declining welfare in dry regions? Labor migration is the most common strategy, but attributing this to desertification (or to excessive population growth) is problematic. In the Sahel, for example, large numbers of individuals participate in long- and short-term migration to coastal countries, where historically they have sought employment (see Painter 1985; Caldwell 1975). The historical depth to this pattern and the difficulty of attributing causality to a single factor, however, blur the relationship between desertification in the Sahel and labor migration. Poverty is prevalent and opportunities for nonfarm employment are minimal in dry regions, and so migration may be more of a response to social and economic conditions than to environmental change. Evidence on the links between migration and desertification, on the one hand, and between migration and reduced incomes reflected in lower herd or land productivity, on the other, is far from conclusive (Watts 1987).

The population/desertification controversy, then, needs to be situated properly in its historical context. While the commonly used variable, the land/people ratio, shows a recent reduction in some dry regions, this may be more the result of loss of land than of excessive demographic growth. It may also reflect the inmigration of poor farmers to these areas, in response to increasing shortages of land in other parts of a country. In Kenya alone, more than 25 percent of arid and semi-arid lands in certain districts has been removed from the local economy to accommodate wildlife parks, which drastically worsen land/people ratios in the affected communities (Little 1984; Sperling 1987). In most semi-arid regions of Africa (and Asia to some extent, see Spooner and Mann 1982), farmers also have increasingly pushed the limits of rainfed agriculture by migrating into drier areas to farm (Bernus 1981; Ibrahim 1984). As indicated earlier, the state facilitates these processes by expropriating lands from dry regions and by encouraging agricultural settlements.

The Effects of Labor Shortages. In discussing the relationship between labor (and population) and land degradation, the effects of labor shortages need to be examined with care (Collins 1987). Often a relationship among labor shortages, poverty, and environmental decline is assumed, absent of any empirical data.

Documenting the relationships between local labor use and land management practices requires considerable micro-level data, which should be collected from samples stratified by one or more indicators of wealth (livestock ownership or income, for example). Recent work by Scoones on Zimbabwe (1990), Jodha on India (1986), and Watts on Nigeria (1987) demonstrates the importance of using stratified samples for exploring the variation in local management practices, as well as the relationships between poverty and land degradation. In the semiarid areas of Baringo District, for instance, research shows that differences in herd and land management do correlate with household labor availability and economic welfare (for a similar finding in Botswana, see Kooijman 1978). Table 7.1 shows that labor constraints in this area are most acute among poor (IV) and very poor (V) homesteads, who usually do not herd their goats and sheep separately, do not seasonally move their herds outside of a six-km radius of their settlement, and often do not herd their cattle during the wet season when grazing is relatively plentiful. This pattern contrasts sharply with that of wealthier households (strata I and II), which have sufficient labor to herd goats and sheep separately, to move cattle and sheep seasonally, and to herd cattle during the wet season.

These different herding practices have implications for resource use in the area. First, because animals are frequently unsupervised in the wet season, they enter grazing zones that are normally left "fallow" until the dry season; they wander onto cultivated fields where they consume crops, instigating conflict between their owners and farmers; they consume vegetation that has not had a chance to grow back after the dry season; and they (especially goats) often browse on vegetation along steep slopes and in riverine areas already vulnerable to erosion. Second, as noted above, labor shortages result in less mobility for many homesteads, which, in turn, leads to the overuse of certain zones around population settlements. While this may not result in irreversible environmental

TABLE 7.1 Herd Management Strategies, Baringo, Kenya[a]

Management	Category of Homesteads[b]		
Characteristics	I and II	III-H and III-L	IV and V
Average Labor Size (in Adult Units)[c]	7.73	5.74	3.95
% of homesteads that herd G and S separately	63.00	15.00	0.00
% of homesteads that leave goats unherded[d]	13.00	17.00	67.00
% of homesteads that do not herd goats in wet season	37.00	58.00	100.00
% of homesteads that do not herd cattle in wet season[e]	40.00	50.00	86.00
% of homesteads that move cattle or sheep more than 15 km	80+	25-30	5-8

Source: Author's field notes. See note "a".

[a] The table is based on Little (1992:153). It draws on a sample of 38 homesteads that owned at least some cattle, goats (G), and sheep (S).

[b] Category (I) and (II) homesteads are the wealthiest with average herds of 137 and 49 livestock units (LSU), respectively; while category (IV) and (V) are the poorest with holdings of 12 and 4 LSU. Homesteads of categories (III-H) and (III-L) are in between, but their herds are closer to those of the poor than to the herds of the rich.

[c] This represents the average amount of homestead labor.

[d] Defined to mean that the homestead's goats are not herded at least 75 percent of the time.

[e] This includes homesteads that did not look after their cattle at least 25 percent of the days during the wet season.

damage (degradation), it does mean that range resources are not evenly utilized over space, and that herds may suffer from shortages of fodder when they rely only on village-based pastures.

Finally, with households pursuing a variety of management strategies, it is difficult to reach a consensus on such collective strategies as restricting the use of certain pastures seasonally. This is why much of the "reserved grazing" system, once operative in the Baringo area, has fallen into disuse (Little 1985). To practice a rotational grazing pattern implies that herders share common interests and have sufficient labor to move their animals, which is often no longer the case.

Another dry area where local labor and land management strategies are documented is the communal areas of Zimbabwe. Using a household sample stratified by wealth, Scoones (1990) provides considerable information on household labor availability and its effects on herd and land management strategies. As in the Baringo case, Scoones is working in an area identified as seriously degraded. The Zimbabwe work documents that wealthier households have enough labor to move their cattle to distant pastures, while poorer households do not control the resources to do so. The latter group suffers disproportionately during drought because it does not control the labor to take advantage of the full range of grazing available (Scoones:408-409). Another important finding of the Zimbabwe research is the extent to which labor availability allows certain households to exploit strategic grazing resources, such as swamps. These key resources are critical for allowing herds to survive in an environment as heterogeneous and "patchy" (Scoones 1989a) as the communal areas of Zimbabwe. As in the Baringo case, mobility is the key to the sustainable use of Zimbabwe's dry areas, where, in Scoones's words, "Restrictions on such flexible and locally specific responses imposed by fenced grazing schemes or veterinary controls should be carefully examined . . . lest they reduce the sustainability of the system" (1989a:20).

Gender and the Division of Labor. Local labor use is mediated through customary patterns of division of labor. Recent transformations in labor relationships, especially the increased dependence on wage labor markets, create additional pressures for certain segments of the household. Women in particular have been compelled to absorb tasks normally carried out by men, many of whom have migrated to towns and other areas of employment. The increased workload of females in herding economies is at least partially a response to this loss of labor (Dahl 1987). As Sperling shows for the Samburu of northern Kenya, "Male emigration intensifies the female workload. . . They become more directly involved in many aspects of herd care, such as fencing, watering, curative regimes, and forage collection" (1987:179). These additional demands, however, are not always accepted passively. In some areas, pastoral women have refused to contribute to certain tasks (e.g., moving animals to remote pastures) that require long absences and excessive labor beyond their already heavy burdens (Little 1992; Sperling 1987). Where women have accepted the increased labor demands accompanied by male emigration, little is known about what payments or other benefits, if any, women have received from the extra work. Local ideologies and patterns of gender division are not always easily modified to accommodate new labor arrangements (Talle 1987). Research to date has assumed: (1) tasks can be easily reallocated--in terms of both knowledge and technique--among household members; (2) household members share common interests and motives and thus will be willing to contribute additional labor for tasks deemed beneficial for the collective household; and (3) beliefs about what are appropriate labor roles for household members can be easily changed. The restrictions that Islam places on women's public display of work is an obvious example of an ideological constraint. Another is the seasonal herding camp of eastern Africa, which often assumes cultural importance in the development and identity of young males (see Spencer 1988). Cultural constraints on substituting labor in certain tasks and by certain categories of individuals can strongly influence how land and other resources are used in dry regions.

Women of poor households are most seriously impacted by labor shortages and the decreased mobility and environmental problems that ensue. Not only do they absorb additional tasks, but because of localized degradation resulting from decreased mobility, they must search farther from their homes for firewood and other natural products (e.g., wild plants) required for cooking (Monimart 1989). Ensminger, for example, estimates that because of sedentarization in northern Kenya, some Orma women now allocate 13 times more labor for collecting fuelwood than women who remain in nomadic households (1987:37).

Income Relationships

As indicated earlier, with several important exceptions (e.g., the oil-exporting Middle East), incomes in dry regions often are low for political, economic, and historical reasons (Kates and Haarmann 1992). Moreover, available assets and revenue are usually very unequally distributed. Data from dry regions of Senegal (Sutter 1987), Niger (Starr 1987), Kenya (Grandin 1983; Little 1992), Tanzania (Kjaerby 1979), and Somalia (Little 1989) demonstrate that for one of the most important assets, livestock, distribution is highly segmented. While as much as 45 percent of livestock units may be owned by only 10 percent of herders, the bulk of households (up to 50 percent) may own 15 percent or less of the total herd. Such inequities in wealth, as was pointed out, imply different capacities to allocate labor to resource management activities and to forego immediate benefits for longer-term conservation concerns. The increased disparities in the distribution of wealth and the related phenomenon of "semi-proletarianized" workers are among the most important changes occurring in dry regions. They, in turn, have strong implications for how resources are conserved or abused in these areas. To blame low-income households for environmental problems in dry areas is, to use O'Leary's phrase, like "blaming the economic victims" (1984). It avoids examining the structural/political economic processes that determine the distribution of income, and it does not lend itself to finding constructive policy solutions for land management problems.

Case Studies of Land Degradation ("Desertification")

In selecting case studies for this paper, I focused on materials that (1) were well documented in the literature; (2) provided both ecological and social/political data; (3) had information on both structural/political economic (macro) factors and local (micro) variables, particularly relating to land, labor, and income activities; (4) depicted an area defined as threatened by desertification; and (5) provided some longitudinal depth to the analysis. Many studies meet at least some of these criteria, but are weighted toward either social or ecological aspects. For example, some of the most interesting ecological work is coming out of southern Africa (Biot, Sessay, and Stocking 1989; Abel and Blaikie 1990), but it is not always supplemented with enough social and economic data to explore labor use and other processes highlighted in the previous section (the Zimbabwe materials may be an exception, but as they are summarized in earlier sections they are not included as a case study). The Baringo case, which I know quite well, is omitted because it is cited throughout the paper. Other studies either emphasize the structural/political economic (and historical) side of the debate (Cliffe 1988; Prah 1989) or the local behavioral side (Bovin and Manger 1990; McCabe 1990; Coughenour et al. 1985), and do not attempt to integrate the two sets of questions. The following examples, then, represent some of the better-documented cases of land degradation in dry regions that meet the specified criteria, but obviously they do not represent the full range of the literature.

The Integrated Project in Arid Lands (IPAL), Marsabit, Kenya

The Integrated Project in Arid Lands (IPAL) research is one of the most detailed, long-term studies of land degradation in a dry region of eastern Africa. The IPAL project also had a component that focused on the applications of the research for rural development, and it attempted several different types of interventions as well. This dimension of IPAL is not reviewed here, but readers should consult Fratkin for an assessment of it (1991). Initiated in 1977 under the direction of Hugh Lamprey, an avid believer in desertification (1975), the

IPAL study of Marsabit District, Kenya, was a multidisciplinary effort by anthropologists, geologists, historians, ecologists, range management specialists, and hydrologists that generated masses of biophysical and socioeconomic data and reports (see, for example, the bibliographic list in IPAL 1984). It spanned a period of approximately eight years and was strongly motivated by the twofold premise that desertification was measurable and that it was occurring in northern Kenya. The project, which was part of UNESCO's "Man and the Biosphere Program," received substantial amounts of funding from UNEP and often reported its findings in UNEP's *Desertification Control Bulletin* (see O'Leary 1984).

The social science research carried out by IPAL was competent and resulted in several important publications on social change and land management among Rendille and Gabbra pastoralists (O'Leary 1985). These provide detailed data on land, labor, and income relationships, including information on tenure institutions, local resource practices and knowledge, labor organization and migration, and herder and gender differentiation. The work of O'Leary (1984; 1989; 1990) and Sobania (1980), both of them social scientists who worked for IPAL, places Rendille and Gabbra pastoralism in its larger political, economic, and historical contexts, showing how colonial grazing policies restricted population movements and how the Rendille and Gabbra have recently entered wage labor markets to supplement their incomes. While the natural science research of IPAL often claimed that desertification was a widespread problem in northern Kenya caused by local management practices (Lamprey and Yussuf 1981; IPAL 1984), the project's social scientists cautioned against sweeping generalizations about the role of herders in generating land degradation there. Thus, O'Leary admits to the strong presence of degradation around large settlements, but he does not attribute it to herder practice (1984). Rendille herders recognize the existence of overgrazing, but note that it is a reversible process, and they can identify several overgrazed areas that have recovered after being left fallow (O'Leary 1985:81). Instead, O'Leary argues that these problems are the result of faulty policies that encourage sedentarization and poverty (1984, 1989).

Herders in Marsabit increasingly pursue a sedentary form of pastoralism that overuses areas around settlements but leaves large parts of the range underutilized and underpopulated (IPAL 1984). Insecurity is another reason why herders of Marsabit now prefer to reside in large settlements. For a recent analysis of the relationship between political insecurity and environmental degradation (including desertification), see Hjort af Ornas and Salih (1989). It is estimated that about 45 percent of herders in the district reside around four major population settlements, and that large parts of the surrounding range areas are underutilized, including 11 percent of usable rangeland that for more than three years contained no domestic livestock (or people) at all (O'Leary 1984:17). Around the towns population density is likely to exceed 50-60 per km², while the density for the entire district (including the areas covered by IPAL) is less than 2 per km² (IPAL 1984). Clearly, if there is a population problem in Marsabit it is caused by poor distribution rather than excessive growth and density.

The settlements of Marsabit District are pockets of economic and ecological impoverishment, marked by increased economic differentiation, dependence on food aid, and fragmented domestic units where women increasingly assume herding duties for absent males. As in many of the dryland economies of eastern Africa (see Coughenour et al. 1985; Homewood and Rogers 1987; Hogg 1987), environmental problems are not caused by customary land use that relies on population mobility, but by social, political, and economic processes that restrict movement and lead to the overutilization of certain zones. In the case of the IPAL project area, this is further aggravated by the recent tendency of wealthy and well-connected ranchers, who often are residents of local towns, to acquire private land titles in the more favorable range zones and to fence "immense areas" of communal grazing grounds (Schlee 1991:154). Confronted with such problems, it is not surprising that local herders do not view either overgrazing or soil erosion as particularly significant issues (IPAL 1984:457-463).

In spite of the relatively lengthy period of data collection on environmental processes (1977-1985), the ecological arguments of IPAL are far from convincing. While IPAL's environmental studies rate much of Marsabit District as degraded or under threat of degradation, the project's data show that most degradation has been in the vicinity of a few large settlements and water points, an observation that is hardly novel. Even in the case of settlements, IPAL admits that much of the damage is probably reversible:

> The present situation is a complete absence of vegetation around the settlements in an ever-increasing circle, which, at Korr, is estimated as having a 9 km radius . . . Some degree of regeneration has proved possible by the IPAL enclosures in Korr and Kargi. The successional trend appears to be from annual grasses and herbs, which are the first colonizers, to dwarf shrubs with, hopefully, after long periods of protection, a return to the original vegetation (IPAL 1984:227).

Notwithstanding the massive amount of natural science research, the IPAL scientists rely on very subjective indicators of range condition: "The range condition for the range sites contained here is a *subjective judgement* [my emphasis] based on the following attributes: soil stability, composition of desirable and non-desirable plants, bare ground and litter cover and the state of erosion. It is based on four categories: excellent, good, fair and poor" (IPAL 1984:106). With such general categories and with their own perceptions of "desirable" plants and erosion, it is not surprising that the IPAL ecologists found much of the area to be degraded: "With the exception of one mountain range type, none of the range types had a condition better than 'fair', indicating the seriousness of the degradation of the grazing resource condition in the study area" (IPAL 1984). The range specialists see a lack of sound management strategies by the herders as the major cause of land degradation. (This is not a problem, according to IPAL's social science work: see O'Leary 1985.)

In sum, although IPAL's work provides important information on range ecology and pastoral practice, the socioeconomic and biophysical components of the study are not well integrated, so that contradictory statements are made about the causes of land degradation. The IPAL ecologists are willing to make sweeping generalizations on data and criteria that they themselves admit are subjective. With the exception of some good data on ecological change around settlements--although the changes are probably not irreversible and, therefore, not really degradation by our earlier definition--they provide little convincing evidence that land degradation is severe in the rangelands of Marsabit. Because their program was funded in part by organizations (e.g., UNEP) with a vested interest in the reality of desertification, and because they did not work with a consistent definition of what degradation entailed, IPAL seemed willing to assume desertification in cases where its own data and analysis indicated otherwise. (For a good discussion of the "institutionalization" of the desertification debate in development bureaucracies, see Warren and Agnew 1988 and Spooner 1989).

More recent ecological research in the same region, also supported by UNEP, reveals similar inconsistencies and contradictions regarding the extent of degradation. Compare, for example, these two quotations from the same document:

> The general picture for Marsabit is that no significant degradation occurred during the 16 years [1956-1972] except for Logologo and a little for Illaut. . . Because the degradation was apparent only at Illaut and Logologo, it would not be fair to derive a rate and generalize it to the whole study area (Government of Kenya/UNEP 1990:152).

> The results of this study show that desertification which is land degradation in arid, semi-arid and sub-humid areas is a major problem in the study areas [also including Baringo, Kenya]. The main forms of desertification identified were soil and vegetation degradation. The soils are being degraded through water and wind erosion, while vegetation is through tree and shrub cutting and overgrazing (Government of Kenya/UNEP 1990:157).

As in the IPAL study, this research raises doubts but, again, when the environmental data are inconclusive, desertification is assumed.

An even more recent study contradicts both the IPAL and the Government of Kenya/UNEP findings by suggesting that land degradation in Marsabit is not a problem: "Most of the district's rangelands are in good condition . . . . This primarily means that the rangelands are not degraded . . . . less than 1% of the district's rangelands are in poor condition" (Herlocker and Walther 1991:52). Regarding this last assessment, conducted by the Government of Kenya with support from German technical assistance, Herlocker and Walther are not willing to claim desertification when serious questions remained. Considered together, the three studies raise troubling questions about the measurement and interpretation of degradation even when "hard" evidence is offered.

"Desertification" in Western Sudan

Several local studies of environmental degradation, also motivated by a concern for desertification, have been conducted in the Sudan during the past two decades (Ahlcrona 1988; Haaland 1980; Manger 1981; Ibrahim 1984). In fact, the alarmism in the 1970s about desertification in Africa was based initially on research results from the Sudan (see Lamprey 1975; Rapp, Le Houérou, and Lundholm 1976), and the country remains an important focus of dryland studies.

The case study in this section is based on fieldwork conducted in Darfur, in western Sudan (Ibrahim 1984, 1987; Behnke 1985), but where appropriate it draws on similar work in central Sudan (Ahlcrona 1988). In many respects, the Sudan case is more informative than the IPAL work because it addresses resource management problems at the boundary between crop and livestock production activities, where competition and conflict are inherent (Bovin and Manger 1990; Little 1987). Because rainfall is sufficient to support dryland farming, Sudanese herders themselves are cultivating, which adds to the complexity and potential problems of land use in the area. It is at this interface of different production activities and interest groups (farmers, pastoralists, and commercial ranchers) that ecological problems in dry regions seem most serious.

Like the IPAL project, the objectives of Ibrahim's work in Darfur was to assess the extent and causes of desertification (1984, 1987). The study covers most of the salient social and economic parameters related to land, labor, and income, but it does not give explicit attention to the effect that differential access to these factors has on land use. For Ibrahim, the major source of land degradation is the expansion of cultivation into formerly pastoral areas, where the intensification of land use--including the mechanization of agriculture--leads to soil erosion and the destruction of perennial vegetation. Pastoralists are forced into even more marginal habitats, where overgrazing is now occurring (Little and Horowitz 1987:10). While the data support the IPAL position that sedentary forms of production are more destructive of the environment than nomadic pastoralism, the Sudan case is more complex because rainfed cultivation, unlike in Marsabit, Kenya, is widespread. With material welfare and food security having deteriorated in recent years, herders themselves are compelled to plant millet beyond the agronomic boundary, the very thing that will degrade their resources and reduce their future yields (Little and Horowitz

1987; for a West African example of this, see Bernus 1981). The state supports this tendency to cultivate in dry regions, and among large-scale capitalist farmers it has subsidized the establishment of mechanized farming schemes--an action that strongly degrades soils and vegetation (Ibrahim 1987).

Without providing explicit data on labor use, Ibrahim's work shows how pastoralism has become increasingly sedentary, partly in response to the emigration of individuals to work for wages outside of Darfur. Along with crop production, this form of sedentary animal husbandry results in localized degradation around villages (the major locations of desertification) and further restricts nomadic herders into less and lower quality rangelands. The "concentric rings of degradation" around settlements that Ibrahim discusses at length reminds one of the land problems around large villages in northern Kenya recorded by the IPAL team. As in other studies in the Sudan (Manger 1988; Haaland 1980), Ibrahim shows that livestock ownership in these settlements is increasingly controlled by absentee herd owners (often merchants), who use their political connections to insure access to grazing and water. They, in turn, hire impoverished pastoralists to herd their cattle.

Other research from Darfur documents a spontaneous "enclosure movement" among farmers and sedentary livestock owners that reduces the land available for extensive pastoralism, increases densities of population, and attentuates environmental problems (Behnke 1985). By carefully documenting the various usufructuary rights associated with different types of land (e.g., flooded arable land), the study reveals how wealthy farmers and sedentary herders are able to claim the most valuable land without reprisal from the state or local community. As Behnke points out, "The local communities in the area of heaviest grazing, on the other hand, supported enclosure because it gave them exclusive control over what had traditionally been a resource shared with nomadic herders" (1985:21). The customary tenure system did not acknowledge "the right to enclose," but it did recognize that more valuable land had more restrictive claims to its use and, therefore, it seems to have accommodated this recent process without major controversy among sedentary residents.

Conflict has emerged, however, between the new "owners" and the nomadic herders who used to rely on these lands for seasonal use: "The range enclosure movement is . . . based on competition between transient and permanently resident livestock keepers for control of a diminishing range resource" (Behnke 1985:18). Pastoralists, who now are fined by local courts if they graze on enclosed lands, must alter their seasonal grazing patterns to avoid conflict. While the environmental effects of this transformation are not explored in detail, Behnke provides some indication that the impacts are not favorable: "Short-term declines in regional livestock productivity may therefore be a common feature of the shift from open to enclosed systems of range ownership in nomadic areas" (Ibid. 1985:24). Research from elsewhere in Africa and Asia confirms that drop-offs both in herd and in environmental productivity often result from "enclosure" movements in dry regions (Jodha 1986; Scoones 1990; Sandford 1983).

The Darfur materials reveal in some detail how the increased pressure on land around settlements differentially affects women (Ibrahim 1984). It does so by increasing their labor requirements for collecting fuelwood and water, reducing their yields of crop and dairy products, and denying them access to income from new activities like mechanized farming (1984:137-144). Ironically, they are also called on to contribute labor to conservation schemes and to local forestry efforts that "combat" resource degradation.

The Darfur research is far more tolerant of the concept of desertification than other recent work in dry parts of Africa (see Bernus 1981; Abel and Blaikie 1990; Sandford 1982; Watts 1987), including O'Leary's research under the IPAL program (1985). For Ibrahim, there is little question that desertification is a real phenomenon that threatens Sudanese villages and environments; he uses photos, for example, to graphically depict the effects of desertification (1984). While he demonstrates the effects that labor shortages and reduced access to land have on resident households, he views increased labor migration as a response to desertification rather than to low incomes, and writes of the contribution of human activities to processes of land degradation. His own statements seem to imply that the "overpopulation" problem and its relationship to degradation is caused by political processes that have alienated large tracts of land for agricultural schemes and heightened insecurity in surrounding areas, which are then left unoccupied (Ibrahim 1984). This connection, however, is not made; therefore, unlike the IPAL case, we do not learn if the "population" problem is a result of distorted distribution particularly around large settlements or excessive growth rates. Because the population density of the region is noted to be only $3.5/km^2$ (Ibrahim 1984:99), it is likely that the former--uneven population distribution--is the case.

The work of Ahlcrona (1988) in a nearby region of the Sudan challenges some of Ibrahim's findings about the extent and causes of desertification. Relying on field surveys, aerial photos, and Landsat imagery, Ahlcrona shows that, in contrast to Ibrahim, "It was not possible to verify any transformation of sand dunes or any systematic expansion of desertified perimeters around villages during 1961-1983" (1988:69). What "degradation" occurred was reversible and was probably more a response to rainfall variation than to human activities. This finding, which needs to be critically tested in other regions, raises questions about the results of the IPAL and other research (Coughenour et al. 1985; Hogg 1987) concerning the positive relationship between degradation and population settlements (peri-urban zones). It should be noted that even the most skeptical critics of the desertification concept often accept this correlation.

Ahlcrona's work also questions Ibrahim's findings on the links between cultivation and land degradation. It argues that when observations are taken over a period of more than 20 years there is no evidence of a permanent transgression of agriculture into the range areas, and, therefore, it is doubtful that excessive cultivation is a major cause of desertification in the Sudan (with the exception of large-scale, mechanized agricultural schemes). Moreover, this study shows that, in contrast to Ibrahim's work, it is declining rainfall that accounts for declines in crop yields, not land degradation: "It was concluded that rainfall was highly correlated to the yield of millet and sorghum . . . and that man-made degradation has not been the main cause behind the declining yields in the region studied" (Ahlcrona 1988:40). While Ahlcrona's work lacks the social and economic depth and detail of Ibrahim's, it does raise serious doubts about many of his findings regarding environmental trends. In more general terms, what is instructive here is that the presence of poverty, excessive population, and short-term losses in vegetation and environmental productivity are not sufficient conditions to assume irreversible ecological change ("desertification"). Land management problems and wrenching socioeconomic hardships do exist in western Sudan, as Ibrahim skillfully demonstrates, but these do not imply permanent environmental damage.

Land Degradation in Rajasthan, India

Studies of land degradation in the arid areas of Rajasthan, India, provide interesting parallels as well as contrasts to the African materials. The sheer volume of publications attests to the interest shown in the region, and the area has been integral to discussions of desertification in Asia (Spooner and Mann 1982; Jodha 1983, 1986, 1991). While changes similar to those of the African examples are found in Rajasthan, it differs significantly in one respect: the region's population growth and density is relatively high--in excess of 100 per km^2 in some locations. Even in Indian terms, Rajasthan's rate of population growth, particularly in its arid districts, is very high (Caldwell 1984:24).

In addition to demographic factors, changes in land tenure and market relations are frequently identified as contributing to desertification in Rajasthan. A major land reform program was instituted in the late 1940s, with the objective of replacing the exploitive feudal pattern of land ownership through the redistribution of land to smallholders. The state enforced the reallocation, but because it did not account for local differences in land quality and ecology much of the land distributed under the program was unsuitable for agriculture (Jodha 1982). Instead, large parts of the arid zone should have been left under forest and pasture cover. As in the Sudan case, the increased cultivation of the drylands facilitated the loss of vegetative cover and increasingly exposed the soils to wind and water erosion. This resulted in declining crop yields and an increase in labor emigration, both processes that further impoverished the region (Jodha 1982:345).

Jodha shows that while the land reform program reduced inequities in landholdings, it also created considerable ambiguity over the responsibility for common pasture and forest management. Theoretically, under the reforms local villages were allocated responsibility for managing common rangelands and forests, but they were given neither the resources nor the authority to enforce regulations. The state, in turn, frequently undermined their efforts by subdividing and titling communal lands for private use. With few controls on access to common lands, mismanagement ensued and wealthy members of the community began to carve off large parcels from the communal lands. The state has laws that provide for minor fines for those who expropriate public lands, but the penalties are well below the "returns from private use of CPRs" [Common Property Resources] (Jodha 1983:268). After three to five years of illegal occupation, the land usually ends up effectively being under the ownership of the occupant.

Jodha provides excellent time series data on the loss of CPRs, which resulted both from inappropriate policies and private encroachment. In a recent survey of 82 villages the areas of CPRs have declined by at least 30 percent, and in some cases by 50 percent, since 1950-1952 (Jodha 1991:6). This loss has encouraged overutilization of the remaining CPRs and of the surrounding range areas that are increasingly congested. Mismanagement is further provoked by the lack of either state or communal controls on the remaining CPRs, a situation created by the initial government reforms. Relying on detailed local data, Jodha shows how the transformation of CPRs has most seriously impacted poorer households of the community, who depend on wild products (for subsistence and sale) and grazing from the CPRs. Unlike richer members of the community, they have not benefited from the "privatization" of the CPRs and do not own large parcels of grazing land to compensate for the decline in communal grazing. It is estimated that CPRs contribute more than 20 percent of the poorest households' total income (Jodha 1986:1177).

Another factor, mentioned earlier, that is frequently cited as contributing to desertification in Rajasthan is excessive human population. This variable is underplayed by Jodha, but is emphasized by others:

> The increase in population in only 10 years (1961-1971) was 63%. . . As a consequence there is a high dependency ratio. . . The man-land ratio is declining fast. . . The spread of new technologies has increased human and livestock pressure on resources and led to social disintegration (Malhotra and Mann 1982:309-310.)

Clearly population has grown rapidly in the last several years, but it is unclear whether this was the variable that initiated land management problems or was a response to local conditions of underdevelopment (including ecological problems) and poverty. As has been shown elsewhere, population growth can be stimulated by conditions of impoverishment that motivate households to increase the factor of production--labor--that they control (Collins 1988).

Earlier discussion has shown how the loss of CPRs has resulted in overutilization of existing lands, and there is evidence that many of the grazing zones that Rajasthan pastoralists seasonally moved to are no longer open to them. Not only has this forced herders to restrict their pattern of movements and to overutilize available pastures (Jodha 1982:309; also see African examples discussed earlier), but it has exaggerated population pressure as well. Because of the sizable amount of irrigation and opportunities for nonfarm employment, an emphasis on population densities (land/person ratios) may be inappropriate for Rajasthan. In addition, the growth in population pressure and demographic rates may stem from the immigration of families from more favorable agricultural areas, where land inequities and shortages are likely to be more severe. The "pushing out" of poorer families to more marginal areas, including arid rangelands, is a process that has been observed in several different regions and has been shown to have severe ecological implications for receiving locations (Bernus 1981; Little 1992).

The final social variable highlighted in arguments about land degradation in Rajasthan is the market. During the past two decades the region has been opened up to trade through the creation of roads and considerable market infrastructure (Jodha 1980, 1985). As a result, marketplaces have grown considerably, the number of merchants working in the area has increased, the value of many local products (e.g., meat and wool) has grown rapidly, and local land values have risen. According to Jodha, these processes are at least partly responsible for the decline in CPRs and the degradation that has taken place. With the penetration of commercial markets, land from the CPRs has been converted to the production of market-oriented products, and land speculation has taken place as well. In addition, such CPR products as fuelwood, charcoal, and wild foods are now collected for sale, and "in place of occasional visits of traders with bullock carts and caravans of camels to collect supplies, now trucks and tractors frequent villages to collect supplies during all seasons" (Jodha 1985:273). Wealthier merchants and farmers induce overexploitation by undermining whatever limited regulations on CPRs exist and by entering into buying contracts for CPR products with local people. Poorer households, in turn, are increasingly dependent on the market for subsistence requirements and are said to be excessively using the land (including CPRs) in order to earn cash (Jodha 1980).

The physical environment of Rajasthan has been well studied, but nonetheless several questions about the extent of "desertification" remain. For example, what often is interpreted as degradation may actually be reversible within a feasible time frame. Jodha lists several ecological changes as indicators of desertification. These include: "the deterioration of fertile lands due to the removal of top soil or the submersion of fertile lands under shifting sands; the conversion of fertile lands into patches of saline wasteland in the low lying areas near seasonal streams; the drying up of wells or the increased salinity of well water; the replacement of superior perennials by inferior ones or annual grasses including non-edibles in the grasslands; the increased population of malformed or stunted trees; and finally the increased human misery reflected by increased seasonal migration and accentuated pauperization through recurrent famine" (1982:345). Undoubtedly, physical changes have occurred and perhaps even longer-term declines in crop yields, as some data indicate (Jodha 1991); but without consideration of climatic and other natural factors it is questionable whether these changes are solely man-induced or irreversible. Dhir, for example, while acknowledging resource problems in Rajasthan notes, "Fortunately, the process is gradual and the stage of irreversibility has not yet been reached" (1982:331). This does not mean that the situation in the region is not serious, but without additional empirical confirmation and analyses the alarm over permanent ecological change may be premature.

Agenda for Research and Policy

This chapter has tried to broaden the discussion of population growth and environmental change by looking at the social processes that contribute to land degradation in dry regions. Such issues as poverty, labor shortages, and wealth inequities reveal greater significance in explaining environmental problems than do demographic factors. If there is a population "crisis" in the dry areas, it is one of uneven distribution rather than accelerated growth. The case studies printed here highlight the uncertainties in current understandings of dryland ecologies and demonstrate the importance of time-series data for assessing environmental degradation.

What is striking, in light of recent research findings, is how little government and donor programs working in dry regions have been affected. Even today, major donors of Europe were formulating strategies to "roll back" the desert, and many of the archaic assumptions about common property systems and herder behavior were being invoked. Countries with large expanses of dry regions, such as Botswana, Nigeria, Kenya, and the Sudan, are still encouraging sedentarization, individual land titling, and other interventions that create rather than solve problems of land degradation. Clearly there are strong political interests for encouraging land reform in dry regions--for example, African and Indian elites have done well under private land-titling programs--that benefit from the desertification scenario. These incentives are reinforced by Western governments and organizations, sometimes pressured by their citizens to respond to global environmental problems, which fund inappropriate programs under the auspices of confronting desertification.

The Forum for Research

Part of the problem has been that most research findings are published in scientific journals or presented at academic conferences to which policymakers are unlikely to have access (publications like *Haramata*, from the International Institute for Environment and Development and *Pastoral Development Network*, by the Overseas Development Institute, are important exceptions). Because there are such strong political motivations and ideologies underlying the desertification debate, it is easy to avoid information that is not presented in a policy-oriented forum. It may be discarded regardless, but possibilities of influencing policy are strengthened when results are presented in an appropriate context. One example that may have a general impact on how governments and donors view dry regions is the conference "Technical Meeting on Savanna Development and Pasture Production," conducted by ODI and IIED and funded by the Commonwealth Secretariat. The workshop's papers are summarized in Behnke and Scoones (1991) and provide considerable insights into dryland ecologies and appropriate development actions, while using a policy forum to present their findings. In addition, this meeting is being followed up with case studies of range management and degradation in three countries (Kenya, Nigeria, and Zimbabwe)--most of them carried out by local researchers-- that will facilitate policy dialogues in those countries.

Because concerns about environmental degradation have dominated dialogues about the development of dry regions, programs for these areas have been driven by environmental rather than social and developmental concerns. And most of these, in turn, have been couched in negative or defensive (military-like) terms--"combating environmental degradation" or "fighting against degradation." The government ministry responsible for arid and semi-arid lands in Kenya, for example, is called the Ministry for the Development and Rehabilitation of Arid, Semi-Arid and Wastelands--not exactly a title that espouses a positive perception of dry regions. Yet, are the environmental problems of arid areas any worse than for cities like Nairobi or Delhi? Social science research in these areas is influenced by the association of dry ecosystems with environmental concerns, since funding agencies are also influenced by larger ecological issues. Until the environmental negativism about dry regions is replaced by a more positive awareness, "development" in these areas will be focused more on physical resources and wildlife than on people.

Research Directions

Several priorities for social science research have been identified in previous sections and need not be repeated here. Instead, I point to two other areas that could benefit from additional information. The first concerns local perceptions of environmental degradation, those that particularly influence behavior and welfare. In dry regions very little is known about how degradation is defined locally, how environmental problems are perceived by different social groups (based on gender, class, age, etc.) who may not share the same concept of degradation, and how beliefs about degradation are translated into action. For example, a pastoral woman may define degradation by declining yields from dairy herds, or by the amount of time spent in collecting fuelwood. Whether or not the land or environmental problem resulting in these changes is irreversible (degradation) or not--a concern to a physical scientist--may be irrelevant to her. Another member of the same household may perceive degradation in a very different manner. Since physical degradation in dry regions is difficult to measure and to define in a locally meaningful way, efforts to uncover local definitions of degradation may lead to appropriate development actions more quickly than lengthy, elaborate ecological studies. It means, however, that in-depth fieldwork (at least one to two years) is required to learn how local perceptions of degradation are translated into behavior and how these viewpoints may vary by different social and economic categories.

A second and related area for future research concerns the restructuring of labor relationships and its effects on resource use. While some interesting recent data on intrahousehold income distribution exist, I am unaware of studies that link intrahousehold differences to local land management or environmental strategies. These relationships nonetheless are being transformed. It appears tautologic to suggest that actors (male/female, young/old) will contribute more effort to activities, including conservation efforts, that reap the most income. Yet, do women in dry regions more carefully manage environmental resources that particularly enhance their material welfare? Can the same be said of men? If affirmative answers can be given to both questions, then are changes in labor relations that blur, on the one hand, the responsibility for labor and, on the other, the allocation of income benefits likely to result in land management problems? For example, if women do not reap much income from cattle production--as opposed to other activities--are they less likely to be concerned with herd and pasture management than are other workers? These kinds of questions are difficult to answer with present data, but need to be unraveled in studies of the relationship between income and resource management. General questions of poverty (low income) and its impact on the environment need to be recast in terms of "poverty for whom."

To conclude, serious problems of social and economic underdevelopment persist in dry areas, but it may be inappropriate to rally action and research on the basis of "desertification." Social science research can be more effective if it treats environmental problems as epiphenomena of underdevelopment and poverty, and directs attention to dry regions on the basis of human, not ecological and alarmist, concerns about overpopulation.

References

Abel, N. O. J., and P. M. Blaikie (1989). "Land Degradation, Stocking Rates and Conservation Policies in the Communal Rangelands of Botswana and Zimbabwe." *Land Degradation and Rehabilitation*. 1:101-123.

———— (1990). "Land Degradation, Stocking Rates and Conservation Policies in the Communal Rangelands of Botswana and Zimbabwe." *ODI Pastoral Development Network Paper 29a*.

Ahlcrona, Eva (1988). *The Impact of Climate and Man on Land Transformation in Central Sudan*. Lund, Sweden: Lund University Press.

Anderson, David (1984). "Depression, Dust Bowl, Demography, and Drought: The Colonial State and Soil Conservation in East Africa During the 1930s." *African Affairs*. 83:321-343.

Anderson, David, and Richard Grove (eds) (1987). *Conservation in Africa: People, Policies, and Practice*. New York: Cambridge University Press.

Bartels, G., G. Perrier, and B. Norton (1991). "The Applicability of the Carrying Capacity Concept in Africa: A Comment on the Thesis of de Leeuw and Tothill." in *New Concepts in International Rangeland Development: Theories and Applications*. R. Cincotta, C. Gay, and G. Perrier, (eds). Logan, UT: Department of Range Science, Utah State University, 25-32.

Bassett, Thomas J. (1988). "The Political Ecology of Peasant-Herder Conflicts in the Northern Ivory Coast." *Annals of the Association of American Geographers*. 78(3):453-472.

Behnke, Roy H. (1985). "Open-Range and Property Rights in Pastoral Africa: A Case of Spontaneous Range Enclosure in South Darfur, Sudan." *ODI Pastoral Development Network Paper No. 20f*.

Behnke, R. H., and I. Scoones (1991). "Rethinking Range Ecology: Implications for Rangeland Management in Africa." Paper presented at the Technical Meeting on Savanna Development and Pasture Production, 19-21 November 1990, Woburn, UK.

Beinart, W. (1984). "Soil Erosion, Conservation, and Ideas about Development in Southern Africa." *Journal of Southern African Studies*. 11:52-83.

Bernus, Edmond (1981). "Touregs nigériens: Unité culturelle et diversité régionale d'un peuple pasteur." Paris, France: ORSTOM (Mémoires ORSTOM No. 94).

Binns, Tony (1990). "Is Desertification a Myth?" *Geography*. 75: 106-13.

Biot, Y. (1990). "Evaluating the Sustainability of Agricultural Production. An Example from Botswana." Part 1: The Rangeland Ecosystem, Part 2: The Forecast. *Pedologie*.

Biot, Y., M. Sessay, and M. Stocking (1989). "Assessing the Suitability of Agricultural Land in Botswana and Sierra Leone." *Land Degradation and Rehabilitation*. 1:263-278.

Bishop, Joshua (1988). *Indigenous Social Structure, Formal Institutions, and the Management of Renewable Natural Resources in Mali*. Washington, DC: The World Bank.

Blaikie, Piers, and Harold Brookfield (1987). *Land Degradation and Society*. New York: Methuen and Company.

Bovin, Mette, and Leif Manger (eds) (1990). *Adaptive Strategies in African Arid Lands*. Uppsala, Sweden: The Scandinavian Institute of African Studies.

Caldwell, J. C.(1975). *The African Drought and Its Demographic Implications*. New York: The Population Council.

---------- (1984). "Desertification. Demographic Evidence 1973-1983." *Australian National University Occasional Paper No. 37*. Canberra, Australia: Australian National University Development Studies Centre.

Cincotta, Richard P., Z. Yanqing, and Z. Xingmin (1992). "Transhumant Alpine Pastoralism in Northeastern Qinghai Province: An Evaluation of Livestock Population Response during China's Agrarian Economic Reform." *Nomadic Peoples*. 30: 3-25.

Cliff, Lionel (1988). "The Conservation Issue in Zimbabwe." *Review of African Political Economy*. 42:48-58.

Cloudsley-Thompson, John (1988). "Desertification or Sustainable Yields from Arid Environments." *Environmental Conservation*. 15(3):197-204.

Collins, Jane (1987). "Labor Scarcity and Ecological Change." in: *Lands at Risk in the Third World: Local-Level Perspectives*. Peter D. Little and Michael M. Horowitz eds. p.19-37. Boulder, CO: Westview Press.

--------, (1988). *Unseasonal Migrations: The Effects of Rural Labor Scarcity in Peru*. Princeton, NJ: Princeton University Press.

Cook, H. J. (1983). "The Struggle Against Environmental Degradation - Botswana's Experience." *Desertification Control Bulletin*. 8:9-15.

Coughenour, M. B., J. Ellis, D. Swift, D. Coppock, K. Galvin, J. T. McCabe, and T. Hart (1985). "Energy Extraction and Use in a Nomadic Pastoral Ecosystem." *Science*. 230:619-624.

Dahl, Gudrun (1987). "Women in Pastoral Production: Some Theoretical Notes on Roles and Resources." *Ethnos*. 52(1-2):246-279.

de Leeuw, P. N., and J. C. Tothill (1990). "The Concept of Rangeland Carrying Capacity in Sub-Saharan Africa - Myth or Reality?" *ODI Pastoral Development Network Paper 29b*.

de Ridder, N. and Wagenaar, K.T. (1984). "A Comparison Between the Productivity of Traditional Livestock Systems and Ranching in Eastern Botswana," *ILCA Newsletter* 3(3):3-5.

Dhir, R. P. (1982). "The Human Factor in Ecological History." in: *Desertification and Development: Dryland Ecology in Social Perspective*. B. Spooner and H.S. Mann, eds. New York: Academic Press. Pp. 311-332.

Downs, R. E., and S. Reyna (eds) (1988). *Land and Society in Contemporary Africa*. Hanover, NH: New England Universities Press.

Dregne, H. E. (1983). *Desertification of Arid Lands*. New York: Harwood Academic Publishers.

Drinkwater, Michael (1990). "Rangeland Management in Africa. Further Comments on Pastoral Development Network Paper 28b." *ODI Pastoral Development Network Paper 30e*.

Eckholm, Erik, and Lester Brown (1977). "Spreading Deserts: The Hand of Man." *Worldwatch Paper No.13*. Washington, DC: Worldwatch Institute.

Ellis, J. E., and D. M. Swift (1988). "Stability of African Pastoral Ecosystems: Alternate Paradigms and Implications for Development." *Journal of Range Management*. 41(6):450-459.

Ensminger, Jean (1987). "Economic and Political Differentiation among Galole Orma Women." *Ethnos*. 52(1-2):28-49.

Fratkin, Elliot (1991). *Surviving Drought and Development: Ariaal Pastoralists of Northern Kenya*. Boulder, CO: Westview Press.

Galaty, J., D. Aronson, D. Salzman, and A. Chouinard (eds) (1981). *The Future of Pastoral Peoples*. Ottawa, Canada: International Development Research Center.

Glantz, Michael H. (1977). *Desertification: Environmental Degradation in and around Arid Lands*. Boulder, CO: Westview Press.

—————, (1987). *Drought and Hunger in Africa: Denying Famine a Future*. New York: Cambridge University Press.

Glantz, Michael H., and Nicolai S. Orlovsky (1986). "Desertification: Anatomy of a Complex Environmental Process." in: *Natural Resources and People: Conceptual Issues in Interdisciplinary Research*. Kenneth A. Dahlberg and John W. Bennett eds. Boulder, CO: Westview Press. Pp.213-229.

Goodland, R. J. A., and H. S. Irwin (1975). *Amazon Jungle: Green Hell to Red Desert?* Amsterdam, The Netherlands: Elsevier.

Gorse, Jean Eugene, and David R. Steeds (1987). "Desertification in the Sahelian and Sudanian Zones of West Africa." *World Bank Technical Paper No. 61*. Washington, DC: The World Bank.

Government of Kenya (1990). "Report of the Kenya Pilot Study (FP/6201-87-04) Using the FAO/UNEP Methodology for Assessment and Mapping of Desertification." in: *Desertification Revisited: Proceedings of an Ad-Hoc Consultative Committee on the Assessment of Desertification*. R.S. Odingo, ed. Nairobi: UNEP. Pp 123-177.

Grandin, B. E. (1983). *The Importance of Wealth Effects on Pastoral Production: A Rapid Method for Wealth Ranking*. ILCA

————— (1986). Land Tenure, "Sub-Division, and Residential Change on a Maasai Group Ranch." *Development Anthropology Network*. 4(2):9-13. Binghamton, NY: Institute for Development Anthropology.

Grove, A. T. (1977). "Desertification in the African Environment." in: *Drought in Africa 2*. African Environment Special Report 6. David Dalby, R.J. Harrison Church, and Fatima Bezzaz, eds. London: International African Institute. Pp.54-64.

Gulbrandsen, Ornulf (1980). *Agro-Pastoral Production and Communal Land Use: A Socio-Economic Study of the Bangwaketse*. Gaborone, Botswana: Government Printer.

Haaland, Gunnar (1980). "Problems in Savannah Development." in: *Problems of Savannah Development*. G. Haaland, ed. Bergen, Norway: Department of Sociology, University of Bergen. Pp.1-37.

Hardin, G. (1968). "The Tragedy of the Commons." *Science*, 162:1243-1248.

Hare, F. Kenneth (1977). "The Making of Deserts: Climate, Ecology and Society." *Economic Geography*. 53:332-345.

Helldén, Ulf (1990). "Desertification -- Facts and Myths. Report from the SAREC/Lund International Meeting on Natural Desertification and Man Induced Land Degradation". *Orenas Slott*, 5-7 December.

Herlocker, Dennis, and Dierk Walther (1991). "Condition of Marsabit District Range Lands." in: *Marsabit District: Range Management Handbook of Kenya*. Vol II, No. 1 Pp. 51-52. H. Schwartz, S. Shaabani, and D. Walther, eds. Nairobi, Kenya: Ministry of Livestock Development.

Hill, Alan G. (ed) (1985). *Population, Health and Nutrition in the Sahel. Issues in the Welfare of Selected West African Communities*. London, England: Routledge and Kegan Paul.

Hitchcock, Robert (1980). "Tradition, Social Justice and Land Reform in Central Botswana." *Journal of African Law*. 24:1-34.

Hjort af Ornas, Anders, and M. A. Mohamed Salih (eds) (1989). *Ecology and Politics: Environmental Stress and Security in Africa*. Uppsala, Sweden: Scandinavian Institute of African Studies.

Hogg, Richard (1986). "The New Pastoralism: Poverty and Dependency in Northern Kenya." *Africa*. 56(3):319-333.

--------, (1987). "Settlement, Pastoralism, and the Commons: The Ideology and Practice of Irrigation Development in Northern Kenya." in: *Conservation in Africa: People, Policies, and Practice*. David Anderson and Richard Grove, eds. New York: Cambridge University Press. Pp.293-306.

Homewood, Katherine, and W. A. Rodgers (1987). "Pastoralism, Conservation, and the Overgrazing Controversy." in: *Conservation in Africa: People, Policies, and Practice*. David Anderson and Richard Grove, eds. New York: Cambridge University Press. Pp.111-128.

Hopcraft, P. (1981). "Economic Institutions and Pastoral Resource Management: Consideration for a Development Strategy." in: *The Future of Pastoral Peoples*. J. Galaty, D. Aronson, D. Salzman, and A. Chouinard, eds. Ottawa, Canada: IDRC. Pp. 224-243.

Horowitz, Michael M (1976). "Sahelian Pastoral Adaptive Strategies Before and After Drought." in: *Proceedings of the West Africa Conference*, Tucson, Arizona, 1976. P. Paylore and R. Haney, eds. Tucson, AZ: University of Arizona. Pp.26-35.

--------, (1979). "The Sociology of Pastoralism and African Livestock Projects." *AID Program Evaluation Discussion Paper*. Washington, DC: USAID.

Horowitz, Michael M, and Peter D. Little (1987). "African Pastoralism and Poverty: Some Implications for Drought and Famine." in: *Drought and Hunger in Africa: Denying Famine a Future*. M. Glantz, ed. Cambridge: Cambridge University Press. Pp.59-82.

Ibrahim, Fouad N. (1984). *Ecological Imbalance in the Republic of the Sudan - With Reference to Desertification in Darfur*. Bayreuth, Germany: Bayreuther Geowissenschaftliche Arbeiten.

--------, (1987). "Ecology and Land Use Changes in the Semiarid Zone of the Sudan." in: *Lands at Risk in the Third World*. Peter Little and Michael Horowitz eds. Boulder, CO: Westview Press. Pp. 213-229.

IPAL (Integrated Project in Arid Lands) (1984). *Integrated Resource Assessment and Management Plan for Western Marsabit District, Northern Kenya. Vols. 1 & 2.* Walter Lusigi, ed. Nairobi, Kenya: UNESCO.

Jodha, N. S. (1980). "Agricultural Tenancy in Semi-Arid Tropical Villages of India." *Progress Report Economics Program 17.* Andhra Pradesh, India: ICRISAT.

————, (1982). "The Role of Administration in Desertification: Land Tenure as a Factor in the Historical Ecology of Western Rajasthan." in: *Desertification and Development: Dryland Ecology in Social Perspective.* B. Spooner and H.S. Mann eds. New York: Academic Press. Pp. 333-350.

————, (1983). *Market Forces and Erosion of Common Property Resources.* Patancheru, Andhra Pradesh: ICRISAT Center.

————, (1985). "Population Growth and the Decline of Common Property Resources in Rajasthan, India." *Population and Development Review.* 11(2).

————, (1986). "The Decline of Common Property Resources in Rajasthan, India." *ODI Pastoral Development Network Paper 22c.*

————, (1991). "Rural Common Property Resources: A Growing Crisis." *IIED Gatekeeper Series No. SA24.* London, England: International Institute for Environment and Development.

Kates, Robert, and Viola Haarmann (1992). "Where the Poor Live: Are the Assumptions Correct?" *Environment.* 34 (4): 4-11, 25-28.

Kenya, Republic of (1991). *Economic Survey: 1991.* Nairobi: Central Bureau of Statistics.

Kjaerby, F. (1979). "The Development of Agro-Pastoralism among the Barabaig of Hanang District." *BRALUP Research Paper No. 56.* Dar es Salaam, Tanzania: University of Dar es Salaam.

Kooijman, K (1978). *Social and Economic Change in a Tswana Village.* Leiden: Afrika-Studiecentrum.

Lamprey, H. F. (1975). *Report on the Desert Encroachment Reconnaissance in Northern Sudan, 21st October to 10 November 1975.* UNESCO/UNDP, mimeo, 16pp.

Lamprey, H., and H. Yussuf (1981). "Pastoral and Desert Encroachment in Northern Kenya." *Ambio.* 10(2).

Le Houérou, H. N. (1989). "The Grazing Land Ecosystems of the African Sahel." *Ecological Studies 75.* Berlin: Springer-Verlag.

Leslie, Paul, P. Fry, K. Galvin, and J. T. McCabe (1988). "Biological, Behavioral and Ecological Influences on Fertility in Turkana." in: *Arid Lands Today and Tomorrow: Proceeding of an International Research and Development Conference.* E. Whitehead, C. Hutchinson, B. Timmermann, and R. Varady, eds. Pp. 705-712. Boulder, CO.: Westview Press.

Little, Michael A. (1989). "Human Biology of African Pastoralists." *Yearbook of Physical Anthropology.* 32:215-247.

Little, Peter D.(1984). "Land and Pastoralists." *Cultural Survival Quarterly.* 8(1):46-47.

————, (1985). "Absentee Herd Owners and Part-Time Pastoralists: The Political Economy of Resource Use in Northern Kenya." *Human Ecology.* 13:131-151.

————, (1987). "Women as Old Payian (Elder): The Status of Women among the Il Chamus (Njemps) of Kenya." *Ethnos.* 52(1-2):81-102.

----------, (1989). "The Cattle Commodity System of the Kismayo Region, Somalia: Rural and Urban Dimensions." *Working Paper No. 51.* Binghamton, NY: Institute for Development Anthropology.

----------, (1992). *Elusive Granary: Herder, Farmer, and State in Northern Kenya.* Cambridge: Cambridge University Press.

Little, Peter D., and David Brokensha (1987). "Local Institutions, Tenure, and Resource Management in East Africa." in: *Conservation in Africa: People, Policies, and Practice.* David Anderson and Richard Grove, eds. Pp. 193-209. New York: Cambridge University Press.

Little, Peter D., and Michael M Horowitz, eds. (1987). *Lands at Risk in the Third World.* Boulder, CO: Westview Press.

MAB (1975). "The Sahel. Ecological Approaches to Land Use." *MAB Technical Note 1.* Paris, France: United Nations Educational, Scientific and Cultural Organization.

Mabbutt, J. A. (1989). "Desertification: The Public Record." in: *African Food Systems in Crisis Part One: Microperspectives.* Rebecca Huss-Ashmore and Solomon H. Katz, eds. Pp. 73-109. New York: Gordon and Breach Science Publishers.

Maher, C. (1937). *Soil Erosion and Land Utilization in the Kamasia, Njemps, and East Suk Reserves.* Nairobi, Kenya Ministry of Agriculture.

Malhotra, S. P., and H. S. Mann (1982). "Desertification and the Organization of Society." in: *Desertification and Development: Dryland Ecology in Social Perspective.* B. Spooner and H.S. Mann eds. New York: Academic Press. Pp. 305-310.

Manger, Leif O. (1981). *The Sand Swallows Our Land.* Bergen, Norway: Department of Social Anthropology, University of Bergen.

----------, (1988). "Traders, Farmers and Pastoralists: Economic Adaptations and Environmental Problems in the Southern Nuba Mountains of the Sudan." in: *The Ecology of Survival.* Douglas H. Johnson and David M. Anderson eds. Pp. 155-172. Boulder, CO.: Westview Press.

McCabe, J. Terrence (1990). "Turkana Pastoralism: A Case Against the Tragedy of the Commons." *Human Ecology.* 18:81-104.

McCracken, J. (1982). "Experts and Expertise in Colonial Malawi." *African Affairs.* 81:101-116.

Merry, Douglas (1987). "The Local Impact of Centralized Irrigation Control in Pakistan: A Socioeconomic Perspective." in: *Lands at Risk in the Third World: Local-Level Perspectives.* Peter D. Little and Michael M. Horowitz eds. Pp.352-372. Boulder, CO: Westview Press.

Monimart, Marie (1989). *Femmes du Sahel: La Desertification au Quotidien.* Paris: Editions Karthala et OCDE.

Mortimore, Michael (1989). *Adapting to Drought: Farmers, Famines and Desertification in West Africa.* New York: Cambridge University Press.

National Academy of Sciences (NAS) (1975). *Arid Lands of Sub-Saharan Africa.* Washington, DC: National Academy of Sciences.

National Research Council (1986). *Proceedings of the Conference on Common Property Resource Management.* Washington, DC: National Academy Press.

Nelson, Ridley (1988). "Dryland Management: The 'Desertification' Problem." *Environmental Department Working Paper No. 8.* Washington, DC: The World Bank.

Nicholson, Sharon E. (1982). "The Sahel: A Climatic Perspective." *Club du Sahel Paper D(82)187.* Paris, France: Organisation for Economic Co-operation and Development - Permanent Interstate Committee for Drought Control in the Sahel (CILSS).

Odingo, Richard S. (1990a). "Report of the Kenya Pilot Study (FP/6201-87-04) Using the FAO/UNEP Methodology for Assessment and Mapping of Desertification." in: *Desertification Revisited: Proceedings of an Ad-Hoc Consultative Committee on the Assessment of Desertification.* Nairobi, Kenya: UNEP.

————, (1990b). "The Definition of Desertification: Its Programmatic Consequences for UNEP and the International Community." *Desertification Control Bulletin 18:31-50.*

O'Leary, Michael (1984). "Ecological Villains or Economic Victims: The Case of the Rendille of Northern Kenya." *Desertification Control Bulletin 11:* 17-21.'

————, (1985). "The Economics of Pastoralism in Northern Kenya: The Rendille and Gabbra." *Technical Report No. F-3.* Nairobi, Kenya: UNESCO.

————, (1989). "Changing Responses to Drought in Northern Kenya: The Rendille and Gabra Livestock Producers." in: *Property, Poverty and People: Changing Rights in Property and Problems of Pastoral Development.* P.T.W. Baxter and Richard Hogg eds. Manchester, UK: University of Manchester. Pp.55-79.

————, (1990). "Drought and Change Amongst Northern Kenya Nomadic Pastoralists: The Case of the Rendille and Gabra." in: *From Water to World Making.* Gisli Palsson, ed. Uppsala, Sweden: The Scandinavian Institute of African Studies. Pp.151-174.

Painter, Thomas (1985). *Peasant Migrations and Rural Transformations in Niger: A Study of Incorporation within a West African Capitalist Regional Economy, c. 1875 to c. 1982.* Ph.D. Thesis. Binghamton, NY: State University of New York at Binghamton.

Palmer, E.(1977). *The Desert of the Exodus, vols. I and II.* New York: Arno Press.

Penning de Vries, F. W. T., and M. A. Djiteye, eds. (1982). *La Productivité des Pâturages Sahéliens.* Wageningen, Germany: Centre for Agricultural Publishing and Documentation.

Peters, Pauline (1988). "The Ideology and Practice of Tswana Borehole Syndicates: Co-operative or Corporation?" in: *Who Shares: Co-operatives and Rural Development.* D. Attwood and B. Baviskar, eds. New Delhi: Oxford University Press. Pp. 23-45.

Pickup, G. (1985). "The Erosion Cell — A Geomorphic Approach to Landscape Classification in Range Assessment." *Australian Rangeland Journal.* 7(2):114-121.

Prah, Kwesi K. (1989). "Land Degradation and Class Struggle in Rural Lesotho." in: *Ecology and Politics: Environmental Stress and Security in Africa.* Anders Hjort af Ornas and M.A. Mohamed Salih, eds. Uppsala, Sweden: Scandinavian Institute of African Studies. Pp. 117-129.

Rapp, A., H. N. Le Houérou, and B. Lundholm (1976). *Can Desert Encroachment Be Stopped? A Study with Emphasis on Africa.* Stockholm: Secretariat for International Ecology and UNEP.

Reining, Patricia (1978). *Handbook on Desertification Indicators.* Washington, DC: American Association for the Advancement of Science.

Rosanov, B. (1990). "Global Assessment of Desertification: Status and Methodologies." in: *Desertification Revisited: Proceedings of an Ad-Hoc Consultative Meeting on the Assessment of Desertification.* Nairobi, Kenya: UNEP.

Runge, C. F. (1981). "Common Property Externalities: Isolation Assurance and Resource Depletion in a Traditional Grazing Context." *American Journal of Agricultural Economics.* 62:595-606.

--------, (1986). "Common Property and Collective Action in Economic Development." *World Development.* 14(5): 623-635.

Sandford, Stephen (1976). "Pastoralism under Pressure." *Overseas Development Institute Review.* 2:45-68

--------, (1982). "Pastoral Strategies and Desertification: Opportunism and Conservatism in Dry Lands." in: *Desertification and Development: Dryland Ecology in Social Perspective.* B. Spooner and H.S. Mann eds. New York: Academic Press. Pp.61-80.

--------, (1983). *Management of Pastoral Development in the Third World.* Chichester, England: John Wiley and Sons.

Schlee, Gunter (1991). "Traditional Pastoralists -- Land Use Strategies." in: *Marsabit District: Range Management Handbook of Kenya, Vol II,* No. 1 Pp. 130-164. H.Schwartz, S. Shaabani, and D. Walther, eds. Nairobi, Kenya: Ministry of Livestock Development.

Schmink, Marianne, and Charles H. Wood (1987). The "Political Ecology" of Amazonia. in: Lands at Risk in the Third World: Local-Level Perspectives. Peter D. Little and Michael M. Horowitz eds., Pp. 38-57. Boulder, CO: Westview Press.

Scoones, Ian (1989a). "Economic and Ecological Carrying Capacity Implications for Livestock Development in the Dryland Communal Areas of Zimbabwe." *ODI Pastoral Development Network paper 27b.*

--------, (1989b), "Patch Use by Cattle in Dryland Zimbabwe: Farmer Knowledge and Ecological Theory." *ODI Pastoral Development Network paper 28b.*

--------, (1990). *Livestock Populations and the Household Economy: A Case Study from Southern Zimbabwe.* Ph.D. Thesis. University of London.

Sobania, N. (1980). *The Historical Tradition of the Peoples of the Eastern Lake Turkana Basin.* Ph. D. Thesis. London, England: University of London.

Spencer, Paul (1988). *The Maasai of Matapato.* Bloomington, IN: Indiana University Press.

Sperling, Louise (1987). "Wage Employment Among Samburu Pastoralists of Northcentral Kenya." *Research in Economic Anthropology.* 9:167-190.

Spooner, Brian (1989). "Desertification: The Historical Significance." in: *African Food Systems in Crisis Part One: Microperspectives.* Rebecca Huss-Ashmore and Solomon H. Katz, eds. Pp. 111-162. New York: Gordon and Breach Science Publishers.

Spooner, Brian, and H. S. Mann (1982). *Desertification and Development: Dryland Ecology in Social Perspective.* New York: Academic Press.

Starr, Martha (1987). "Risk, Environmental Variability and Drought-Induced Impoverishment: The Pastoral Economy of Central Niger." *Africa*. 57(1):29-50.

Stebbing, E. P. (1935). "The Encroaching Sahara: The Threat to the West African Colonies." *The Geographical Journal*. 85:506-24.

————, (1938). "The Man-Made Desert in Africa. Erosion and Drought." *Journal of the Royal African Society*, Supplement (January).

Steeds, David (1985). *Desertification in the Sahelian and Sudanian Zones of West Africa*. Washington, DC: World Bank.

Sutter, John W. (1987). "Cattle and Inequality: Herd Size Differences and Pastoral Production Among the Fulani of Northeastern Senegal." *Africa*. 57: 196-217.

Talle, Aud (1987). "Women as Heads of Houses: The Organization of Production and the Role of Women among the Pastoral Maasai in Kenya." *Ethnos*. 52 (1-2):50-80.

Tucker, Compton J., Harold E. Dregne, and Wilbur W. Newcomb (1991). "Expansion and Contraction of the Sahara Desert from 1980 to 1990." *Science*. 253:299-301.

UNCOD (1977a). *World Map of Desertification*. Nairobi, Kenya: United Nations Environment Program.

————, (1977b). *Report of the United Nations Conference on Desertification, Nairobi, 29 August-9 September 1977*. Nairobi, Kenya: United Nations Environment Program.

UNEP (United Nations Environment Program) (1984). *Activities of the United Nations Environment Program in the Combat against Desertification*. Nairobi, Kenya: United Nations Envrionment Program.

————, (1991). *Status of Desertification and Implementation of the UN Plan of Action to Combat Desertification*. Nairobi, Kenya: United Nations Environment Program.

Warren, A., and C. T. Agnew (1988). *An Assessment of Desertification and Land Degradation in Arid and Semi-Arid Areas*. London: International Institute for Environment and Development Paper. No. 2.

Warren, A., and J. K. Maizels (1977). "Ecological Change and Desertification." in: *Desertification: Its Causes and Consequences*. United Nations Desertification Secretariat, ed. Oxford: Pergamon. Pp. 169-260.

Watts, Michael (1987). "Drought, Environment and Food Security: Some Reflections on Peasants, Pastoralists, and Commoditization in Dryland West Africa." in: *Drought and Hunger in Africa*. Michael H. Glantz, ed. New York: Cambridge University Press. Pp.171-211.

Westoby, M., B. Walker, and I. Noy-Meir (1989). "Opportunistic Management for Rangelands Not at Equilibrium." *Journal of Range Management*. 42(4):266-274.

White, Cynthia (1987). *Poor Pastoralists are Poor Herd Managers: Changing Animal Ownership and Access to Land Among the WoDaaBe (Fulani) of Central Niger*. Paper presented at workshop on "Changing Rights in Property and Pastoral Development," University of Manchester, England, 23-25 April.

8

The Socioeconomic Matrix of Deforestation

Marianne Schmink

This chapter takes a critical look at how deforestation is defined and explained. It begins by adopting a social definition of deforestation, one that links it to the livelihoods of local populations in forested areas. The behavior of forest users is analyzed within the socioeconomic matrix that shapes forest conversion patterns. A common framework of analysis is used to compare deforestation in Amazonia and in India.

The main objectives of the chapter are: 1) to reorient thinking about deforestation to a focus on multiple users, especially the local resident population; 2) to analyze international and national market trends and policy actions, migration, and land tenure as primary elements of the socioeconomic matrix of deforestation; and 3) to emphasize the importance of social dynamics (household and interest group strategies, conflict and cooperation) that contribute to deforestation.

Deforestation: The Socioeconomic Issues

Definitions

Deforestation is surprisingly complex to define and measure, even with the benefit of modern satellite technology. Differences in concepts and measures of deforestation significantly affect the analysis of rates and trends, the assessment of the seriousness of the problem, and the focus of suggestions for policy. The most restricted definition of deforestation, used by the FAO (1981b), refers to the wholesale conversion of forests (over 40 percent) to other uses (1981b). Others such as Myers (1984) and Johnson (1991) define deforestation to include the modification of forest structure and composition through such activities as selective commercial logging that impoverish the resource base without completely clearing the forest. The cases of deforestation considered in this paper include not only wholesale forest conversion but also significant modification due to logging and fuelwood collection.

Following Blaikie and Brookfield's (1987) conceptualization of land degradation, I argue in this paper that definitions of deforestation are not technical, but social. Forests are always being modified, and even burned, by natural forces as well as human interference. Deforestation has been occurring over millennia and has long been considered a major manifestation of the expansion of civilization (ICIHI 1986; Tucker 1983:179). It becomes a problem only when perceived as such by society.

For our purposes, deforestation might be defined as: "the reduction in the capability of a forest to fulfill a particular function." This definition would have one meaning for a biologist concerned with the forest's function as habitat for specific animal species, another for a U.S. conservationist concerned with the role of the forest in affecting global climate patterns, a third for the government of a tropical country where the economic productivity of the forest is the main concern, and a fourth for the local communities accustomed to free access to the forest resources.

In this paper, local land managers are the point of departure for analysis. The loss of forest resources is often perceived as a problem by the local populations who depend on forest resources, but these same social groups may benefit from the conversion of forests to other, more economically productive uses. The working definition of deforestation used here emphasizes degradation of tropical forests to the detriment of their use by local populations who depend on them for a significant portion of their livelihood.

Only recently has deforestation been perceived as a global problem. This is because of the perception that planetary resources are reaching the limits for supporting the world's population and economic systems. The global dimensions of the deforestation problem are perceived most clearly in the developed world, most of whose forests were long since cleared (Mather 1990; Williams 1990:189-190). The increased pressures on tropical forests are particularly alarming because, in general, the soils that support tropical forests are more sensitive to changes induced by human interference, and less able to recover their productive capability without human assistance (Blaikie and Brookfield 1987:12). Deforestation in the tropics also affects a large population of people who depend on the forest for a living.

Deforestation History

Estimates of the decline in the world's forest areas throughout history vary. Problems of measurement and definition have led to disagreements about the magnitude of deforestation. Moreover, forest conversion cannot always be considered permanent due to the dynamics of forest regeneration (Williams 1990:180). One recent authoritative source estimated that from pre-agricultural times to the present, the world's forest areas declined by about one-fifth, from 5 to 4 billion hectares (Repetto 1988:2). Forests and woods still cover more than two-fifths of the earth's surface and account for 60 percent of the net biomass productivity of the planet's ecosystems, according to this source. In temperate zones, forested areas are stable, not declining, because of the reversion of previously cultivated land to woods.

The historical roots of deforestation can be traced to the process of world economic development, beginning in the late fifteenth century with the integration of the world market and the reorientation of land use in already settled areas (Mather 1990; Williams 1990:180). From the seventeenth century on, European expansion posed the greatest threat to the world's forests as the colonial powers introduced new crops and new forms of exploitation in the tropics and subtropics.

The greatest declines in forest cover have been in temperate forests (32-35%) and in subtropical woody forests (24-25%), less for tropical climax forests (15-20%) and tropical evergreen forests (4-6%) (Repetto 1988:2). But since World War II, pressures leading to deforestation have shifted from temperate to tropical zones. Rates of forest conversion in developing countries have been increasing since the "development decades" of the 1950s and 1960s. For example, from 1950 to 1983, Central America lost 38 percent of its forest area, while 24 percent of Africa's forests were cleared during the same period (Repetto 1988:6).

Of particular concern is the current rate of tropical deforestation and the perception that it is increasing. FAO estimates in 1981 showed an annual rate of forest loss of 11 million hectares per year, but by 1990 the figure was revised upwards to 17 million hectares per year (Johnson 1991:7). Annual clearing removes an estimated 56,000 km^2 of forests in tropical America, 37,000 km^2 in tropical Africa, and 20,000 km^2 in Asia (Williams 1990:196). These aggregate rates include widely varying patterns in different countries and regions. Local deforestation is severe in Central America, Western and Sahelian Africa, and Southeast Asia. In some countries (e.g. Madagascar, Sierra Leone, Nigeria, Cote d'Ivoire, Bangladesh, India, Sri Lanka, Malaysia, Thailand, Philippines), if current rates of deforestation continue, virtually all forests will disappear in a matter of a few decades (Johnson 1991:8-9).

Macro Policy Statements About Deforestation

The complexity of the socioeconomic factors behind forest modification, and their variability from one setting to another, have made it difficult to go beyond ill-informed aggregate assertions about the causes of deforestation. These, in turn, often have led to poorly conceived policy initiatives.

A positive relationship between population growth and deforestation rates has been observed at the global level. For example, Repetto cites projections based on population growth, increases in demand for food, and declining agricultural yields that predict deforestation of 10-20 percent of tropical forest areas by the year 2020 (1988:6-9). Several reports produced in the 1980s, such as *The Global 2000 Report* (Barney 1980), Worldwatch Institute's 1984 *State of the World*, and World Resources Institute's *World Resources 1986*, used aggregate-level statistics to suggest that population growth was behind the expansion of migratory agriculture that was the main cause of deforestation (Bedoya 1991).

However, these simplistic Malthusian correlations hold only at the most aggregate level. On a country-by-country or region-by-region basis, population growth does not predict deforestation rates (Mather 1990). In some areas, such as those dominated by cattle ranching, deforestation rates increase as population density declines. The distribution of population, and of access to income and productive resources among different social groups, are better predictors of deforestation patterns. Population growth is only one factor among many that make up the "pressure of production on resources" (Blaikie and Brookfield 1987:240) that commonly leads to resource degradation, including deforestation.

The so-called "fuelwood crisis" is a good example of simplistic, unicausal thinking about population and deforestation. During the 1970s, a population-driven gap between demand and supply of fuelwood, especially in Africa, was diagnosed to be the driving force behind deforestation (Anderson and Fishwick 1984; FAO 1981a). This perception led to massive reforestation efforts intended to address the "fuelwood gap," which were largely unsuccessful.

In fact the relationship between fuelwood and deforestation was poorly diagnosed to begin with (Dewees 1989). The fundamental causes of deforestation were more closely related to land clearance for agriculture and livestock. Nor did planners understand that farmers might be able to adapt their fuelwood strategies in the face of deforestation, and therefore saw little need to plant trees for fuel. In general, the dynamics of behavior by forest users is overlooked in macro-level analyses. Yet the rationality of individuals and groups, and the interactions between them, play a major role in forest conversion patterns.

Like reforestation for fuelwood, many major policy efforts have focused on massive tree-planting efforts rather than addressing the factors that lead to deforestation. The Tropical Forestry Action Plan (TFAP), announced in October 1985, was an international initiative jointly developed by the FAO, the

World Bank, the World Resources Institute, and the United Nations Development Programme. The plan called for US$8 billion to be invested in tropical countries over a five-year period beginning in 1987. The focus was on timber and fuelwood plantations and on industrial forestry. This emphasis on industry needs, and the lack of consultation with nongovernmental organizations, led to criticisms by the environmental community that the plan neglected essential conservation efforts. Nor did the TFAP include attention to the needs of local forest-dependent peoples (Johnson 1991).

Forestry sector policies such as the TFAP and the projects supported by the International Tropical Timber Organization take a narrow view of the deforestation problem and its solution. Industrial forestry and fuelwood plantations are to be expanded to supply future demand for tropical timber. These proposals show relatively little concern for natural forest conservation and management, much less for the broader development issues that lie behind deforestation (Johnson 1991:26). These solutions would hardly address the deforestation problem that is the focus of this paper in that they would do little to improve the livelihood of local populations.

While most of the blame for tropical deforestation is placed on small farmers, and secondarily on ranchers and loggers, others have focused on the role of government policies in encouraging deforestation. An international research project reported in Repetto and Gillis (1988) identified the negative environmental impacts and high economic costs of government policies, however worthy their goals in stimulating economic growth or poverty alleviation. The factors behind deforestation in tropical countries are deeply rooted in the development patterns of those countries, characterized by rapidly growing populations, the slow growth of employment opportunities, concentration of land holdings, and distortions in land tenure institutions (Repetto 1988:15-16). In most cases, government policies have contributed directly or indirectly to the factors that lead to deforestation.

The Socioeconomic Matrix

The complexity and variability of factors leading to deforestation defy simple models or unicausal explanations. In this paper I follow the approach adopted by Blaikie and Brookfield (1987:3), and focus on the intersection between, on the one hand, the evolving strategies of particular "forest managers," and, on the other hand, the changing social, political, and economic circumstances, or matrix, that frames their behavior. This conceptual framework is adapted from a study of conflicts over resources in Brazilian Amazonia (Schmink and Wood 1992).

The initial focus in this approach is on the individuals or small groups who make management decisions about use of forests in a particular local site. But the analysis broadens to include the interaction of different local groups, and how their actions are shaped by, and may affect, the socioeconomic and political context over time. Thus there are three key aspects to the approach: focus on multiple users and their interactions; contextual analysis at different levels of social structure; and attention to historical dynamics.

There usually is not one, but multiple users of forest resources. Each different actor or social group has a particular "rationality" for forest use, and they are often in conflict (Schmink 1987; Schmink and Wood 1987). Relatively few case studies focus on the interaction of social groups in the deforestation process. Two recent studies of Amazonia have analyzed this interaction: one focuses on conflicts between different social groups over resources in Brazilian Amazonia (Schmink and Wood 1992), and the other emphasizes coalition strategies among small farmers in Ecuador (Rudel 1993).

The multiple forest users who are the protagonists in the case studies of deforestation analyzed here respond to particular situations from the standpoint of their own objectives, constraints, and perceptions. These are defined by the characteristics of each individual or social group (age, gender, ethnicity, education, social class) and their access to resources, including property, markets, and technologies. This access flows from the socioeconomic structures of the society in question (class and property relations, market systems, and macroeconomic policies). It is these socioeconomic structures that make up the matrix of forest management decisions.

The history of forest use in a particular site is the product of the interaction of multiple forest users responding to changing circumstances at different scales of analysis (local, regional, national, and international). The "political ecology" approach seeks to take these different levels into account by focusing on key variables that constitute the socioeconomic matrix of deforestation. The most important population characteristics of this socioeconomic matrix are migration patterns, land distribution and settlement patterns, and household economic strategies. The matrix itself changes over time, in part due to the actions and interactions of specific land users (Schmink and Wood 1992).

The complexity of interactions among these levels complicates the tasks of predicting outcomes and designing policy. This is likely one of the reasons for the simplicity of many analyses and prescriptions.

Case Studies of Deforestation

The definition of deforestation adopted here takes the point of view of local residents as land managers. Several authors have argued persuasively in favor of focusing on the resource "user" (Blaikie and Brookfield 1987; Vayda 1983; Rocheleau 1987). The political ecology approach used in this paper looks at individuals within their social context, that is, within the complex and changing socioeconomic and political matrix that shapes their perceptions and behavior (Schmink and Wood 1987). There is a broad literature that points to the importance of global markets and public policies in driving deforestation (e.g., Repetto and Gillis 1988). The framework developed below links the study of individual decisions about forest use with the strategies adopted in situations of conflict or cooperation among social groups, and how these strategies respond to a changing market and policy environment.

In addition to linking these different levels of analysis, what is unique in this approach is the focus on the social dynamics that lead to indeterminate outcomes and that can either accelerate or retard pressures for deforestation. There are two sources of indeterminacy: one is the interaction over time between different social groups occupying a forest area; and the other is the modification of the socioeconomic matrix through historical change.

The socioeconomic matrix is the set of structural conditions that frames decisions to clear forested areas. The basic elements outlined in Figure 8.1 make up the context within which particular interest groups, domestic units, and individuals participate in forest conversion. These conditions derive in part from international market demands and policy instruments, which interact with national development policies (especially road and railway construction, resettlement schemes, and industrial incentives) and land tenure patterns to determine how individuals and groups will behave with respect to the forest resources they wish to exploit. Population pressure, due to high fertility rates or to migration, tends to accelerate the pressures to deforest. So does the interaction among different social groups competing for the resources in frontier areas.

The content of each level of analysis (global, national, local/regional, or household/community) will vary for each case, and the variables in Figure 8.1 interact in ways too complex to permit easy synthesis. Change takes place at all levels represented in the figure. It is not just structural conditions that are important, but also the strategies of groups and individuals in responding to those conditions over time. The matrix is intended to focus attention on the principal variables that must be considered in analyzing the social causes of deforestation, but it does not constitute a predictive model.

I will draw on information from two cases to illustrate the usefulness of the framework presented in Figure 8.1, and the importance of analyzing social dynamics. I will begin by applying the framework to Amazonia, a region where I have firsthand experience in Brazil (see Schmink 1987; Schmink and Wood 1987; Schmink and Wood 1992), and there are other useful studies of deforestation in Peru and Ecuador (Bedoya 1991; Rudel n.d.). The Amazonian case is interesting because of this cross-national comparative data. Next the framework is applied to India, drawing together the common elements from studies in several regions of that country. I will conclude by briefly comparing the two cases and drawing out the implications for research and policy.

FIGURE 8.1 The Socioeconomic Matrix of Deforestation

Global Context

Markets
Demand for forest goods
Foreign investment

International Aid Policies
Development lending
Structural adjustment
Environmental conditionality

National Context

Markets
Transportation
Prices
Financial markets

Policy
Roads and infrastructure
Price supports and subsidies
Extension services

Migration
Population pressures
Frontier expansion

Land Tenure
Land distribution
Property regimes

Regional/Local Context

Settlement Patters
Localized population pressures
Resource distribution and access

Interest Groups
Conflicts over resources
Coalitions and alliances

Household/Community Context

Gender Relations
Division of labor
Family size and composition

Family/Community Strategies
Access to resources
Income sources and employment
Temporary migration

Deforestation in Amazonia

Figure 8.2 is an elaboration of the socioeconomic matrix specific to
Amazonia (based on information from Brazil, Peru, and Ecuador). The
information in the chart refers to cross-national similarities in Amazonian deve-

FIGURE 8.2 The Socioeconomic Matrix of Deforestation: Amazonia

Global Context

Commodity Markets
Demand for rubber, coffee, timber

Foreign oil, mining, and timber cos.

International Aid Policies
World Bank and IDB support for roads and colonization
Environmental conditionality
Geopolitics

National Context

Markets
Transportation precarious
Export orientation

Policy
Developmentalism:
   Road-building prog. for food prod.
   Fiscal incentives for ranching & logging
   Population redistribution

Migration
Migration from Andes
   northeastern and southern Brazil
Frontier expansion
Economic crises and migration

Land Tenure
Skewed land distribution
Amazon "safety valve"
Competing property regimes
   (titles, reserves, informal)

Regional/Local Context

Settlement Patterns
Population pressures along roads
Contested frontiers

Interest Groups
Conflicts over land, timber, minerals; loggers, ranchers, miners, migrants, natives
State and large investors
vs. small producers

Household/Community Context

Gender Relations
Men: agriculture
Women: domestic and off-farm

Family/Community Strategies
Conditions of access to land
Off-farm income
Temporary migration: mining; land clearing
Resistence movements

lopment trends. Distinctions, especially of the Brazilian case, will be mentioned briefly.

Although Amazonia's vast forests are generally viewed as primordial, there is growing evidence that significant portions of the basin contain forests whose vegetation reflects past human manipulation (Baleé 1989; Sponsel 1986). Numerous sites have been documented of Indian "black earth," rich, anthropogenic soils preferred by present-day natives due to their high fertility (Eden et. al. 1984; Smith 1980). At least 11.8 percent of the region's upland (*terra firme*) areas have been documented as anthropogenic forests dominated by palms, bamboo, Brazil nuts, lianas, and other resources especially useful to human inhabitants (Baleé 1989 1989). As more evidence accumulates of the management practices of contemporary populations and their ancestors, this figure will most likely be adjusted upward.

Anthropologists who wrote in the 1950s extrapolated from the modern ethnographic record to describe precolonial Amazonia in terms of small bands of Indians who lived in temporary settlements and subsisted by fishing, hunting, and shifting cultivation (Steward and Faron 1959; Meggers 1954). In contrast to this familiar image, current reconstructions of the preconquest vindicate the chronicles written by Carvajal and other explorers who registered the existence of permanent settlements that held from several thousand to tens of thousands of individuals, possibly more (Bush, Piperno, and Colinvaux 1989; Denevan 1976; Gibbon 1990; Moran 1981, 1989; Roosevelt 1987). There is growing evidence that current forms of indigenous adaptation are the result of de-evolutionary change due to the depopulation and wholesale disruption wrought by the European conquest (Hemming 1978, 1987).

During the colonial period, Amazonian forests were exploited for key extractive products such as clove, sarsaparilla, cacao, cinnamon, aromatic roots, and palm oils. During the late nineteenth and early twentieth centuries the region experienced a boom in exploitation of natural rubber. The penetration of world market demand for these commodities led to the violent transformation of indigenous systems of resource management and their replacement by a loosely knit society of dispersed settlers linked by river traders. But these extractive activities entailed little deforestation, and agriculture remained small in scale until recent years. In Brazil, production of export products, such as sugar, coffee, and gold, was concentrated along the eastern coast during the colonial period. The Atlantic coastal forest was almost entirely removed by the mid-twentieth century (Dean 1983).

It was with the expansion of the Western market economy during the period following World War II that a period of rapid deforestation began in Amazonia. Demand for such commodities as coffee and oil stimulated investments in infrastructure in Peru and Ecuador's Amazonian regions. Bilateral development assistance and loans from the multilateral development banks (World Bank and Inter-American Development Bank) helped to finance ambitious road-building and migrant farmer colonization projects intended to absorb land-poor populations from other regions of each country, and to increase production of food crops. These macro-level factors unleashed an unprecedented wave of deforestation in Amazonia.

Amazonian policies were infused with the euphoric "developmentalism" of the 1950s and 1960s. Plans to develop the region during this period were part of national policies to address the pressures for land reform. Land distribution is highly skewed in the Amazonian nations, and most of the population is concentrated in highland and coastal areas. The lowland Amazonian region was viewed as the "safety valve" for populations elsewhere and as a reservoir of relatively untouched natural resources to fuel development. Road-building programs were instituted in order to open up the region. In Brazil, generous programs of fiscal incentives were designed to attract the excess capital of investors from the country's more developed Center-South. Both migrant colonists and wealthy investors were active deforesters in different parts of the region.

At the time, there was little concern about the environmental consequences of these policies. In keeping with the developmental precepts, the value of land was calculated in terms of its production for the market. Land with standing natural forest was considered "naked" until improved by clearing and planting. Land rights and titles were contingent on demonstrating this kind of "improvement" or productive use. As a market for land developed in Amazonia, it provided a continuing stimulus to deforestation.

The waves of migration into Amazonia stimulated by these policies, and by good prices for tropical commodities such as coffee, penetrated into the region along new highways built by governments and by private timber and oil companies. The migrants came from different areas for different reasons: from the overcrowded Andean regions; from the modernizing agricultural areas of southern Brazil; and from the stagnant Brazilian northeast. Arriving in the lowland colonization areas, they encountered a chaotic land tenure situation in which informal use rights competed with titled property and traditional collective ownership regimes of native Amazonians. As the pioneer frontier pushed forward, the colonists intruded on indigenous territories on the frontier's edge, while they risked being swallowed up by wealthy investors following in the deforested path of their labor investments.

Population pressures built up along the penetration highways as migrant farmers and investors, including small-scale miners, engaged in contests over access to each new resource opened up by the roads. Ranchers, colonists, indians, miners, loggers, and others came together on the ground in a pitched battle over each new resource "front" opened up by the roads. In some cases different groups worked together in what Rudel (1993:25) has called cross-class "growth coalitions" for the purpose of opening up and deforesting new areas. For example, the roads built by loggers provided a means of entry for small farmers, who returned the favor by supplying timber from their newly cleared plots of land. More frequently, different interest groups engaged in prolonged disputes, sometimes violent, over access to land, minerals, and other resources (Schmink and Wood 1992). In both cases, the interactions between groups were likely to increase the pressures for deforestation.

In western Brazil, these pressures stimulated the growth of a resistance movement among Amazonian rubber tappers based in the state of Acre (Alegretti 1990; Campbell 1990; Schwartzman 1989). The organization began in the mid-1970s in response to the expulsion of traditional forest-dwelling peoples by newcomers who had been encouraged by government policies to purchase land for ranching. The movement developed nonviolent protest techniques to oppose forest clearing and successfully pressured the government to create "extractive reserves" where the indigenous peoples would retain their rights to selective forest exploitation.

Indigenous groups suffering pressure on their land also were induced to deforest more as a means to assure their rights to land, but they still cleared significantly less forest than did their colonist neighbors (Bedoya 1991:93-94; Rudel 1993:40, 110-131). Cultural differences accounted for the native groups' reluctance to clear forest, compared to colonists who sought to clear as much as possible. The overall economic rationality of the two ethnic groups differed significantly: whereas natives were more concerned with the overall return from all crops for subsistence purposes, the colonists focused on the crops intended for the market (Bedoya 1991:96-100).

Household-level decisions about resource use, and deforestation in particular, were linked to access to land for subsistence and to other resources. Access to credit was a significant stimulus to deforestation (Rudel 1993:118). In the frontier setting there were few employment options, especially for women. Men worked primarily in agriculture and mining, and women carried out domestic chores and participated in menial service and commerce activities in town. Households with access to land were the fortunate ones, but they depended on available family labor to carry out land clearing and other agricultural tasks. Lack of rural schools meant that families often lived in town, while men worked on the agricultural plot and returned to town only on weekends. Labor constraints contributed to poor management capabilities of small producers (Collins 1986). The need for family labor provided a strong incentive for large rural families.

Most households had only uncertain access to land, which contributed to their instability. The pressure to establish land rights stimulated them to clear as much land as possible, although the uncertainty of land tenure provided little incentive for long-term management strategies. Even migrant farmers who took great care in managing their traditional highland resources did not do the same in the lowlands (Bedoya 1991:75).

Migrant farmer settlements in Amazonia were not very stable over time. Within a generation or less, many settlers abandoned their lots for one reason or another, or children of the settlers moved to new areas to open up and deforest their own land (Bedoya 1991:61-62; Rudel 1993:137). The availability of extensive land areas, and disputes over land, encouraged this instability of settlements, which stimulated further deforestation. In Peru, the coca economy contributed to population movement as demand for coca stimulated deforestation for planting, and eradication programs caused producers to shift their fields to newer, less accessible areas (Bedoya 1991:110-111). Secure land titles were associated with lower rates of deforestation in Peru, because farmers with titles planted more permanent crops, but only the highly lucrative coca was cultivated intensively using modern inputs (Bedoya 1991:84-86).

Complex internal migration patterns, including seasonal and temporary migration for work in mining areas or on large estates, tended to separate families for long periods. When men migrated, wives were often left with the care of the family's property and livelihood while awaiting remittances or the return of their menfolk. In Brazil, many migrants took work in small-scale mining when they were unable to claim land. The fluidity of internal migration patterns undermined the development of sustainable resource management strategies.

In large forested areas such as Amazonia, colonists must be able to muster coalitions to provide the economic resources necessary to settle and deforest blocks of land (Rudel 1993:25). Since World War II, traditional patterns of pioneer frontier expansion increasingly have given way to cross-class coalitions and to "free rides" provided by the road-building and other policies sponsored by national or international institutions. Under these conditions, small producers can be the most important agents of deforestation, as in Peru and Ecuador's Amazon region (Bedoya 1991:42). In Brazil, on the other hand, clearing for cattle ranches accounted for far more deforestation than did migrant farmers, except in the western state of Rondonia--site of the World Bank-financed Northwest Brazil Integrated Development Program (POLONOROESTE) colonization project (Browder 1988). Moreover, conflicts over land between cattle ranchers and small producers contributed to deforestation, since legal land rights rewarded forest clearing.

Much of the frontier pattern described above is typical of other parts of Latin America where small farmers, often following logging roads, penetrate into forested areas to clear small plots and plant subsistence crops. After a few years, their land is swallowed up by larger ranchers seeking to expand their areas of pasture (Collins and Painter 1986; Rudel 1993). The importance of these frontier dynamics, in areas where population density is so low, belies Malthusian explanations. The pervasive pattern stems from a common legacy of land tenure concentration and population pressure in more settled areas. In settled areas, too, population growth often is not the main cause of deforestation.

For example, a study in Honduras documented how the expansion of capitalist agriculture for export, stimulated by the government at the expense of food production, exacerbated land concentration and encouraged deforestation as land was reallocated from forest, fallow, and food crops to export crops and livestock (Stonich 1989:282).

The evidence from Amazonia and Central America points to the importance of national development initiatives in laying out the terms of the battle over tropical forests during the postwar period. With their strong ties to international lending and assistance, these policies favored access by industrial producers to forested areas. In some situations, small farmer migrants responded with complex sustenance strategies, including multiple migrations. In others they developed group strategies to resist attempts by others to take over their land. Migration, interactions among groups, resistance movements, and an evolving political and economic context were the dynamic elements that led to specific outcomes in each local situation.

Deforestation in India

India has experienced two main eras of deforestation: the first at the turn of the century just after the British took over direct rule, and the second during the 1940s and following independence in 1947 (Tucker 1988:91). According to official government figures, forest cover decreased from 40 to 20 percent of India's territory between 1947 and 1977, leaving only about 11 percent under adequate tree cover (Shiva, Sharatchandra, and Bandyopadhyay 1983:48).

Official government policies have placed much of the blame for deforestation on forest-dwelling populations, but the origins of the current forest crisis in India can be traced to the forest policies initiated by the British and continued after independence. These policies have emphasized commercial exploitation and the generation of revenues for government, at the expense of the use rights of local populations for their own subsistence.

By contrast with the colonial experience in Amazonia, where indigenous populations and resource use systems suffered wholesale transformation, in India the British legacy left many aspects of rural communities intact. In parts of India, tribal populations that depended on forest resources for their livelihood continued a relatively unbroken tradition based on informal, collective management systems. Until 1865, these populations had complete freedom to exploit the forest products they depended upon for shelter, food, tools, and medicines.

These policies changed little after India's independence in 1947. The new Forest Act passed in 1952 stated categorically that natural resources were for the benefit of the nation, not of a village "accidentally" located nearby (Anderson and Huber 1988:37-43). National interests in forested lands primarily were defined in terms of supplying raw materials for wood-based industries (pulp,

FIGURE 8.3 The Socioeconomic Matrix of Deforestation in India

Global Context

Commodity Markets
Demand for tea, timber
Foreign tea and timber cos.

International Aid Policies
British colonial policies
World Bank lending

National Context

Markets
Fuelwood
Export orientation

Policy
Post-independence nationalism
Orientation to forest revenues
State forest ownership

Migration
Increasing pop. growth rates
Internal migration

Expulsion with use prohibitions

Land Tenure
Repression of traditional land rights
Fragmentation of traditional
   landholdings
Exclusion of tribals from reserved
   areas

Regional/Local Context

Settlement Patterns
Movement into marginal areas

Interest Groups
National vs. state govern., allied with
   loggers
Traditional communities and
   resistance movements

Household/Community Context

Gender Relations
Women: main subsist. producers
   and fuelwood gatherers
Conflicts over women's part.
   in resistance movements

Family/Community Strategies
Dominance of panchayats
Inter-household competition
Male out-migration
Resistance movements

paper, and plywood) and generating revenues for state coffers (Joshi 1983:27; Kannan 1983; Kulkarni 1983:89). Although the policy stressed the need to satisfy the needs of rural populations, it portrayed as the enemy the forest-dwelling populations, about 90 percent of which are tribals whose economy is essentially forest-based (Fernandes 1983). The Forest Act left these populations little choice but to clear forest illegally or to seek wage work with the logging companies (Agarwal, Chopra, and Sharma 1982:54).

During the 1960s, India's principal developmental goal was to increase food production through adoption of Green Revolution technologies, which led to wholesale conversion of areas under control of the Forest Department. The eradication of malaria in the 1950s, and the vast road networks that were built in the state of Uttar Pradesh, for example, decimated the forests of the Himalayan foothills and facilitated the felling of timber in remote areas. One result was soil erosion, floods, and landslides.

Of the 3.4 million hectares of forest destroyed from 1951 to 1973, 71.5 percent was converted to croplands and another 11.8 percent to river valley projects, mainly for irrigation (Shiva, Sharatchandra, and Bandyopadhyay 1983:48). But these official figures refer only to land officially transferred from the Forest Department's control, not to the widespread deforestation that also is occurring on department lands (Agarwal, Chopra, and Sharma 1982:33). In the Himalayan forest states, unsupervised commercial logging and resin tapping and road construction without adequate soil conservation measures are the main causes of deforestation.

The gradual encroachment on local villagers' rights to use forest resources led to localized confrontations with state forestry authorities as early as the late nineteenth century (Guha 1989). Distinct conceptions of property and ownership lay at the root of conflicts between villagers and the state, and rebellion took different forms in different areas. These peasant revolts later were strengthened by nationalist movements that legitimized the forest protests. Gandhi's Non-Cooperation Movement inspired a 1910 rebellion against the forest labor camps where villagers were forced to work without pay (Anderson and Huber 1988:38) and a 1913 revolt against the reservation of chir pine forests (Ahmed n.d.).

Population growth rates began to increase in India during the 1920s, and internal migration increased the pressure on resources, pushing settlers into marginal areas. Construction of roads and railways, and timber concessions granted to contractors, continued to place pressure on forest resources after independence, and especially after 1972 with World Bank financing (Guha 1989:138). In the western Himalayas, male out-migration since 1947 left women as the main subsistence providers (Ahmed n.d.; Guha 1989). Agricultural output dropped due to ecological deterioration, and many families were forced to purchase food (Guha 1989:145).

Deforestation exacerbated the fuelwood scarcity, forcing women to spend increasing amounts of time collecting fuel. As in Amazonia, family fragmentation and the erosion of local forms of organization were exacerbated by environmental change. In some areas, communal village forests traditionally were entrusted to the *panchayats* which consisted of higher-status village men who dominated decision making (Blaikie, Harriss, and Pain 1985; Kelkar and Nathan 1991). Households sometimes competed among themselves for access to the common property resources, and conflicts began to emerge within families. In the village of Reni in 1974, women confronted their own menfolk who sought to sell forest resources to contractors (Agarwal, Chopra, and Sharma 1982:42). Women have a particular stake in retaining access to forest resources in regions where they have autonomous control over income from gathering but not from agriculture (Kelkar and Nathan 1991:162).

The Reni incident was a key episode in the emergence of the Chipko protest movement, in which women have been especially active. Chipko (in which people hug trees to keep them from being cut down) is a nonviolent, grass-roots movement that developed from local peoples' efforts to defend their source of livelihood in response to state efforts to promote commercial forestry (Guha 1989:152-186). It began after a massive flood in 1970, which local villagers attributed to the ecological effects of deforestation. By 1981, Chipko protests had halted commercial logging in Uttarakhand Himalaya.

Most tribal people live in India's Central Forest Belt, the largest forested area after the Himalayas. Forest-based movements also have emerged in this region. A project to convert natural forest to plantations met with resistance from villagers who destroyed millions of seedlings of eucalyptus, known as the "ecological terrorist" because it steals soil moisture and nutrients needed for other crops (Anderson and Huber 1988:53). The project was terminated in 1981.

Policies designed to respond to international market demands, with the support of international lending agencies, are features of the global context shared by both cases reviewed here. In both cases, national development policies continued where international demand left off in the postcolonial period, targeting forest resources to address national development goals. Policies to encourage large-scale logging operations, extensive cattle ranching, and export agriculture were fostered at the expense of the undervalued natural forest resources. Encouragement of food crop production and population redistribution were secondary objectives of government policies in some places. These goals were to be achieved, if necessary, at the expense of the local populations, who depended on the forests for their livelihood. These people found themselves pitted against state agencies, company workers, and migrant farmers in sometimes violent confrontations.

In India, forest conflicts took place between local villagers and state agencies, allied with the logging industry, that sought to exploit local forests for tax revenues and profits. The Indian example has some elements in common with other Asian cases (Hurst 1990). The imposition of colonial land tenure concepts and institutions at the turn of the century caused the deterioration of the communal management that villages previously had practiced. In long-settled areas, direct state intervention led to grass-roots resistance from ethnic minorities. After independence, the same policies were continued with the goal of promoting the national good, even at the expense of local villagers.

In Latin America, by contrast, most conflicts took place between different social groups, with the state serving as mediator. The state was not neutral, however (Schmink and Wood 1992). State mediation usually favored elite classes over the interests of relatively powerless local populations whose claims threatened to impede the government's development plans. As Anderson and Huber (1988:18) described it:

> The process of international development must respond to the relentless pressures of competing interests. The state pursues its insatiable appetite for more revenue and thus for more control; the poor people maintain a continuous search for necessities (such as firewood) or for sources of minor benefits; the private sector and multilateral agencies continue to pursue new sources of profitability--no matter how distant or dubious--to adjust to the trend toward declining profitability and the potential exhaustion of current supplies. Expertise tends to assist the state and the private sector in the pursuit of their interests: in the Bastar case, there is little evidence supporting the interests of the poor people in the forest.

On the other hand, some local protest movements in India and in Brazil have succeeded in changing development plans and in halting deforestation. These grass-roots resistance movements gained significantly in strength after environmentalists and human rights activists took up their cause internationally (Miller 1990; Schmink and Wood 1992). The nonviolent tactics of some, such as the Chipko in India and the rubber tappers' movement in Brazilian Amazonia, have earned them a wide range of political allies and a place in the international policy debates over sustainable tropical forest development. The Chipko movement also has been supported by feminists because of women's active involvement.

Conclusions for Research and Policy

This essay has drawn on two relatively well-documented cases of deforestation to apply a framework that explains deforestation as the outcome of social processes at nested levels of analysis, from the global to the household level. The analysis focused on deforestation as defined from the point of view of local populations who depend on forest resources for a major portion of their livelihood. This perspective in itself illuminates aspects of deforestation that would be overlooked were a different definition of deforestation used. Instead of unicausal explanations, this approach recognizes the complexity of factors that lead to the outcome of deforestation in different settings. The framework presented has the virtue of defining a relatively limited number of variables that interact to shape the socioeconomic matrix of deforestation.

International and national market forces are an important necessary condition for expansion into relatively inaccessible forest areas. Population pressure in already settled areas causes internal migration that increases deforestation. Government policies, especially those favoring road building and industrial resource use, intervene in the distribution of land use and population through migration and land tenure patterns.

The interaction of policies, market factors, and land-population relations in particular local/regional settings sets the stage for the various players with an interest in the forest resources there. How the play will come out in the local contest, however, depends on the interactions among social groups and the resources and power sources they draw on to confront or cooperate with others in the pursuit of their interests. The strategies of individuals, households, communities, and interest groups in responding to changing structural conditions ultimately determine local patterns of deforestation.

These observations suggest that future research focus on the social dynamics of local-level interactions related to deforestation, where very little information currently exists. Such research would help illuminate policy "failures" and the competing agendas of government agencies and their constituencies. It would also direct our attention to the local populations who are most likely to have a stake in sustainable forest management, if their own future livelihood can be assured.

References

Agarwal, A., R. Chopra, and K. Sharma (eds.) (1982). *The State of India's Environment, 1982: A Citizens' Report*. New Delhi: Centre for Science and Environment.

Ahmed, S. (n.d.). *The Socio-Political Economy of Deforestation in India: An Analysis of Conflict Concerning the Use of Forest Resources between Local Communities in the Utterakhand and the State of Uttar Pradesh*.

Alegretti, M. H. (1990). "Extractive reserves: An alternative for reconciling development and environmental conservation in Amazonia." in A. Anderson (ed.), *Alternatives to Deforestation*. New York: Columbia University Press, 252-264.

Anderson, D. and R. Fishwick (1984). *Fuelwood Consumption and Deforestation in African Countries*. Washington, D.C.: The World Bank, Staff Working Paper No. 704.

Anderson, R. S. and W. Huber (1988). *Tropical Forests, the World Bank, and Indigenous People in Central India.* Seattle: University of Washington Press.

Baleé 1989, W. (1989). "The culture of Amazonian forests." in D.A. Posey and W. Baleé (eds.), *Resource Management in Amazonia: Indigenous and Folk Strategies.* New York: The New York Botanical Garden, Advances in Economic Botany 7:1-21.

Barney, G. (1980). *The Global 2000 Report.* Oxford: Pergamon Press.

Bedoya G., E. (1991). *Las causas de la deforestación en la amazona peruana: un problema estructural.* Lima: CIPA, Documento 12.

Blaikie, P. and H. Brookfield (1987). *Land Degradation and Society.* London: Methuen.

Blaikie, P., J.C. Harriss and A. N. Pain (1985). "The management and use of common property resources in Tamil Nadu, India." in *Proceedings of the Conference on Common Property Resource Management,* April 21-26. Washington, D.C.: National Academy Press, 481-504.

Browder, J. O. (1988). "Public policy and deforestation in the Brazilian Amazon." in R. Repetto and M. Gillis (eds.), *Public Policies and the Misuse of Forest Resources.* Cambridge: Cambridge University Press, 247-298.

Bush, M. B., D. R. Piperno and P. A Colinvaux (1989). "A 6,000 year history of Amazonian maize cultivation." *Nature.* 340:303-305.

Campbell, C. (1990). *The Role of Popular Education in the Mobilization of a Rural Community: A Case Study of the Rubber Tappers in Acre, Brazil.* M.A. Thesis, University of Florida.

Collins, J.L. (1986). "Smallholder settlement of tropical South America: The social causes of ecological destruction." *Human Organization.* 45,1(Spring):1-10.

Collins, J. L. and M. Painter (1986). "Settlement and deforestation in Central America: A discussion of development issues." Binghamton, N.Y.: *Institute for Development Anthropology Working Paper.* No. 31.

Dean, W. (1983). "Deforestation in southeastern Brazil."in R.P. Tucker and J.F. Richards (eds.), *Global Deforestation and the Nineteenth-Century World Economy.* Durham and London: Duke University Press, 50-67.

Denevan, William (1976). "The aboriginal population of Amazonia," in William Denevan (ed.), *The Native Population of the Americas in 1492,* (Madison: University of Wisconsin Press), 205-234.

Dewees, P. A. (1989). "The woodfuel crisis reconsidered: observations on the dynamics of abundance and scarcity." *World Development.* 17,8: 1159-1172.

Eden, M.J., W. Braz, L. Herrera, and C. McEwan (1984). *"Terra preta* soils and their archaeological context in the Caquetá Basin of southeast Colombia." *American Antiquity.* 49,1: 124-140.

Fernandes, W. (1983). "Towards a New Forest Policy: An introduction." in W. Fernandes and S. Kulkarni (eds.), *Towards a New Forest Policy: People's Rights and Environmental Needs.* New Delhi: Indian Social Institute, 1-22.

Fernandes, W. and S. Kulkarni (eds.) (1983). *Towards a New Forest Policy: People's Rights and Environmental Needs.* New Delhi: Indian Social Institute.

FAO (United Nations Food and Agriculture Organization) (1981a). *The Fuelwood Situation in the Developing Countries*, map prepared by the Forestry Department. Rome: FAO.

---------- (1981b). *Tropical Forest Resources Assessment Document*, 3 vols.(in Spanish). Rome: FAO 1981.

Gibbon, A. (1990). "New view of early Amazonia." *Science*. 248:1488-1490.

Guha, R. (1989). *The Unquiet Woods: Ecological Change and Peasant Resistance in the Himalaya*. Delhi: Oxford University Press.

Hemming, J. (1978). *Red Gold. The Conquest of the Brazilian Indians, 1500-1760*. Cambridge: Harvard University Press.

---------- (1987). *Amazon Frontier. The Defeat of the Brazilian Indians*. Cambridge: Harvard University Press.

Hurst, P. (1990). *Rainforest Politics: Ecological Destruction in South-East Asia*. London: Zed Books.

ICIHI (Independent Commission on International Humanitarian Issues) (1986). *The Vanishing Forest. The Human Consequences of Deforestation*. London: Zed Books.

Johnson, B. (1991). *Responding to Tropical Deforestation*. Washington, D.C.: WWF, Osborn Center Research Paper.

Joshi, G. (1983). "Forest policy and tribal development: Problems of implementation, ecology and exploitation." in W. Fernandes and S. Kulkarni (eds.), *Towards a New Forest Policy: People's Rights and Environmental Needs*. New Delhi: Indian Social Institute, 25-47.

Kannan, K.P. (1983). "Forestry legislation in India: Its evolution in the light of the Forest Bill, 1980." in W. Fernandes and S. Kulkarni (eds.), *Towards a New Forest Policy: People's Rights and Environmental Needs*. New Delhi: Indian Social Institute, 75-83.

Kelkar, G. and D. Nathan (1991). *Gender & Tribe. Women, Land and Forests*. London: Zed Books..

Kulkarni, S. (1983). "The Forest Policy and the Forest Bill: A critique and suggestions for change." in W. Fernandes and S. Kulkarni (eds.), *Towards a New Forest Policy: People's Rights and Environmental Needs*. New Delhi: Indian Social Institute, 84-101.

Mather, A. S. (1990). *Global Forest Resources*. London: Belhaven.

Myers, Norman (1984). *The Primary Source: Tropical Forests and our Future Earth*. (New York: W.W. Norton & Company).

Meggers, Betty J. (1954). "Environmental limitations on the development of culture," *American Anthropologist*, 56:801-824.

Morán, E. (1981). *Developing the Amazon*. Bloomington: University of Indiana Press.

---------- (1989). "Models of native and folk adaptation in the Amazon". *Advances in Economic Botany* . 7:22-29.

Miller, M. (1990). *International Coalition Building: A Case Study of U. S.-Based Conservation Organizations in the Amazon*. M.A. Thesis, University of Florida.

Repetto, R. (1988). *Overview*. in R. Repetto and M. Gillis (eds.), *Public Policies and the Misuse of Forest Resources*. Cambridge: Cambridge University Press, 1-41.

Repetto, R. and M. Gillis (eds.) (1988). *Public Policies and the Misuse of Forest Resources*. Cambridge: Cambridge University Press.

Rocheleau, D. (1987). "The user perspective and the agroforestry research and action agenda." in H.L. Gholz (ed.), *Agroforestry: Realities, Possibilities and Potentials*. Dordrecht: Martinus Nijhoff, 59-87.

Roosevelt, A. C. (1987). "Chiefdoms in the Amazon and Orinoco." In R.D. Drennan and C.A. Uribe (eds.), *Chiefdoms in the Americas*. Lanham, MD: University Press of America, 153-185.

Rudel, T. K. (with B. Horowitz) (1993) *Tropical Deforestation: Small Farmers and Forest Clearing in the Ecuadorian Amazon*. New York: Columbia University Press.

Schmink, M. (1987). "The 'rationality' of tropical forest destruction." in J.C. Figueroa Coln, F.H. Wadsworth and S. Branham (eds.), *Management of the Forests of Tropical America: Prospects and Technologies*. Rio Piedras, Puerto Rico: Institute of Tropical Forestry, Southern Forest Experiment Station, U.S.D.A. Forest Services, 11-30.

Schmink, M. and C. H. Wood (1987). "The 'political ecology' of Amazonia." in Peter D. Little and M.M. Horowitz (eds.), *Lands at Risk in the Third World: Local Level Perspectives*. Boulder: Westview, 38-57.

---------- (1992). *Contested Frontiers in Amazonia*. New York: Columbia University Press.

Schwartzman, S. (1989). "Deforestation and popular resistance in Acre: from local movement to global network." Presented at the American Anthropological Association meeting, Washington, D.C..

Shiva, V., H.C. Sharatchandra, and J. Bandyopadhyay (1983). "The challenge of Social Forestry." in W. Fernandes and S. Kulkarni (eds.), *Towards a New Forest Policy: People's Rights and Environmental Needs*. New Delhi: Indian Social Institute, 48-72.

Smith, N. (1980). "Anthrosols and human carrying capacity in Amazonia." *Annals of the Association of American Geographers* 70 (4): 553-566.

Sponsel, L. (1986). "Amazon ecology and adaptation." *Annual Review of Anthropology* 15: 67-97.

Steward, Julian H. and Louis C. Faron (1959). *Native Peoples of South America*. (New York: McGraw Hill).

Stonich, S. (1989). "The dynamics of social processes and environmental destruction: A central American case study." *Population and Development Review*. 15: 2(June): 269-296.

Tucker, R. P. (1983). "The British Colonial System and the Forests of the Western Himalayas, 1815-1914." in R.P. Tucker and J.F. Richards (eds.), *Global Deforestation and the Nineteenth-Century World Economy*. Durham and London: Duke University Press, 146-166.

Tucker, R. P. (1988). "The British Empire and India's forest resources: The timberlands of Assam and KuMaon, 1914-1950." in J.F. Richards and R.P. Tucker (eds.), *World Deforestation in the Twentieth Century*. Durham and London: Duke University Press, 91-111.

Vayda, A. (1983). "Progressive contextualization: Methods for research in human ecology." *Human Ecology*. 11(3):265-281.

Williams, M. (1990). "Forests." in B.L. Turner, W.C. Clark, R.W. Kates, J.F. Richards, J.T. Mathews and W. B. Meyer (eds.), *The Earth as Transformed by Human Action. Global and Regional Changes in the Biosphere over the Past 300 Years.* Cambridge: Cambridge University Press, 179-201.

World Resources Institute (1986). *World Resources 1986.* Washington, D.C.: World Resources Institute.

Worldwatch Institute (1984). *State of the World.* New York: Norton Co.

9

Problems of Population and Environment in Extractive Economies

Stephen G. Bunker

Resource-based models of population and environment have been spatially and locationally misspecified. The best-known debates emerge from a blend of Malthusian pessimism with simplified notions of carrying capacity (Keyfitz 1991). This mixture of outdated agricultural economics and biological ecology underlies Paul Ehrlich's worries about the population bomb and Garrett Hardin's prescription of lifeboat ethics to save us all from the tragedy of the commons. Their assumption that exponentially expanding human populations will exceed the available supply of natural resources posits a global population exploiting a global environment. The Club of Rome report on *The Limits to Growth* (Meadows et al, 1972) states, and Ehrlich has belatedly acknowledged, that consumption rates vary greatly between different classes and nations. This qualification sophisticates, but does not change, this assumption. Proponents of this view overlook substitutions between different raw materials, the widely varying amounts of reserves of different natural resources, and the different technologies required for the extraction and processing of different material forms.

The oil shocks of the 1970s gave great popular validity to ideas of direct causal relations between population growth and resource depletion, but has not saved them from being easy targets for rebuttals such as Julian Simon's (1981), based on a demonstration that the prices of the most used raw materials tend to secular decline or that known reserves of most major minerals have increased. Simon makes the economistic mistake of assuming that price reflects relative scarcity of natural resources. Raw materials prices are not set directly by the naturally given abundance or scarcity of the resources on which they draw but by competition between suppliers for market share or market control, between consumers to secure cheap, stable access, by rates of consumption, and by technological changes (Marx 1967, vol. III; Girvan 1976, 1980; Keohane 1984). Diversity of sources; extractive, transport, and storage capacities; and the strategies of states and firms far more directly determine raw materials prices than do absolute amounts of reserves. Excess capacity for extracting a physically rare material drives prices down, and insufficient capacity for extracting a physically abundant material drives prices up. Simon's assertion that price measures physical abundance is specious.

The simplistic debate between cornucopians and doomsayers perverts our understandings of both ecology and economy. The basic assumptions of each are indisputable; material resources are indeed finite, and the more of them we use, the sooner they will be exhausted. On the other hand, many raw materials can be substituted by other raw materials or by composites of raw materials. These global assumptions, and the global level at which the argument is cast, however, tell us nothing about the environmental and social consequences of extracting different raw materials from different environments.

The appropriate response to advocates on both sides of this argument is that raw materials are traded globally, but they must be extracted locally; that the local costs, environmental and economic, of extraction and processing vary between materials and technologies; and that the pollution and ecosystemic disruption that attends much extraction may be a greater problem, globally and locally, than the diminution of reserves in the extracted good. Keyfitz's (1991:11-12) point--that even though the supplies of most minerals are robust, the so-called renewable resources are the ones of which we are most likely to run short--is an appropriate corrective, but still neglects the local effects--demographic, environmental, and social organizational--of the ways that labor and capital are deployed and labor and nature are exploited and destroyed in and around mineral extraction projects.

Advocates on both sides of this debate fail to differentiate systematically between the environmental effects of the ways that different populations exploit different natural resources and use different raw materials. These effects can only be understood if we analyze the interactions between population, environment, and development at specific locations and in terms of the specific resources extracted. We can only understand these local effects, however, if we consider the political and economic actions of firms that extract, process, market, and transform these raw materials, the institutions that finance these operations, and the state agencies--local, regional, and national--that regulate and promote extraction and commerce.

Toward a General Model of Natural Resource Extraction

Secure access to an expanding supply of raw materials is critical to economic growth and stability under industrial capitalism. Industrial firms and states of industrial societies therefore act strategically as well as economically to assure access. These strategies may contravene the sovereignty of nations and the environmental and social well-being of communities in resource-rich areas. An adequately framed discussion of how resource extraction affects population and environment must take into account the complex interdependencies, unequal power, and processual and spatial differences between resource-extractive and industrially transformative economies. These have changed over time and space in ways that require conjoint ecological and political economic explanation.

Historically, increased mass and diversity of materials consumed, economies of scale in extraction and processing, and the progressive depletion of the sources most accessible to industrial centers have combined with the absolute spatial fixity of most mineral resources to increase mean distances between natural resource extraction and industrial production. These increased mean distances have heightened the potential scale economies in transport. These scale economies in turn reinforce the technologically driven increases in the scale of extraction, as larger ships, larger ports, and longer rail lines can only return sunk investments--frequently dedicated to a single extractive enterprise--with larger shipments sustained over longer periods of time. These dynamics restrict greenfield mining projects to large deposits, of which there are relatively few. This further reinforces the tendency toward increased distance between extraction and consumption. It also increases the proportion of raw materials transported across national boundaries prior to transformation, and the likelihood that extractive enterprises dominate the economy and politics of the regions, and sometime the nations, in which they locate.

As distance and scale increase, mines tend to locate in areas with sparse populations and little effective integration into capitalist political, economic, and legal systems and with limited access to technical information required for effective rent bargains or for environmental or social regulation. Isolated exporting nations compete against each other in contracts with informed importing firms and states, so raw materials rents and prices remain low and damages to environments are omitted from contract costs. Increased scale and distance, however, also raise the strategic stakes for consuming firms and the states of industrial societies. Particularly in periods of shifting hierarchy between dominant industrial nations, competitive strategies may induce excess extractive capacity, destabilizing markets and increasing environmental impact beyond the technological minimum required to satisfy world demand.

The costs these states and firms are willing to assume, and their capacity to impose some of those costs on the economies and environments of the resource-holding nations, vary with the organization of state-firm relations in dominant industrial countries, with the political organization of specific resource-holding nations, and with the political and economic organization and condition of the world system of nations. Free trade, imperialism or colonialism, foreign direct investment, and joint ventures or shared responsibility are modes of resource access that have characterized different periods of world resource trade. They involved quite different strategies and had quite different effects on the resource-exporting societies. Within these general types, however, the historical data make it quite clear that strategies were also directly affected by the characteristics and uses of particular commodities, by their position within commercial and military technologies and products, by the volume of demand for them and industrial country perceptions of their relative scarcity, by their location in space (both physical and national-territorial), and by the infrastructural requirements that the combination of location, ore grade and type, energetic requirements, and scale impose. These same characteristics directly affect the demographic, environmental, and economic results of extractive enterprises in the surrounding region. The problem of analysis is to combine the physical and political economic aspects of these processes.

Close attention to the chemical and physical characteristics of materials can provide important insights into the relations and differences between extractive and transformative economies. Within the physical constraints imposed by chemistry, space, and topography, market demand and strategic behaviors critically affect volume, location, and technological options. These are essentially social processes. Attention to material processes is also crucial to the study of the uneven political, economic, and environmental outcomes of natural resource extraction and trade, because it permits analysis of the material frame, or constraints, within which political and economic strategies are played out. In the extractive economy, the chemical and physical characteristics of a raw material constrain the technologies capable of extracting and processing it. These technologies, in combination with scale, determine labor absorption and direct environmental impact. Infrastructural requirements for transport and energy combine with surrounding social and economic forces to set the amount of migration and additional economic activity in the area around the mine and thus affect indirect environmental impacts. In the transformative economy, chemical and physical characteristics determine end uses, and thus indirectly the strategic needs of firms and states.

Actual strategic behavior, however, is also driven by space and location. The absolute physical scarcity of a natural resource, and its relative scarcity (the number of places from which it can be economically extracted), interact with the perceived importance, or criticality, of a raw material in industry to determine the costs that firms and states will incur to secure access to it. Volume of demand is a major factor, and so is the cost of substitution. Strategies must take into account and turn to advantage the political and economic conditions and aspirations of the host countries as well.

All of this requires analysis precisely specified by commodity and by historical period. In this paper I use examples from the oil, coal, iron, copper, aluminum, manganese, cobalt, and gold markets of this century, with additional reference to nonmineral markets for comparison.

An Application of the Model to
Population and Environment in Resource Extraction

Global markets, technologies, transport, and finance for a wide range of raw materials affect local populations' uses of their own environments in ways specific to the natural products extracted, the rents paid for their extraction, the kinds of environments from which they are extracted, and the scale of capital and technology required for competitive participation in raw materials markets. Public policy may be able to mediate these effects somewhat, but the state is constrained by the physical characteristics of the extractive process and by the political and economic forces of external markets and externally controlled capital. The state is also constrained by class and sectoral interests and by the distribution of national population in unevenly developed regions and economic sectors. Understanding the relation between population and environment requires attention to the interaction of these complex, and highly variable, forces.

Global trade is driven by a thermodynamic imperative. Industrial societies can only sustain themselves and continue to grow by importing increased amounts of matter and energy. As proximate deposits of raw materials are consumed, they must be extracted and exported from more remote sites. Economically viable sites diminish in number as they are depleted and as the increased scale of technology in extraction, transport, processing, and production renders smaller deposits unprofitable. For most raw materials, technological advances in extraction, reduction, shipping, and processing have offset the additional cost created by the growing average distance between site of extraction and site of industrial transformation.

These savings buffer the industrialized economies from raw materials scarcities and rising prices, but they enhance the demographic, ecological, and economic disruptions of natural resource extraction in raw materials-exporting societies. Technological solutions to resource depletion allow populations of industrial societies to draw on raw materials whose costs in relation to the other costs of production generally decline. These populations suffer the consequences of energy and matter lost or emitted into the industrial society's environment through its incomplete transformation into commodities, and eventually from its degradation into waste, but as long as new sources can be found to replace the depleted ones, or technical or natural substitutes found to offset scarcity of particular raw materials, the impact of raw materials scarcity on industrial populations tends to be temporary crisis rather than economic and ecological collapse. The introduction of new technologies may displace a portion of total employment in an industrial society and may deeply affect some regions, as occurred with Pittsburgh and, earlier, with the Ruhr Valley, but the resulting depressions are sectoral and regional, rather than societal and national. They are not directly related to environmental processes, but tend to be more technologically, politically, and market driven. For an example on the political and economic antecedents to the crisis in the U.S. steel industry see Prechel (1990).

The situation is quite different in the specific locations from which natural resources are extracted. For populations whose primary ties to industrial economies reside in the extraction and export of raw materials, the experience of depletion or substitution is usually keenly felt, and is likely to endure. The critical relations between population and environment, however, typically occur not at the moment of depletion, but at the time of discovery and development of new extractive locations. As the scale of extraction and processing has increased, and as the number of profitable deposits has diminished, extraction of most mineral and of many vegetable resources more frequently involves very large, capital intensive, highly mechanized projects located in very remote areas. Economies of scale in both processing and shipping depend on heavy and extensive infrastructure, both for the transport of the natural commodity and also for the energy required for its extraction and primary processing. In many extractive industries, increase in scale has enhanced the demographic disruptions and the environmental destruction that resource extraction often occasions. At the same time, the enhanced scale has increased the power, and the vested interests, of the companies that manage the extraction, and their ability to subordinate the local communities and governments of the often remote areas in which they operate. These general tendencies, however, must be disaggregated into the specific raw materials and the specific locations from which they enter global trade.

The relation of infrastructural scale to raw materials exports varies with the physical characteristics of the commodity, particularly with the amount of sterile or overburden per unit of ore and with the kinds of molecular bonds between ore and sterile, as well as with the value per volume of the reduced ore. The world average ore grade of mined copper, for example, in 1976 was 1.03 percent, with a range from .47 in Papua New Guinea to 3.9 in Zaire (Nwoke 1987:173). Iron ore grades, in contrast, range from 33 to over 66 percent. On this basis alone one would expect far greater economic pressure for extensive volume reduction through chemically more complex processes near the mouth of the mine in copper than in iron. One rough indicator of this difference is the relative cost of complying with a single set of pollution control regulations; pollution control capital expenditures as a percentage of total capital expenditure in the U.S. between 1973 and 1979 was 41 percent for copper, 16 for steel, and

13 for all nonferrous metals, even including copper (Sousa 1981:55). Copper reduction and smelting become especially critical as ore grade declines, both because of the need to reduce transport costs and because of the increased need to use the nickel, molybdenum, or gold often found in association with low-grade ores.

The preliminary processing of bauxite, in contrast, is relatively simple; crushing, washing, and drying can result in up to 50 percent available alumina. Bauxite containing the preferred trihydrate aluminum tends to be in lower, flatter areas than most of the richer copper deposits, and more often accessible to waterborne transport. Subsequent stages of processing become more complex and costly, but the effectiveness of bauxite volume reduction and the ease of transport allow far greater flexibility in the location of processing than is the case for copper. Aluminum processing does, however, require massive amounts of electricity, most economical through large hydroelectric installations since the oil shocks of the 1970s, and so imposes large infrastructural costs further along the chain of production.

Iron and coal, with low value to volume and consumed in large quantities, usually require heavy land transport to dedicated large ports. Alluvial gold and cassiterite, in contrast, with very high value to volume, may leave an area in small airplanes taking off from numerous, dispersed runways. Manganese, critical to steel making and 6 to 14 times as valuable as iron by volume, is seldom exploited in sufficient quantity to justify the construction of long dedicated railroads, so tends to be exploited either near a port or in conjunction with iron extraction. Lumbering tends to engender numerous feeder roads maintained only as long as supply lasts in particular areas. The length, density, and duration of these roads varies with the type of forestry--temporary roads can be driven further into jungle to extract dispersed individual teak or mahogany logs than would be economical for less valuable species cut for plywood or pulp.

The physical characteristics and location in space, topography, and ecosystem of particular minerals and mineral deposits have equally varied impacts on surrounding environments. The low ore grade of copper means that huge amounts of earth and rock must be removed to gain access to ore, and the molecular structure of copper and its association with multiple toxic heavy metals and chemicals--sulfurs, oxides, lead, arsenic, cadmium, and bismuth--means that there are multiple toxic residues from extraction and processing. The U.S. Bureau of Mines estimated that 40 percent of the waste materials generated in the history of the United States are from mining, and half of that figure from copper alone. Separating aluminum from its surrounding body is a more mechanical process and involves proportionally far less waste. The effluents come primarily from processing inputs such as fluorides and carbon, but these are far more easily controlled than those actually embedded in copper ore. In comparison to the wide diffusion through multiple river systems of mercury used to aggregate placer gold, though, effluents from both copper and aluminum processing are relatively localized and thus susceptible to regulation.

Within these physical and economic constraints, firms may vary their technological choices according to the cost and availability of labor, subsidies and incentives from the national state, the cost of credit, and the structure of taxes, rents, and royalties (Tanzer 1980; Mikesell 1978, 1980; Godoy 1985). Nonetheless, for most of the major minerals, the scale required for competitive operation has increased rapidly over the last three decades. The introduction of open pit, stripmining, and other capital-intensive techniques in iron and copper increases output per mine, but it also forces expansion in the associated transport, storage, loading, and unloading facilities. In similar fashion, the introduction of economies of scale in transport favors the exploitation of larger deposits at higher rates of extraction. In other words, scale increase in one sector of an extractive industry tends to drive scale increase in other sectors as well, so that there are circular, mutually reinforcing tendencies toward spatial concentration and increase of scale.

Finally, competition between national economies for ascendance in the world system enhances the effects of increasing volume of extraction and scale of operations. Raw materials availability and prices are crucial determinants of the competitive positions of firms and nations in global markets. Scale of operations erects significant capital barriers to entry in most minerals markets. The absolute fixity of the location of resources in the ground inserts questions of national sovereignty into all minerals negotiations; access to minerals has often combined interstate diplomacy with firm to state bargaining. The critical importance of raw materials for both military and industrial strength has encouraged collaboration between states and large minerals firms.

For all of these reasons, minerals markets in industrial countries are prone to control by tightly integrated oligopolies. These oligopolies often restrict new entry by buying up rights to strategic reserves while guaranteeing profits by restricting output. This situation is threatening to ascendant firms and economies attempting to rise in the global economy. The solution for them is usually to attempt to diversify sources, and to do so in a way that lowers the cost of the raw material. This strategy is likely to bring more large projects on-line. This extends the ecological and demographic problems associated with such projects into new areas. The increasing scale and distance from established industrial centers often mean that huge new projects are established in remote areas whose populations are little integrated into capitalist economic and political systems, and have little knowledge of environmental consequences or capacity to protest them. New copper mines in the southern Peruvian Andes (Tintaya, Cerro Verde) and in Papua New Guinea (Ok Tedi, Bougainville), new tin, iron, and bauxite mines in the Brazilian Amazon (Boa Ventura, Carajás, Trombetas), copper, uranium, and aluminum mines and processing plants that impinge on indigenous lands in Australia, and huge hydroelectric projects in subarctic Cree and Inuit lands in Quebec all exemplify this tendency. Increased scale of extraction and processing concentrate and intensify environmental damage. If this strategy is successful in lowering the price of the raw material, it is likely that the administrations around the site of the new projects will have little additional revenue to deal with the social and environmental problems that they face.

The expanding raw materials consumption of Japanese and European industry, and Japanese strategies to lower raw materials costs and stabilize supplies, exemplify all of these tendencies. U.S.-dominated oligopolies in both iron and aluminum threatened Japan's access to the cheap, stable supplies of the raw materials it required for ascent in the world economy. Japan's distance from the major sources of many of the metals it consumed exacerbated its competitive disadvantage. New discoveries of high-grade reserves of bauxite and of iron in Australia, Brazil, and Venezuela, together with these countries' aspirations to capture forward linkages domestically, created an opportunity for the Japanese to break established oligopolies by fomenting joint ventures with the host countries or their direct investments based on promised long-term contracts (Mikdashi 1976). These enterprises rapidly diversified the sources of iron in the 1950s and 1960s (Maull 1984; Tanzer 1980) and the sources of aluminum in the 1970s and 1980s (Bunker and O'Hearn 1992). Japan thus evaded the oligopolies' barriers to entry by manipulating resource nationalism and developmental aspirations in the host countries.

At the same time, Japan led in the construction of large ocean vessels that reduced the cost of raw materials. In Brazil, it became a partner in a joint shipping venture with the Brazilian Companhia Vale do Rio Doce (CVRD), from which it purchased iron. On the basis of the new scale of shipping, the cost of shipping iron ore from Brazil to Japan dropped by 60 percent from 1957 to 1968 (Crandall 1981:23; Walter 1983:23; both cited in Prechel 1990:654). The 60,000-ton maximum cargos that still characterized the Great Lakes trade in the 1950s has given way to 300,000-ton cargos out of Australia and Brazil (USBM 1985:8-9), now the two leading iron ore exporters. Not only do these cargos require larger, deeper port facilities, they also reinforce the economies of larger mining operations. They thus concentrate the environmental and demographic effects of these projects, both in displacing inhabitants at the mine, along the rail line, and at the port as well as in bringing in construction crews, direct employees, and spontaneous migrants. In the process, the Japanese succeeded in creating so much excess capacity in these minerals that the returns to host governments have dropped precipitously. As they were largely debt-financed, these projects are actually draining government revenues and impeding other investments. The state's capacity to regulate the environmental impacts of extractive enterprise or to provide for the social welfare and social order needs of the populations around these projects are therefore reduced.

Demographic and Ecological Effects
of Large Extractive Projects

There are several distinct populations that may be affected by large extractive projects. First are the inhabitants of the areas of the mine itself, of the rail line from mine to port and in some cases of service roads for access to the mine, and of the areas around the port itself. Given the increasing remoteness of new extractive projects from established centers of capitalist industry, these communities often are institutionally, culturally, and linguistically distinct from the controlling interests of the mining operation, as well as from each other. Correctly or not, they are usually perceived as not prepared for the forms of employment that the mine or associated enterprises require. They are effectively marginal to the project, or more precisely, in the way. In some cases, such as the construction of hydroelectric dams for smelters, local inhabitants must be relocated if they are to survive physically. In other cases, large companies simply prefer to control completely the areas surrounding their operations, so attempt to expel local residents.

Most significant in many cases is that the remoteness from institutions of capitalist property means that original inhabitants do not have title or claim to their land in any form recognized by the institutions of the economy and society that have so abruptly been imposed on them. As the national state is usually involved in promoting the exploitation of the minerals, and is anxious to keep costs as low as possible, it tends to favor the property claims of the minerals enterprises. Resettlement allowances are kept to a minimum, both on a per household basis and in the consideration of the areas recognized as affected by the new enterprises. The problems do not end with actual resettlement, however. Hunting, fishing, and agriculture around the areas of mine, rail, and port may all be disturbed by pollution. Hydroelectric dams and settlement ponds may provide habitats for new disease vectors, and migrants may contaminate these new habitats. Rotting vegetation and elements in soils such as mercury under newly formed reservoirs may pollute water downstream of the dam. Air and water carry pollutants from reduction and processing. In one of the most notorious cases, Cerro de Pasco Corporation's emissions of sulphur dioxide and arsenic at its smelter at La Oroya in Peru so contaminated soils in the surrounding area that crops and herds were severely reduced in surrounding haciendas and communities (Flores Galindo 1974; DeWind 1987; Dore 1988). The Corporation eventually agreed to buy some of the damaged land. There is some evidence that it had already determined that the land would recover its fertility if sulphur dioxide emissions were reduced. In any event, after it bought the land, it installed Cottrell scrubbers, reduced its emissions, and started raising sheep on its new lands. In the meantime, however, it had effectively expelled local peasants from their lands and greatly increased the labor reserve available for employment in its copper mines.

Gases and particulate matter, some highly noxious, are released into water or air in mining or in processing iron, copper, coal, and tin. Mercury used to aggregate metals or simply released from disturbed soils or processed ores works its way up the entire food chain. Strip mining accelerates water and wind erosion. Drainage systems built to allow deeper pit mining lower water tables in the surrounding areas. Dust and red mud from preliminary processing of bauxite degrade surrounding waters and vegetation. There are technical fixes for the substantial reduction of many of these emissions, but they tend to be costly and are resisted by investors until pushed by stringent regulations or by international protest (Nwoke 1987; Tanzer 1980). These systems are also costly to maintain, and may be neglected during periods of financial crisis. The sedimentation tanks for Tintaya, a copper mine in the southern Peruvian highlands, are no longer adequate for the accelerated rate of extraction imposed by the newly created regional government that now controls the mine. The same fiscal crisis that motivates the accelerated extraction also impedes further investment in the mine, and the tanks now threaten to rupture or overflow, spilling toxic wastes into the river. It should be pointed out that since pollution control standards have become more stringent and new techniques have been developed, direct degradation from mining and reduction has in many places been greatly reduced.

The second population involved is comprised of workers brought in for the construction of extractive infrastructure. Large construction projects, for railroads, roads, ports, docks, dams, and for the mine and associated housing typically require large work crews. As roads, docks, and towns are usually exclusively dedicated to the extractive enterprise, no return on them can be expected until the entire complex is complete or nearly complete. In order to reduce interest costs as much as possible, firms and states generally attempt to reduce construction time and the period between completion of any component and the initiation of the total plant as much as possible. One way to do this is to make construction of the various components of a project as nearly simultaneous as possible. This strategy enhances the short-term demographic impact of the construction and may seriously undermine the capacity of the local government to provide for social welfare and social order. This shortfall in local government administration compounds the difficulties confronted by the displaced original inhabitants of the affected regions, who are also dependent on the local states. The problem is even worse if the crews bring in violence, disease, or addictions that affect the original population.

The numbers of crews may be quite large: 12,000 at Tucurui in the Brazilian Amazon, with an additional 25,000 directly employed and subcontracted workers at the Carajás mine and 10,000 at the aluminum smelter and port project in Barcarena, all within a 300-mile radius over roughly the same period; 17,000 at Hydro-Quebec's projects in James Bay; 24,000 in Ciudad Guyana for dam and mill construction. Direct and subcontracted employment may drop by over 80 percent at the completion of the project. Effects range from direct violence against local inhabitants through the introduction of drugs, alcohol, or prostitution to the fouling of water sources used for drinking or bathing.

If the project is located in a remote, sparsely populated area not endowed with the institutions of social control, social welfare, and the regulation of property rights usually associated with industrial economies, the temporary influx of large crews, often predominantly male, is likely to so overburden the local social organization that the disciplining of the work force remains the almost exclusive domain of the enterprise itself. Such an arrangement makes no provision, however, for other populations that follow the new roads into the area, searching for work, establishing small businesses, or hoping to claim newly opened land. Furthermore, the socially random location of mineral deposits and the accidental nature of their discovery militate against sequential large projects in the same country. As a result, the termination of the infrastructural projects may leave large populations with little knowledge of the local environment to fend for themselves. Reversion to subsistence or commercial agriculture in an unknown environment, especially where multiple households are making the attempt at the same time, may stress the local environment without providing much income for these households.

The construction crews directly employed are only one, and often the least numerous, of the migratory contingents attracted by large extractive projects. Providers of services for what may be a work force earning, however briefly, well above the regional average are likely to follow the work crews, as are individuals and households looking for employment or for land. The speed with which these projects develop, and the dramatic impact of their announcement, may stimulate large and rapid migratory flows that run ahead of any return information about the actual availability of jobs or land. To the extent that new lands are opened for settlement or for resource extraction, there is likely to be extensive conflict between different classes over access to the land and resources, whether between peasant, rancher, and logger or between lone prospector and large mining company (Schmink and Wood, 1992; Robinson 1986; Roberts 1991).

Occupation and exploitation of lands newly opened by rail and road to extraction projects may have greater environmental impacts than the project itself. In the Brazilian Amazon, massive and conflictual migration for land has resulted in extensive deforestation along the roads and line of rail to the Carajás mines. In Ecuador, official colonization, spontaneous migration, palm oil plantations, and cattle raising have followed roads built for oil exploration and drilling and led to significant deforestation. Migrants from the highlands have displaced indigenous groups in some of these areas. In Brazil, the discovery of gold and cassiterite in areas opened up by mining projects and along lumbering roads has led to dredging and blasting of waterways and pollution of rivers with mercury used to agglomerate the gold dust. In Pará, as in Minas Gerais earlier, pig iron plants have consumed large areas of forest in the form of charcoal. There are strong pressures from mining companies and the institutions that finance them to disconnect the environmental and social costs of the exploitation and occupation of surrounding lands from the accountability of the extractive project strictly defined, but in fact there is no other direct explanation for the opening of these lands. The company itself does not, in most instances, directly create the social and economic conditions that foster migration and structure economic opportunities there, but it does break down the physical barriers to migration into areas the local state may not be adequate to administer.

To the extent that the extractive economy itself generates linked industry, some migrants and unemployed construction workers will eventually find work. Without complex and costly state support, however, there is likely to be a significant lag between the completion of the infrastructure for extraction and the growth of diversified forward users of the raw material. Such a time lag is, in fact, a minor problem in relation to the general difficulties of developing subsidiary industry around resource extraction. Forward linkages from extraction have become increasingly rare as the scale of extraction increases and the cost of transporting raw materials diminishes. There are relatively few incentives to establish forward linkages in remote areas; to the extent that they occur, they usually result from fiscal policies that require a strong state with access to capital or credit. Even under these conditions, sustained development is difficult. Despite extensive planning, massive state subsidies, and political decisions to center national heavy industry in Ciudad Guyana, there are still relatively few enterprises based on fabricating the raw materials extracted and processed in the area (Rakowski 1987; García 1987). Furthermore, unemployed construction workers in Ciudad Guyana seldom have the skills required in industry, and are more likely to engage informally in service sector work. The fiscal incentives and tax holidays justified by the same economic models applied in Venezuela have had even less success in Brazil's Programa Grande Carajás (Bunker 1989a 1989b), and the populations drawn to Parauapebas, Marabá, Tucurui, and Barcarena suffer far higher levels of unemployment.

The unemployed or underemployed spontaneous migrants and former construction workers, whether they seek land or engage in artisanal extraction in areas opened by the transport systems for the larger extractive enterprises or whether they form part of swelling urban agglomerations, are likely to exacerbate the problems of original populations displaced by construction of mines and dams, either as competitors for scarce state or enterprise social welfare funds or resettlement sites or because of invasion of the rest of their lands. In some cases, the established economies of the original inhabitants and the potential economies of the migrants are undermined by the cheap return traffic of raw goods exporting transport, which almost always has excess capacity. The exceptions are in those extractive economies with very high value to volume ratios, such as fur, gold, and cassiterite, where imports of supplies are often bulkier than the exports of raw materials (see Innis 1956). Where the extractive commodity makes up a large share of both GDP and exports, the influx of foreign revenue into the country can also severely disrupt sectoral balance, leading to the so-called dutch disease. The rapid decline of small-scale agriculture in Venezuela and Nigeria after the rapid rises in oil prices are well-known examples of how transitory extractive bonanzas may prejudice major segments of the national population. Less known but equally consequential may be the environmental effects of the national population's having to return to more direct dependence on agriculture again at a later period. To the extent that agricultural intensification requires modifications of the environment maintained by constant inputs of labor, the carrying capacity of the agricultural area may decline under labor shortage. Reintroducing agriculture may result in less environmental care and greater degradation. Similar problems have been documented when migrants to mines have returned to their home areas when falling world demand has reduced mineral output.

Labor intensity interacts with the physical distribution of the mineral itself to affect both demographic distributions and ecological effects. Gold and cassiterite are sometimes diffuse, and susceptible to prospecting and panning by petty entrepreneurs. This kind of extraction may spread over large areas. It also may conflict with corporate claims to exploit the same areas with much more highly capitalized technologies. Copper, iron, and coal allow for strip mining, pit mining, or deep vein mining; the last requires far more labor than the others. In copper, for example, labor to output ratios in different countries varied by a function of 3 to 13 (Nwoke 1987:174). The geological processes that form trihydrate aluminum impose varieties of strip mining, so that there is far less variation in the organic composition of capital in this industry, which tends to be highly mechanized everywhere. This homogeneity is probably enhanced by the aluminum industry's short history. Iron and copper mining and processing are far older, and they have passed through a far greater number of technological innovations or phases than those of aluminum. There are more variants of these different technologies still in use in more places than is the case with aluminum. Total amounts of labor and its rate of remuneration, labor's relative concentration and diffusion in space, and the ability of petty extractors to compete with large corporations all mediate environmental consequences of extraction and the types and sizes of populations that develop around the extractive enterprise. Successful bids by large corporations to displace placers and prospectors may create large contingents of unemployed miners, stimulate their migration to new areas, or provoke violence. Successful attempts by large, land-using enterprises to displace peasant settlers have similar effects. These issues primarily depend on state policy and on its local implementation.

Systematizing these different interactions between resource extraction, population, and environment is only possible through the specification of the variables particular to minerals demand, technologies, and the world trading regime at particular periods or epochs of capital, and to the particular commodities, regimes, environments, demographics, and social organizations relevant to particular extractive sites. In many instances, the relations between central, state, and local government are also critical. The advantages of capturing foreign revenues or of providing cheap raw materials for domestic industries accrue at the national level, while the ecological, political, and social disruptions that attend extraction are highly localized. Again, the relative success of capturing forward and backward linkages around Ciudad Guyana and the comparative failure of the Projeto Grande Carajás reflect the regional focus of the Corporación Venezolana de Guyana and the national control over PGC.

The organic composition of capital in mining varies with the technology available to firms at different periods of time, with the cost of labor (itself a reflection of economic organization and of political regime), with the size and quality of the deposit, and with its transport requirements. Historical comparisons clarify these differences. The strip and open pit mining operations that increased rapidly after World War II are highly capital intensive. The deep vein mining that prevailed until World War II required far more labor. Colonial mining endeavors in Latin America, and later in Africa, were initially faced with significant labor shortages. These resulted in large measure from the absence of a free labor force and were exacerbated by mining entrepreneurs' reluctance to pay wages high enough to attract men who had access to sufficient agricultural land for subsistence. The solution in Peru and Bolivia was found in various forms of forced labor. This was achieved first through the sale of labor and food for the mines in order to collect money to pay colonial taxes. Low wages, the rigors of the journey to Potosi and to Huancavelica, as well as to lesser mining centers, and the dangers of mining itself meant that many of these men did not return, either dying or remaining at the mine site. Potosi's population reached 20,000 by 1550; the exhaustion of the richest and most accessible veins after 1560 required more labor but lowered wages. Viceroy Francisco Toledo introduced in the 1570s a system of direct labor tribute called *mita*, under which communities were obligated to send one-seventh of their able-bodied men to work in the mines at any one time. With this expanded power of recruitment, laborers at Potosi increased to over 10,000, and production peaked there in 1592, after which it declined steadily (Dore 1988). Colonial policies in Africa coercing migration through taxation and restricting commercial agriculture to white settlers lowered wages, and local linkage functioned similarly (see review by Godoy 1985). Both facilitated the mining firms' access to labor at wage rates below the cost of social reproduction. Both achieved this by breaking the relation between indigenous communities and their lands. Both resulted in high levels of migration toward the mines, and both allowed the firms to operate with low levels of capitalization. Low levels of capitalization facilitate output reduction or shutdown when demand is weak, so tend to be associated with fluctuating levels of employment and therefore with high incidence of circular migration.

This kind of system often results in the rapid growth of cities around mines and in the destabilization of populations and of human relations to the environment in the areas from which labor is recruited. The Zambian Copperbelt opened production in the late 1920s. The mines there employed 8,000 Africans in 1927, and 32,000 by 1930. Employment dropped sharply during the Depression, but rose to over 30,000 by 1940. By 1946 the total African population of Copperbelt towns was estimated at over 200,000 (Muntemba 1977:350). Northern Rhodesian colonial policy was to protect European estate agriculture by removing African peasants from proximity to towns and rail lines, so that these towns were little integrated into local agriculture. There was little forward processing or industry either. These urban populations were therefore sustained by services to a low-waged mass labor force and were therefore as unstable as the mining employment itself. Fluctuations in employment exacerbated problems of overcrowding and soil degradation in native reserves or migrant sending areas.

In one well-documented case, migration to mines vastly reduced the labor available to maintain the complex Barotse system of draining a fertile but flood-prone plain. The drainage system deteriorated, but the population that remained was able for a time to exploit the less fertile uplands. Shortened fallow there eventually led to diminished yield. The entire system was tremendously stressed when migrants returned during a recession and attempted to intensify cropping in the upland areas (Van Horne 1977). Labor impressment to Peruvian and Bolivian mines centuries earlier contributed to the decline of terracing and irrigation systems that functioned before the Spanish conquest. Outmigration after depletion or market collapse may also affect new frontiers. Recent attempts to resettle unemployed tin miners in the Bolivian lowlands have degraded those environments as well.

Extractive entrepreneurs who attempt to reduce their capital exposure by employing large masses of low-paid labor usually complain about labor scarcity and appeal to the state to assist them in overcoming the reluctance of "immobile" labor. The use of coolie and penal labor in Peru's guano extraction (Méndez 1987), the transport of workers from the Brazilian northeast to work in the Amazon rubber trade, the stimulation of migration to extract lumber from Canada (Innis 1956), Australia and New Zealand (Blainey 1982), India (Tucker 1989), and Indonesia and Malaya (Brockway 1979) have all been accompanied by elite and colonial worries about the availability and quality of local labor. Collapse of the extractive enterprise has often stranded the migrants in subsistence activities for which they were not prepared in an environment they did not know. Return migration, however, can be equally destructive of the environment. Labor recruitment for mining is likeliest to target relatively dense populations. These populations are likely to have intensified their agriculture and so depend on sustained modifications of the environment such as ditches and terraces. The communal labor needed to maintain this built environment, together with the extra labor required for conservation of soil and water resources, are most at risk when significant portions of the active male population are removed. The return migration of these men and their families is likely to exacerbate, not repair, the effects of neglect.

Collins (1988:133-134) questions the conclusion that outmigration sustains peasant communities by providing access to cash and relieving pressure on land. She cites Maletta's argument that labor off farm can bring about "an irreversible deterioration of the land itself of its precarious infrastructure (irrigation canals, embankments, fences, etc.) which means that many times the decision is equivalent to 'burning the boats'" (1979:64). She also cites Posner and McPherson's (1982) statement that it eliminates the time and the incentive to conserve soil resources. The instability of labor-intensive extraction enhances these problems where labor is drawn under various coercive regimes from intensive subsistence agricultural systems. The relations between population, environment, and extraction are quite different under these conditions than they are under the more recent capital-intensive, generally larger mining projects that have dominated raw materials markets during the past three decades. In these new mines, where 2,500 direct employees may provide for exports of over 35 million tons of iron ore in a year, the major demographic and environmental impacts may come from landlessness and unemployment in the surrounding regions rather than from the disruption of relatively stable subsistence communities. These highly capitalized mines confront a very different problem, however; the large debt loads they carry make it far more difficult to shut them down when minerals prices fall. In some metals industries, this directly exacerbates huge price fluctuations. In order to keep such mines operating during periods of low prices, national states in some cases exonerate companies from their tax obligations, further restricting the local administration's ability to provide social services.

The instability of minerals enterprises is often particularly striking in materials defined as strategic or critical. The United States has maintained stockpiles of such materials. These classifications reflect essential dependencies in industry or defense for minerals found in only a few countries in the world. Cobalt is used in superalloys in jet engines. Fifty-one percent of the world's cobalt production comes from Zaire, and another ten percent from Zambia, and has regularly been stockpiled by the United States. Because stockpiling decisions are based on political perceptions of threat to supplies rather than economically driven demand, purchases and prices can be highly unstable. Between 1978 and 1982, prices for cobalt fluctuated by 1,000 percent, and U.S. purchases by about the same amount. In countries where minerals exports amount to over 90 percent of exports and one-third of GDP, these fluctuations are devastating to state functions and to social well-being (USBM 1985; Shafer

1983). U.S. government use of agricultural surpluses for barter against strategic materials (USOTA 1985) may seriously disrupt local agricultural economies and reduce local self-sufficiency in food.

One of the difficulties of interpreting the demographic effects of mining is that the organic composition of capital in particular mines has a tremendous impact on the degree and type of urbanization. The African cities that grew rapidly around labor-intensive mines were essentially dormitories for a low-paid work force, while a city such as Houston has emerged around the capture and creation of backward, forward, and fiscal linkages around the oil industry. The first city is relatively undifferentiated and dependent, the second has a high degree of divided labor and is technically and financially relatively autonomous. Somewhere in between is Ciudad Guyana, where state corporations have forced forward linkages, primarily around heavy industry, and have created a large planned city of over half a million inhabitants, but where fluctuations in extractive and processing construction continue to have a major impact on the employment and well-being of that population. The point here is that Broken Hill's and Oruro's populations grew completely in function of low paid labor in the mines, while Houston's population has grown around industries linked to oil.

Parauapebas, a completely new town outside of Carajás that grew from nothing to 30,000 between 1980 and 1990--at the same time that employment at the mine itself went from less than a thousand to almost 25,000 and then back to 5,000--is effectively disarticulated from the highly capital-intensive mining operation. These distinctions emerge in part from the commodity extracted, from the organic composition of capital in that industry at a particular time, from decisions by firms in particular locations about the organization of labor and of capital, and from the conditions of world markets for the particular raw materials extracted. They also emerge from the political regime dominant in the extracting area, from policies governing the assignment of costs and the distribution of benefits, and therefore from the allocation of governance between national, regional, and local administrations. The relative success of Ciudad Guyana, for example, is in part due to the establishment of a regional development authority, the Corporación Venezolana de Guyana (CVG), with control over its own investment budget. The CVG is part or full owner of the most significant minerals enterprises in the region, and operates them in part with social objectives rather than maximal economic efficiency in mind. Radetzki (1985) complains that this makes the CVG's iron mines less profitable. In contrast, the mining operations around Carajás are controlled by a profit-seeking state enterprise with a minority of private shareholders, and has no formal link with the Programa Grande Carajás (PGC), the national governmental agency charged with fostering development around the mines. The PGC has no investment budget of its own, but can authorize tax holidays and fiscal incentives. If a mine fosters a new administrative unit, urbanization around it may well include a significant political service sector. If, on the other hand, administrative powers remain distant, the social isolation of the project will be enhanced. The distribution of rents from the mining operation usually reflects the relative powers of local, regional, and national administrations.

The effects of extraction on population, environment, and development are mediated by highly variable but specifiable conditions, including the particular physical characteristics of the natural resource and the topography in which it is located, the biological and hydrological relations of surrounding ecosystems; the technologies dominant in the extraction and processing of the particular natural resource; international product and financial markets that affect its extraction, processing, and eventual industrial transformation; population densities and economic organization of the already existing human settlements in the area levels of welfare and economic activity in the surrounding society; the density of local and national political administration in the area; and national policy and legislation affecting extractive industry.

The relation of infrastructural scale to raw materials exports, for instance, varies with the physical characteristics of the commodity, particularly with the amount of sterile or overburden per unit of ore and with the kinds of molecular bonds between ore and sterile, as well as with the value per volume of the reduced ore. Iron and coal, for example, usually require heavy land transport to dedicated large ports, and aluminum requires large hydroelectric installations, while alluvial gold and cassiterite may leave an area in small airplanes taking off from numerous dispersed small runways. These variations in turn affect the size, composition, settlement patterns, and economic opportunities of the various migratory flows toward the project site and surrounding area. They are also likely to affect struggles for access to land and other resources among the migrants, or between them and earlier inhabitants of the region.

Spatial dispersal or concentration of the raw material will also have variable effects on migration into and settlement in the area. Commodities with low value to volume such as iron and coal are generally exploited in highly concentrated sites; the technologies developed in this century for open pit mining favor the concentration of large-scale machinery and large transport facilities required for economical movement of low-value bulk. This also favors concentration of ownership. Commodities with high value to volume may allow far more spatial dispersal of exploitation, especially when the resource itself is spatially dispersed. Gold and cassiterite, both spread by alluvial action, are examples of resources that may be exploited over wide areas, and are therefore less susceptible to concentrated ownership. In some cases, they may be exploited by households in combination with other activities such as agriculture or hunting. Concentration and dispersal of the resource and its exploitation mediates environmental effects of extraction directly in terms of ecosystemic disturbance and pollution, and indirectly through the different settlement and transport patterns associated with extraction. The relative concentration of population and environmental effects in turn mediates the capacity of the state to regulate extractive activities and the social welfare and order of the populations attracted to the region.

This general set of ideas informs an approach to environmental and demographic change around natural resource extraction that departs from a focus on the commodity itself, working outward from its physical and topographic properties to technologies and labor to world product and financial markets, considering as well the financial and political constraints on states to set and implement policies to ameliorate these effects. Extractive economies are configured by complex interactions between a wide range of social and natural processes. They can be compared across these processes, or dimensions, and so allow us to work toward more precise understandings of how they function, but each configuration will be different.

What these differences mean is that it is impossible to make meaningful general statements about the impact of mineral extraction on population and environment. It is however, quite possible to elaborate propositions about the interactions of multiple processes as these affect population and environment and as they affect the ability of the local communities and states to capture more of the revenues from extraction and to mitigate some of the environmental and social disruptions that extraction induces. These separate processes will configure each particular local economy differently, but they can provide comparable dimensions across multiple regional economies.

These propositions would require assessment of labor requirements in all stages of construction and operation, as well as the other economic opportunities opening in the same area, locational advantages or political incentives for linked industries and agriculture, conditions in the rest of the national society that would promote or hinder migration into the area, and the balance between migration and economic opportunities there. Space and topography, both as barriers to communication and as conditions of production, mediate the spread of these processes around the extractive enterprise. Scale, technology, environmental effects, processing and transport requirements, and pollution are all highly determined by the natural resource itself and the geology and hydrology in which it occurs, as these change over time with technical innovations. Control over the conditions of extraction and the distribution of costs and benefits from extraction reflect a more difficult array of variables, including markets for the commodity, the structure of the local and global firms that extract and trade it, and the number and distribution of economically viable reserves, as well as relations of property and between classes in the regions from which it is extracted.

Examples of propositions susceptible to comparative analysis include:

1)   Populations will change around a new project in function of a) the distance to existing transport infrastructure appropriate for

transporting the raw materials, b) the size of the labor force engaged to construct transport, energy, and mining infrastructure, c) the existence of other construction or other employment opportunities elsewhere at the end of construction, d) the extent of unemployment and landlessness in the larger society, e) the size and organization of the populations displaced by mine and associated infrastructure, and f) the physical characteristics and resource endowments of the surrounding areas.

2)   The economic activities in which these populations engage will reflect a) land tenure and the police power to enforce the security of tenure in land, b) state policies to subsidize or support or repress various kinds of economic activity, and c) the extent to which transport systems around the extractive enterprise serve to cheapen imports or to facilitate access to external markets.

3)   To the extent that national states are willing to allow extraction of minerals under conditions governed by rents set by the marginal producers, societies that export resources available in a wide variety of sources will be able to exert relatively little control over the conditions of extraction, and will be able to provide little protection for the surrounding populations, polities, and environment.

4)   To the extent that minerals firms have far greater expertise and knowledge of the extraction process and of raw materials markets, national states will be at a tremendous disadvantage in attempting to include environmental and social welfare costs in the contracts they write.   See Mikdashi (1976) for eloquent statements by Australian officials about the imbalances of bargaining with imperfect knowledge, and Mezger (1980) for confirmatory statements by corporate officials about their dominance in bargaining with national governments.

Conclusions

Resource-exporting nations face formidable consortia of raw materials-dependent states and firms.   Local systems are usually subordinate to the national state in the writing of minerals contracts.   They are thus at a double disadvantage in attempting to mitigate the consequences of extraction or to be compensated for them.   The well-being of local populations and the integrity of the local environment are thus directly at risk when natural resources are extracted for export.   These local systems are among the most literally peripheral in the world system--they are marginal politically and spatially, despite, and in large measure because of, the critical importance to industrial society of particular forms of matter found in or on their soils. Globally stated, the question about the relation of global resources to global populations simply overlooks these local systems and the effects of extraction on them. Separately, they each comprise tiny portions of the globe, of the world economy, and of the world's population. At any time in history, of course, many of them have suffered the ultimate impoverishment--depletion--and are no longer actively exploited. Collectively and over time, however, these areas make up a significant part of the global environment. The cost of their extraction cannot be reduced to the proportionate reduction of global reserves of a particular resource, because the environmental and social costs of extraction involve complex social and ecological processes that ramify well beyond the natural resource itself.

The relation between the severity of these social and environmental costs and absolute population numbers is highly tenuous. Clearly, consumption rates of any particular raw material affect the scope and scale of the impact of its extraction. Consumption rates are notoriously uneven between societies, however. Indeed, consumption of mineral resources tends to expand more in populations with low growth rates. More directly important may be the regulation of extractive processes and technologies. Most mineral extraction is highly polluting, but much pollution can be contained. Costs of containment are high, and regulation requires a strong and committed state. Extraction-dependent economies are seldom in a position to demand pollution control; their competitive position further weakens their ability to do so. Pollution costs are not incorporated into raw material costs, and prices remain correspondingly low, further stimulating consumption and accelerating depletion. The relations between population and environment, or between population growth and resource depletion, are mediated in complex, crucial ways by the intersection of material processes, technological choices, corporate strategies, and national and international relations. Simple assertions of their direct interdependence at a global level obscure critical processes and impede practical solutions.

References

Blainey, Geoffrey (1982). *The Tyranny of Distance: How Distance Shaped Australia's History*. (South Melbourne: Macmillian).

Brockway, Lucile (1979). *Science and Colonial Expansion: The Role of the British Royal Botanic Gardens* (New York: Academic Press).

Bunker, Stephen G. (1989a). "Staples, Links, and Poles in the Construction of Regional Development Theories", *Sociological Forum*, IV,4:589-609.

--------, (1989b) "The Eternal Conquest." *NACLA* May.

Bunker, Stephen G. and Denis O'Hearn, (1992) "Strategies of Economic Ascendants for Access to Raw Materials: A Comparison of the United States and Japan," in Ravi Arvind Palat (ed.), *Pacific-Asia and the Future of the World-System*, (Westport, CT: Greenwood Press), 87-102.

Crandall, Robert (1981). *The U.S. Steel Industry in Recurrent Crisis: Policy Options in a Competitive World*. Washington, D.C.: The Brookings Institute.

Collins, Jane L. (1988). *Unseasonal Migrations: The Effects of Rural Labor Scarcity in Peru*. Princeton: Princeton University Press.

DeWind, Josh (1987). *Peasants Become Miners: The Evolution of Industrial Mining Systems in Peru, 1902-1974*. New York: Garland Publishing, Inc.

Dore, Elizabeth (1988). *The Peruvian Mining Industry: Growth, Stagnation, and Crisis*. Boulder: Westview Press.

Flores Galindo, Alberto (1974). *Los mineros de la Cerro de Pasco 1900-1930*, (Lima: Pontificia Universidad Católica del Peru).

García, María-Pilar (1987). "La experiencia de la Guayana venezolana como un polo de desarrollo: un fracaso del modelo teórico, de la institución planificadora, o del estilo de planificación?" *Cuadernos de la sociedad venezolana de planificación, Guayana 25 años después: teoría y práctica de la planificación urbana*. Caracas. 17-71.

Godoy, Ricardo (1985). "Mining: Anthropological Perspectives." *Annual Review of Anthropology* 14:199-217.

Girvan, Norman (1980). "Economic Nationalism vs. Multinational Corporations: Revolutionary or Evolutionary Change?" in H. Sklar (ed.), *Trilateralism: the Trilateral Commission and Elite Planning for World Management*. Boston: South End Press, 437-67.

----------, (1976). *Corporate Imperialism: Conflict and Expropriation*. (New York: Monthly Review Press).

Hardin, Garrett (1968). "The Tragedy of the Commons." *Science*. 1162,1:243-8.

Innis, Harold A. (1956). *Problems in Canadian Economic History*. Toronto: University of Toronto Press.

Keohane, Robert O. (1984). *After Hegemony: Cooperation and Conflict in the World Political Economy*. (Princeton: Princeton University Press).

Keyfitz, Nathan (1991) "Population and Development within the Ecosphere: One View of the Literature." *Population Index*. 57,1:5-22.

Maletta, Hector (1979). "Campesinado, precio, y salario." *Apuntes*. 10:53-86.

Marx, Karl (1967). *Capital: A Critique of Political Economy*. Vol III. (New York: International Publishers).

Maull, Hans W. (1984). *Energy, Minerals and Western Security*. Baltimore: The Johns Hopkins University Press.

Meadows, D.H., D.L. Meadows, J. Randers, and W.W. Behrens III (1972). *The Limits to Growth*. New York: Universe Books.

Méndez, Cecilia (1987). "La otra historia del guano: Peru 1840-1879." *Estudios y Debates*. 1:7-46.

Mezger, Dorothea (1980). *Copper in the World Economy*. New York: Monthly Review Press.

Mikesell, Raymond (1978). "Trends in Foreign Investment: Agreements in the Resources Industry." *Resource Policy*. 8:194-99.

------------, (1980). "Mining Agreements and Conflict Resolution." in S. Sideri and S. Johns (eds.), *Mining for Development in the Third World*. New York: Pergamon Press, 198-210.

Mikdashi, Zuhayr (1976). *The International Politics of Natural Resources*. Ithaca: Cornell University Press.

Muntemba, Maud (1977) "Thwarted Development: A Case Study of Economic Change in the Kabwe District of Zambia." in Robin Palmer and Neil Parsons (eds) *The Roots of Rural Poverty in Central and Southern Africa*. Berkeley: University of California Press, 358-364.

Nwoke, Chibuzo (1987). *Third World Minerals and Global Pricing: A New Theory*. London: Zed Press.

Posner, Joshua and M. McPherson (1982) "Agriculture on the Steep Slopes of Latin America." *World Development*. 10:341-353.

Prechel, Harland (1990). "Steel and the State." *American Sociological Review*. 55,5:648-668.

Radetzki, Marian (1985). *State Mineral Enterprises: An Investigation into their Impact on International Mineral Markets*. Washington, D.C. Resources for the Future.

Rakowski, Cathy (1987). "Planificación y la división del trabajo en ciudad guayana." in *Cuadernos de la sociedad venezolana de planificación, Guayana 25 años después: teoría y práctica de la planificación urbana*. Caracas, 81-100.

Roberts, Timmons (1991). "Forging Development, Fragmenting Labor: Subcontracting and Local Response in an Amazon Boomtown." Unpublished Ph.D. dissertation in Sociology, Johns Hopkins University.

Robinson, Kathryn M. (1986). *Stepchildren of Progress: the Political Economy of Development in an Indonesian Mining Town*. Albany: SUNY Press.

Schmink, Marianne and Charles Wood (1992). *Contested Frontiers in Amazonia*. (New York: Columbia University Press).

Shafer, Michael (1983). "Capturing the Mineral Multinationals: Advantage or Disadvantage?" *International Organization*, 37:93-119.

Simon, Julian (1981). *The Ultimate Resource*. Princeton: Princeton University Press.

Sousa, Louis J. (1981). *The U.S. Copper Industry: Problems, Issues, and Outlook*. Washington, D.C.: U.S. Bureau of Mines.

Tanzer, Michael (1980). *The Race for Resources: Continuing Struggles over Minerals and Fuels*. New York: Monthly Review Press.

Tucker, Richard (1989). "The Depletion of India's Forests under British Imperialism: Planters, Foresters, and Peasants in Assam and Kerala." in Donald Worster (ed.) *The Ends of the Earth*. Cambridge: Cambridge University Press, 118-140.

USOTA (1985). *Strategic Materials: Technologies to Reduce U.S. Import Vulnerability*. (Washington, DC: Congress of the United States, Office of Technology Assessment).

Van Horn, Laurel (1977). "The Agricultural History of Barotseland, 1860-1964." in Robin Palmer and Neil Parsons (eds.) *The Roots of Rural Poverty in Central and Southern Africa*. Berkeley: University of California Press, 144-170.

USBM (U.S. Bureau of Mines) (1985). "Iron Ore." Washington, D. C. United States Bureau of the Interior.

Walter, Ingo (1983). "Structural Adjustment and Trade Policy in the International Steel Industry. in William C. Cline (ed.), *Trade Policy in the 1980s*. Washington, D.C. Institute for International Economics, 383-425.

10

Urbanization and the Environment in Developing Countries: Latin America in Comparative Perspective

Bryan R. Roberts

Urbanization is, at one and the same time, both a central and a secondary issue in the population-environment debate. It is central because urbanization is a global process that not only concentrates population in cities but transforms rural areas, and thus has substantial implications for the environment in both. In the last two or three centuries, it has been the major force altering the environment through stimulating massive increases in individual and collective consumption, including that of energy. Urbanization has thus changed land use patterns, depleted natural resources, and polluted the physical environment. Yet, urbanization is a secondary issue since there are no universally valid relations by which urbanization mediates between population and environment. The environmental impact of urbanization cannot be disentangled from that of other, associated factors, such as the pattern of economic growth and population increase. Thus, the pace and type of industrialization or the consumption needs of a growing population, rather than urbanization *per se*, are, in many respects, the environmentally crucial variables. Because of economies of scale, it is, as Lowry points out, unlikely that a dispersed population would consume less energy or other resources than a concentrated one of the same size and level of technology (1991). And the environment may benefit more from making changes within existing cities, through, for instance, substituting collective for private transport, than by promoting a more dispersed pattern of settlement.

From the perspective of this paper, the crucial difficulty is the ambiguity surrounding the term environment. The environment is often understood as the physical environment. In cities, this is the built environment together with its atmosphere and infrastructure, such as sewage disposal, water supply, and other utilities. And outside the cities, it covers the range of habitats that are the subject of other papers in this work--forests, deserts, mining regions, or commercial farming areas. There is also the social environment, and I shall contend that it is essential to consider both the physical and social environment since the two are interdependent. The social environment is the set of social, economic, and political relationships that people use, with different degrees of success, to make a living.

Urbanization fundamentally alters the relation between an individual's social and physical environment in two ways. It differentiates populations spatially by economic activity and by social status, and does this in both city and countryside. Urbanization also decreases a population's control over its immediate environment by intensifying dependence on nonlocal social and economic relationships, as employees of firms, as clients of bureaucracies, and as consumers of products.

The extent to which subgroups of a population segregate themselves spatially and make different demands on resources not only has a direct impact on the physical environment but determines the specific population impact of general changes in that environment. The various forms of pollution in the city and the lack of urban infrastructure are likely to affect people differently depending on where they live and the degree to which wealth or social relationships enable them to mitigate the negative aspects of their immediate environment and lessen their dependence on it. For instance, the aged who are socially isolated or single-parent females without kin or friends for support can be prisoners of their physical environment to a greater extent than those with extensive networks of social support. The interesting challenge in this field is, then, to relate these changes to each other--not just to relate urbanization to the creation of slums, to the increase in pollution, or, for that matter, to deserted villages and abandoned terraces, but to see the way in which changes in household organization and consumption are themselves part of the environmental problem.

The historical context must be kept in mind because it is only recently that urbanization has become a universal phenomenon. Urbanization, as a process of increasing differentiation, economic interdependence, and population concentration on a global scale, began with the Industrial Revolution at the end of the eighteenth century. Though cities, and urban civilizations, have risen and fallen since at least the beginnings of recorded history, the urban proportion of the world's population probably remained relatively constant until the eighteenth century (Lampard, 1965). The limits on urban growth previous to that time were set by the capacity of an urban center to dominate a sufficiently large hinterland to ensure a steady supply of foodstuffs and raw materials for industry. The slowness of transport, the limited reach of markets when most of the population was embedded in subsistence agriculture, and the sheer amount of military force needed to secure the supply routes meant that only with great difficulty could a city dominate enough space over a sufficient span of time to support a large population.

Once sustained urban growth had begun, it helped unleash population growth, but with increasing demands made on the environment. Here the key variables are the resource constraints of different sectors of economic activity as indicated by Keyfitz (1991: 15): "Primary is based essentially on land and the population limit is sharp, secondary is more flexible, but it still requires materials and energy, tertiary is the most flexible of all, requiring little outside of skill and effort."

Industrialization raised productivity in both town and countryside, providing increasing numbers of job opportunities in the cities and enabling food production to rise to feed the growing urban populations. Lowry (1991:160) points to the high productivity of agricultural workers in developed countries based on large expanses of land and access to large amounts of commercial energy (tractors, combines, fuel, and fertilizer). At later stages of urbanization, in both developed and developing countries, urban economies have become increasingly service-oriented and communications-based, but the consequences of this change for the relation of population to the environment is ambiguous. It means, potentially, a greater flexibility in urban spatial organization, permitting the dispersal of work sites, even to the extent of employees using computer-based communications to keep in touch with head offices while working out of their homes. But the move to a service economy has not lessened the consumption of materials and energy. Increasing prosperity, and a dramatic rise in "service" employment is accompanied by an increase in resource-depleting consumer goods. These are often produced, as in the case of electronics, with considerable negative impact on the environment, and the markets for them are made possible by the service economy, particularly finance and advertising (Gershuny and Miles 1983). Certain services, such as high tech medicine, are greedy consumers of resources.

The next section examines the nineteenth-century patterns of urbanization in various countries of the developed world. The aim is to provide some benchmarks for evaluating contemporary urbanization patterns in the industrialized world. Since the nineteenth-century patterns vary in their impact on the environment, they also reinforce the point that the relation between urbanization and environmental change is a specific one that must be abstracted carefully from particular historical and regional contexts.

Urbanization and the Environment in the Nineteenth Century

In the nineteenth century, industrialization, as Landes (1970) puts it, unbound Prometheus, permitting an enormous increase in the production of basic commodities and cheapening their unit price for both rural and urban populations. Increasing commodity production generated income opportunities in the cities, attracting rural migrants, while cheap industrial goods displaced rural handicraft production. Agricultural production changed to meet the demands of the cities both for foodstuffs and for the raw materials of industry. Subsistence farmers were gradually displaced and/or were motivated (at times by coercion) to market their production to meet urban needs. The resultant patterns of urbanization differed, however, according to both the scale and type of industrialization.

I will take the cases of the United States, Britain, and France. By the end of the nineteenth century, these three countries were among the major industrial powers of the day. In 1900, the U.S. was the largest producer of crude steel in the world (10,352,000 metric tons a year), followed by Germany (6,461,000 metric tons), and Britain (4,980,000 metric tons), with France in sixth place, at 1,585,000 metric tons (Mitchell 1976, Table E9, 1985, Table E9). But they exhibited different patterns of urbanization.

Britain had, by far, the largest city population of the three by 1900, with France having the smallest number of city dwellers (Table 10.1). City growth was more concentrated in both France and Britain than in the United States. London and Paris were substantially larger than the combined populations of the next three largest cities of Britain and France respectively (the index of 3 city primacy), and maintained their advantage from 1860 to 1900. New York had a slightly larger population in 1860 than the next three largest U.S. cities combined, but by 1900 this advantage had disappeared. The U.S. urban population was a relatively dispersed one when compared with Britain's. By 1900, when 40 percent of the U.S. population was classed as urban and 78 percent of the British population was so classed, the U.S. had more of its urban population in small towns (under 25,000), and less in large cities (over 100,000) than did Britain (U.S. Bureau of the Census 1975: Series A 57-72; Lawton 1978:Table 3.2).

The different patterns of urbanization indicated by these data had one common environmental impact. Rapid urban growth made urban poverty, overcrowding, and the health dangers of cities major public preoccupations in the three countries. Despite the differences in their urbanization, all three had large cities whose growth was heavily fueled by inmigration and that outpaced the capacity of public authorities to provide adequate infrastructure or to ensure law and order. The scale of these nineteenth-century cities and their rate of growth was unprecedented, and was the major environmental issue of the day. Devising a form of government for these cities that would reconcile economic

TABLE 10.1 Growth of European, United States and Mexican Cities Between Different Time Periods[a]

	Rate of Pop. Growth	*Rate of Urbani- zation*[b]	*Growth Largest City*	*Growth of Next 3 Cities*	*% Living in Cities at start of per.*[c]	*at end of per.*	*Index 3 City Primacy start*	*end*
United States 1860-1900 (15 cities)	2.2	1.1	2.7	3.3	10.0	14.6	1.2	1.0
France 1860-1900 (6 cities)	0.1	1.1	1.2	1.2	7.0	11.0	2.3	2.2
Britain 1860-1900 (11 cities)	1.2	0.5	1.8	1.5	29.8	36.8	2.2	2.5
Mexico 1940-1980 (8 cities)	3.0	1.6	4.9	4.7	21.9	42.0	2.6	2.6

Sources: B. R. Mitchell, *International Historical Statistics, The Americas and Australasia*, Table B4, *European Historical Statistics, 1750-1970*, Table E9. G. Garza and Departamento del Distrito Federal, *Atlas de la Ciudad de México*, Cuadro 4.2.

[a] For France and Britain, these are cities with a population of 250,000 or more in 1900. For Mexico it is cities with more than 250,000 in 1980.
[b] Urbanization is the annual rate of growth of the proportion living in cities with respect the total population.
[c] This is the ratio of the largest city to the next three largest cities.

growth with public order and the general welfare was a challenge in the nineteenth century, and has continued to be so since. In terms of assessing the performance of urban government in developing countries, it is as well to remember that it took almost a century in the cities of the developed world to come to at least partial terms with the problems of overcrowding, sanitation, and law and order.

The differences between the three cases appears more clearly when we consider the impact of urbanization on the hinterlands of the cities. In Britain, nineteenth-century urbanization did little to restructure the rural areas of the

United Kingdom. By the late eighteenth century, Britain already had a highly market-oriented agrarian structure based on a flourishing network of small towns, and peasant farming had all but disappeared. The major changes in the rural environment had occurred prior to the nineteenth century, thus antedating industrialization. The migrants that supplied the labor force of the industrial cities in the north of England in the nineteenth century were often rural craftworkers and their families who had been displaced by factory production. Also, the rapid industrialization of Britain in the nineteenth century, mainly based on textiles, served an external market. The hinterland that British cities transformed through trade in manufactures and agricultural products was global, and included Africa, Asia, and Latin America. This hinterland supported a dense urban population in Britain, but Britain's own rural environment was probably more shaped by the demands of fox hunting than by the growth of cities in the nineteenth century (Hoskins 1970).

French urbanization in the nineteenth century also had little impact on the rural environment, but for somewhat different reasons than in the British case. Though there was a gradual shift of population to the cities (noted in Table 10.1), French population growth was low and France had few large cities. Consequently, their transformative impact on the highly regionalized and peasant-based agrarian structure of France was a slow one. French industrialization, as compared to that of Britain, was based on small-scale manufacture, much of it in luxury items, and oriented to the domestic as well as the export market. Thriving industrial centers, such as Lyon, were basically made up of workshop production.

In contrast to the British and French cases, urbanization in the United States was an integral part of rural development. The spread of towns and cities opened the internal frontier, serving as bridgeheads to the settlement of agricultural regions through channeling international migration to both rural and urban areas. The major part of U.S. manufacturing production by the end of the nineteenth century was based on transforming agricultural, livestock, and forest products, and the economic character of cities was shaped by the specialization of their hinterlands, as in the case of Chicago with the grain and cattle industries of the midwest.

In these three cases, the crucial variable in understanding environmental change is, consequently, industrialization, not urbanization. The pattern of industrial development was quite different in the three cases--large-scale production of mass consumption goods for a world market in the British case, small-scale production of luxury goods in the French case, and industrialization geared to the development of the internal market, both rural and urban, in the U.S. case. Nor, in these three cases, was living in cities the important factor in changing consumption patterns. In the U.S., immigrant farmers were no less dependent on the market for nonfoodstuffs than their urban counterparts, and their reliance on purchased sources of energy was likely to be greater on a *per capita* basis given their needs for heating, transport, and power to cultivate the fields. In Britain, the rural population depended on the market for most basic needs, including a substantial part of its food needs, and the range of goods found in the house of an urban worker was unlikely to be substantially greater than in that of a rural worker. Poverty circumscribed consumption, not rural or urban location. And the same is likely to have been the case in France.

In these three nineteenth-century cases, urbanization and economic development went hand in hand, and the environmental consequences of urbanization at a local, national, or global level were basically the consequences of the particular pattern of economic development. Continuing economic development in the twentieth century had an increased environmental impact through the dramatic rises in individual as well as collective consumption of goods. This impact was to be mediated, however, by the differences in the pattern of development--such as the extensive use of space in U.S. cities and their regional dispersal, leading to higher per capita consumption of petrol and other energy sources than in the European cases (Lowry 1991; Table 6). But economic development combined with citizen action helped to ameliorate some negative environmental impacts. Changing technology and greater affluence permitted the "cleaning-up" of the Western European and U.S. cities in the post-World War II period, substituting cleaner for "dirty" fuels, and enforcing environmental protection regulations.

Urbanization in the Developing World

The situation of the developing world is quite different. Whereas the impact of urbanization on the environment in the developed world is subsumed within that of economic development, in the developing world urbanization appears to have a more independent impact on the environment due to the lack of economic development. Various studies have shown that many countries of the developing world are overurbanized in the sense that their level of agricultural productivity cannot feed the urban population, making them dependent on food imports, while their level of industrialization does not generate enough employment to provide an adequate standard of living for the urban population (Bradshaw 1987).

Population processes are inseparable from these outcomes, but these must be understood in the wider context of the pattern of economic development occurring in developing countries. It was the opening-up of areas for commercial farming and making them safe for Western colonization that helped reduce mortality from diseases such as malaria. The income opportunities created in mines, on plantations, and subsequently in cities, made the restricted land opportunities for peasant farming less of a limit on household formation and procreation than it might otherwise have been. The result has been high rural rates of natural increase. The situation has, however, been little different in the cities. In 1960, urban birth rates in developing countries were high relative to those of developed countries: 37.9 per 1000 population compared with 23.5 in the developed world. The high urban birth rates in the 1960s are accounted for by various factors, including the compositional one that the urban population includes migrants who come at childbearing ages. The birth rate for the rural population of developing countries was higher, 44.1 per 1000 population, but this difference was compensated by the higher rural mortality, 21.7 compared to 25.4 (Lowry 1991; Table 3). The evidence available for the period since 1960 indicates that birth rates in developing countries have dropped substantially. The partial data available for rural and urban locations suggest that this drop has occurred in both urban and rural birth rates (United Nations 1992; Table 9).

These population trends underline the point that the association between urbanization and population growth is a complex one. It is the transformation in the overall economy of developed countries, not just urbanization, that leads to the kind of modernization that, combined with increased access to family planning, results in declining birth rates: higher levels of education, job opportunities for both men and women outside the home, or economies in which children represent a serious economic cost. However, higher rates of illiteracy continue in many urban areas of developing countries, and urban shelter is often self-constructed and overcrowded and thus not a barrier to establishing a new household or accommodating extra children. Many income opportunities can be obtained by working at home with the help of child labor, and, importantly, the absence of social welfare systems means that children may be viewed as providing security against old age and incapacity.

A further consideration casting doubt on the uniform effect of urbanization in reducing population growth is that the current decline in birth rates is likely to be based on different factors among subgroups of a population. Wood and Carvalho (1988:169-177) argue that different class logics are present in the strategies of family limitation of the mid-1970s in Brazil. The urban poor had fewer children because it was difficult to meet basic subsistence; women chose birth control and worked for pay to increase household income. In this situation, children were a short-term liability, but fewer children mean, in the longer term, less support in advanced stages of the household life cycle. For the increasing number of wage workers in the agricultural regions of the North-East of Brazil, restricting family size was necessitated by lack of access to land and the dearth of local income opportunities--children were both a long-term and short-term liability. For the middle classes, restricting family size was less a question of basic subsistence than of quality of life for themselves and their children.

Sources of Variation

The weak positive association between urbanization and economic development in the developing world has deep historical roots. Exploring these roots is crucial to understanding contemporary variations in urbanization. The pattern of urbanization imposed by the colonial powers on what is now the developing world was basically centrifugal, with cities serving as bridgeheads to the control and settlement of the rural areas from which the resources would be extracted to the benefit of the colonial power (Morse 1962). The significance of these colonial developments is that they implanted urban forms designed to tie the colony into the economic and political system of the core country. At one level, this meant the introduction of urban spatial planning based on sharp class and status distinctions--such as between the native city and the city of the "white" colonizer (King 1990). At another level, it encouraged the consumption of manufactures and of luxury goods in general that could only be satisfied by intensifying external trade. And, at yet another level, the colonial city became the focus of a dendritic pattern of urbanization in which economic and political transactions were funneled to a few centers, and then outward to the core (Johnson 1970).

Though the colonial powers varied in their manner of colonization--the Spanish restructured the economy and society of Hispanic America, while the British in India used the existing political and economic structure to extract a surplus--the function of the colonial city was essentially the same. It was an entrepot serving to consolidate the control of the core and intensify interdependence. It had few orthogenetic functions--those functions that enhance the distinctiveness of local society and its culture--but neither did it serve to introduce innovations that would stimulate development. The legacy of the colonial city has endured to the present. Most of the major cities of the developing world are colonial creations. They were built, or rebuilt, to represent the core's idea of the relation of cities to their environment. They became the trojan horses of western consumption values, outward-looking bastions of Western models of modernization and economic growth, sharpening the distinction between the "modern" urban and the backward "rural" so that urban bias would seem a natural tool in the achievement of modernity.

Differences in the colonial experience account, in part, for differences in the contemporary pattern of urbanization in developing countries (Roberts 1989). The type classifications in Table 10.2 are based on contrasts in the timing and mode of incorporation into the world economy. These two variables provide a preliminary understanding of the key historical experiences affecting urbanization in the developing world, particularly the varieties of agrarian transformation. I refer to the transformation of existing social relations through proletarianization and through breaking down subsistence production. Incorporation produced many variations in this process, as Wallerstein (1976) shows, at times reinforcing existing relationships, at others introducing archaic ones, as in the "re-feudalization" of Central Europe. Those areas that were incorporated earliest and through the most radical restructuring of rural social relationships show the highest current levels of urbanization and initially high but declining rates of urbanization, namely the various areas of Latin America (Type 1). Four other types of incorporation can be identified from among the geographical groupings in Table 10.2.

Another type is the early incorporation of countries such as China and India, which had well-developed internal markets, but whose economies were not fundamentally restructured by incorporation into the world system (Type 2). In both these countries, internally generated economic changes were slow, and colonialism or, in the case of China, neocolonialism brought no radically new sets of social and economic relationships (Raza et al. 1981; Skinner 1977). Both the rates and levels of urbanization are low, though the rates of urbanization in China have fluctuated sharply due, in part, to policy changes.

The experiences associated with Type 3, those fitting the countries of North Africa, are similar since they, too, were incorporated early and without fundamental restructuring, and rates of urbanization are low in this type also. The contrast with China and India is that lacking large internal markets, countries of this type have been heavily dependent on external trade, resulting in moderately high levels of urbanization, associated with an old urban culture (Abu-Lughod 1980:30-51). Abu-Lughod points out that the focus of urban growth has alternated between coast and interior in North Africa, depending on the international context and whether interior or Mediterranean trade routes were favored (1971, 1980).

Type 4 is of relatively early incorporation, with uneven restructuring, resulting in "islands" of export agriculture or mining amidst subsistence cultivation. The countries of sub-Saharan Africa and Southeast Asia fit this type, though to varying degrees. The level of urbanization is low, but the rate of urbanization is high. For more detailed accounts, see Gugler and Flanagan (1978) and Hart (1987) for Africa, and, for Asia, Armstrong and McGee (1985:88-110), who provide a typology of urban trajectories, and Nemeth and Smith (1985).

A fifth type also shows high rates of urbanization. It is the most recent form of incorporation into the world system, that associated with the political economy of the cold war. These are the "new" Asian countries, of which South Korea is the example in Table 10.2, which became significant locations of foreign investment in the 1950s and 1960s. Their rapid transformation is closely linked to two phenomena: considerable economic and military aid, and their proximity to Japan as it seeks markets and outlets for investment (Nemeth and Smith 1985:197-204; Deyo 1986).

The Constraints of Dependent Industrialization

There are also certain common features to urbanization in developing countries set by a global economy in which industrialization has been dependent on the technology and capital of the advanced industrial countries. The predominant form of industrialization has been, until recently, that of import substitution oriented to the production of basic goods for the internal market. Tariffs and overvaluing the local currency cheapened the costs of imported technology while protecting local manufacture. This created both a bias towards capital-intensive production of sophisticated consumer goods, increasingly dominated by multinational corporations, and a more labor-intensive production of basic goods that remained in the hands of local capital. One important spatial implication of import-substitution industrialization is its tendency to concentrate population in a few large urban centers, with the rest of the population scattered in villages or in towns poorly endowed with urban infrastructure.

In Latin America, the most urbanized part of the developing world, industrialization was to be rapid under import-substitution, but it concentrated in few cities, and even in these had a limited capacity to absorb labor productively--limited by capital-intensive technology, by size of the domestic market for its goods, and by the increasing supply of labor. The result has been an apparently dualistic labor market, and similar trends have been reported for cities in Africa and Asia (Bromley and Gerry 1979; Rodgers 1989; Standing and Tokman 1991). "Luxury" goods and services for the high-income populations aree provided by formally organized enterprises, and low-quality goods and low-cost convenience services (eg., street vending, neighborhood stores offering credit) are provided by small-scale enterprises (at times through subcontracting) and the self-employed--the "informal" sector (Portes and Walton 1981; Roberts 1978). The low cost is often based on convenience, as when cigarettes are sold by the unit and not by the packet in neighborhood stores or when meals are bought from street vendors.

From the 1970s, a new mode of integration has begun to replace import-substitution--export-oriented industrialization. This type of integration is most marked in the case of the Asian "tigers," but it has become an increasingly widespread model in Latin America and in parts of Southeast Asia, such as Malaysia and Thailand. Mexico, for instance, has experienced a dramatic rise in the importance of industrial exports in its export composition, and in the importance of employment in the export industries. Unlike import-substituting industrialization, the export-oriented model does not depend on proximity to domestic markets, but, rather, on transport routes and the cheapness and availability of labor and of infrastructure. Urbanization under this pattern follows a different spatial logic. It is less likely to be associated with the dominance nationally of any one city, and can lead to a more dispersed pattern of urban growth.

TABLE 10.2 Annual Rates of Urbanization in the Developing World: 1950-1990

Rates of Urbanization	Type 1		Type 2		Type 3	Type 4	Type 5	
	Temperate South America[a]	Tropical & Middle America[b]	China	Southern Asia[c]	North Africa[d]	Tropical Africa[e]	Southeast Asia[f]	South Korea
1950-60	1.2	2.1	5.5	0.9	2.0	3.1	1.7	2.6
1960-70	0.7	1.8	0.0	1.2	1.8	3.1	1.4	3.8
1970-80	0.6	1.4	1.2	1.7	1.0	2.9	1.7	3.3
1980-90	0.4	1.1	5.3	1.7	1.1	2.7	2.2	2.4
Total Population growth, 1950-1990	1.6	2.7	1.8	2.3	2.5	2.7	2.2	1.9
% urban population 1990	86.1	70.7	33.4	27.3	44.6	28.7	29.9	72.0
% in urban agglomerations, 1985[g]	39.1	34.3	27.2	25.9	30.9	6.2	29.7	53.4

(continues)

TABLE 10.2 (continued)

Rates of Urbanization	Type 1		Type 2		Type 3	Type 4	Type 5	
	Temperate South America[a]	Tropical & Middle America[b]	China	Southern Asia[c]	North Africa[d]	Tropical Africa[e]	Southeast Asia	South Korea
% urban growth due to natural increase 1950-1990[h]	69.4	63.4	39.3	62.6	62.4	48.0	56.0	38.0

Sources: United Nations (1991) *World Urbanization Prospects 1990*, Table A-2,A-4; United Nations (1987) *The Prospects of World Urbanization*,TableA-7

aTemperate South America includes Argentina, Chile and Uruguay.
bTropical and Middle America are the other countries of South America, Central America, and Mexico.
cSouthern Asia includes Afghanistan, Bangladesh. Bhutan, India, Iran, Maldives, Nepal, Pakistan, and Sri Lanka.
dNorth Africa includes Algeria, Egypt, Libya, Morocco, Sudan, and Tunisia and Western Sahara.
eTropical Africa is all of Africa excepting North Africa and Southern Africa (Botswana, Lesotho, Namibia, South Africa, Swaziland).
fSoutheast Asia includes Brunei, Cambodia, East Timor, Indonesia, Kampuchea, Laos, Malaysia, Myanmar, Philippines, Singapore, Thailand, Vietnam, and Brunei.
gAgglomerations are capital cities and cities with 2 million inhabitants or more in 1985. Figures are the proportion of the urban population of each area who live in such cities.
hThis is the proportion of urban growth accounted for by vegetative population growth.

I will use data from Latin America to explore some of the implications of these patterns of industrialization for urbanization. The data on Mexico in Table 1 cover the years of Mexico's rapid period of urbanization from 1940 to 1980. These were also years of industrial development, mainly concentrated in Mexico City, aimed at import-substitution and at producing basic goods for the internal market. Mexico starts with a large city population when compared to the United States or France. By 1980, it had a larger city population than Britain did in 1900, and it is concentrated in one big city--Mexico City. Mexico's rate of urbanization is the most rapid of the four countries in the table, indicating large-scale immigration. This is assuming that birth and death rates are similar for the city and noncity populations. In that case, the proportionate gain in the city population (the urbanization rate) is accounted for by migration from smaller places. When the contribution of migration is added to that of the natural increase of the population, the result is city growth rates, such as the 4.9 percent for Mexico City, that have doubled that city's population every 15 years or so. Compared with the three nineteenth-century cases, Mexico's urbanization is much more rapid and more uneven in terms of its concentration.

The differences with the nineteenth-century European pattern of urbanization also appear when we extend the analysis to include Brazil, the largest Latin American country in both area and population, and look at the overall patterns of urban growth, not just that of the large cities. Urbanization and population growth occur at a much greater pace in Brazil and Mexico than in the case of Britain in the nineteenth century. The following data are taken from Lawton (1978; Table 3.2) Garza and DDF (1987:Cuadro 4.2) and estimates from the *Censos do Brasil* for 1940, 1950, 1960, 1970, and 1980. Using population in places of 20,000 or more as the criterion for urban, Brazil urbanized at an annual rate of 2.6% between 1940 and 1980. The equivalent rate for Mexico was also 2.6%, whereas Britain's rate based on places of 20,000 or more was 0.9% between 1860 and 1900. Annual population growth rates in these periods were 2.6% for Brazil, 3.0% for Mexico, and 1.2% for Britain. The cities of both countries consequently grew much faster than did those of Britain at their moments of rapid urban growth. The rates of overall urban growth for intermediate and large cities were, as a result, almost three times higher than in the cases of Britain.

The shape of the urban system also contrasts. By 1980, the population of both Brazil and Mexico is heavily concentrated in the large cities (100,000+), which contain 42.7% and 48.1% respectively of the total population. In Britain, urbanization resulted in a substantial population (17.5% of the total population) living in small and medium-sized towns and cities (between 20,000 and 100,000). This is explained by the type of economic development in Britain, which encouraged some urban de-centralization in smaller industrial and commercial centers. The pattern for Brazil and Mexico shows no such de-concentration, with less than 8% in both countries living in small and medium-sized towns and cities. For Brazil, as for Mexico, estimates of sizes of towns and cities is complicated by differing definitions of the spatial areas of those towns and cities.

In the decade 1980-1990, the growth pattern of Mexican cities appears to have altered. Corona (1991) assesses the reliability of the 1990 census, suggesting a possible undercounting of population. The undercount may have been particularly high for Mexico City, but even accepting the highest (and improbable) estimate of the city's 1990 population--19 million--gives an annual growth rate of 3 percent from 1980 to 1990, indicating a slowing of the metropolitan area's growth, due to less inmigration and lower birth rates. Population growth declined to an annual rate of 1.9%, while the growth rates of cities with more than 100,000 inhabitants declined from 4.2% annually in 1970-1980 to 2.1% in 1980-90, barely exceeding population growth (Garza and DDF 1987: Cuadro 4.2; INEGI 1990: Cuadro 5). Within these overall growth rates, there are marked differences depending on the type of city. The fastest-growing cities are those linked to export-oriented industrialization--the northern border cities, with a growth rate of 3.5% annually, and those interior cities that are specializing in the export economy, with a growth rate of 3.6% annually. In contrast, the reported growth rate of Mexico City has declined to 1.0% annually in contrast to 4.3% from 1970-1980. The growth rate of the two next largest cities, Guadalajara and Monterrey, also shows sharp declines, from 4.3% in 1970-80 to 2.3% in 1980-90.

Implications for the Environment

Though the different rates and levels of urbanization in the developing world are likely to mean diverse outcomes for the rural and urban environment, urbanization in poor economies will have negative impacts on the environment. Hardoy and Satterthwaite point out that the spatial growth patterns of third world cities have combined the worst environmental features of the cities of the developed world with those peculiar to underdevelopment (1985). Thus, per capita automobile ownership is comparable to that of many Western cities, and the range of industrial pollutants is present, but with less strict controls. Then there are the problems resulting from widespread poverty and overcrowding, such as garbage, exposed sewage, scarce and polluted water, and infectious diseases.

Though urban consumption in the developing world is higher and different from rural consumption, the difference is not a great one when the low-income urban population is the point of comparison. Rural consumption increases with urbanization to include industrialized foodstuffs, other basic goods, and consumer durables. A greater part of a rural budget than an urban budget is provided by self-provisioning, exchange, and remittances from migrants. In

Mexico, for instance, about a quarter of the budget of the poorest families, who are mostly rural, appears to be made up by these items. These estimates are my own, but are taken from Cortés and Rubalcava's (1991) study of household income in Mexico based on a national survey of income and consumption. However, low-income urban populations do not differ substantially in that self-provisioning, exchange, and remittances appear to make up about 15 percent of their budgets, and, like their rural counterparts, they are likely to own their accommodation--in both cases mainly through self-construction. What is striking in Mexico is the extent of consumerism, though this is likely to be less true of other developing regions. Benaría reports that among the poorest of her Mexico City sample, 85% had television, 55% had a refrigerator, 45% had a sewing machine, 40% had a washing machine, and 15% had a car (1991: Table 2). Though these are higher levels of consumption than are likely to be found in Mexican rural areas, the difference is explained as much by lower income or factors such as absence of electricity than by any rural cultural disdain for modern consumer goods.

High rates of urbanization, especially when combined with high rates of population growth, create severe problems of urban infrastructure and housing. And these problems will emerge whether or not a country has a high level of urbanization. The urban problems of Calcutta are no less severe than those of Mexico City, though India is predominantly a rural country and Mexico predominantly an urban one. Rapid rates of urbanization, combined with high rates of population growth, result in particularly high comparative rates of contamination in the largest cities. Thus, Jiménez and Velázquez report higher concentrations of sulphur dioxide in the atmosphere of metropolitan Manila than in New York, Los Angeles, and Chicago, and higher concentrations of particulate matter than in Tokyo, New York, and London (1989:55). As Schteingart points out for Mexico City, the issue is one of rapid urban growth, but at low levels of economic development (1989a). The number of vehicles in use in Mexico City has expanded rapidly with its growth, and a large proportion of that expansion is based on the continued use of old vehicles that heavily pollute the environment. Pollution emissions in the city increased by 45 percent between 1972 and 1983, probably less than the population increase in these years.

Also, the impact of large cities on the rural environment can be substantial even in predominantly rural countries. Delhi is reported to have brought in 223,600 tons of firewood by rail from forests some 700 kilometers away, and the expansion of India's cities has taken up 1.5 million hectares of some of the best agricultural land in the last thirty years (Centre for Science and Environment 1989).

It is in these two areas--the impact on the rural environment and on the nature of work and survival in the cities--that urbanization in the developing world has had important consequences for the physical environment. In both areas, it has disrupted planning and placed piecemeal pressures on already fragile systems. In the next two sections, I will look at the impact of urbanization on the rural and urban environments, considering the social as well as the physical environment. The focus will be poverty, which illustrates the importance of considering both the physical and the social environment. In terms of physical environment, poverty includes a range of factors depending on whether the poverty is rural or urban: poor soils; small plots of land; lack of water supply; inadequate and overcrowded housing; lack of sewage or garbage disposal; and poor access to utilities such as water or electricity. The social environment of poverty is often the same in rural and urban environments. It includes lack of employment opportunities. It also includes the relative absence of informal support relationships, such as those with kin, friends, or neighbors, and formal ones, such as with welfare agencies. Support relationships are essential to help those who cannot cope on their own resources, such as the aged, the unemployed, or female-headed, single-parent families.

The Impact on the Rural Environment

Urbanization in countries in which agricultural land is scarce can relieve population pressure on land and make it easier to raise productivity for the benefit of both rural and urban populations (Oberai 1991). This has not happened for two principal reasons. The first is the high rate of natural increase among the rural population in developing countries, which continues to grow in absolute numbers despite high rates of rural-urban migration (Lowry 1991:Table 3). The second reason is the way in which the rural structure is being transformed during urbanization. Urbanization has occurred in the context of what is essentially a dualistic agrarian structure with pockets of often export-oriented commercial agriculture amidst a mass of subsistence producers. These peasant economies diversified with urbanization, with labor migration becoming an increasingly normal part of the household economy in many parts of the developing world. Off-farm earnings enabled peasants to retain their smallholdings, but the economic viability of small-scale farming was undermined as population pressure on land increased, soils became exhausted, and governments sought to cheapen urban food prices by importing cheap foodstuffs and limiting their market price. In this context, urbanization acted to maintain small-scale farming, but not to make it more productive or profitable. It was this diversification that enabled the Mantaro Valley area in Peru to sustain a relatively high population growth for much of the twentieth century despite the average size of landholding declining from three hectares at the beginning of the century to one hectare by 1970 (Long and Roberts 1978).

There were few intermediate places to absorb the flow of migrants seeking off-farm income opportunities since uneven rural development meant that there was no network of market towns to support regional economic growth based, as in the U.S. in the nineteenth century, on the processing of agricultural products and the servicing of the agricultural population. Roads were built connecting the large cities with the rural areas, making it easier for people to migrate, and bringing urban goods to the countryside, including industrialized foodstuffs. In a study of several remote highland villages of Peru, Figueroa found that farming households were spending half their budgets, most of which was earned by labor migration, on purchasing industrialized foodstuffs, such as cooking oil, canned goods, and sodas (1984). Increasingly, in most developing countries, rural-urban migration became based on direct migration from peasant village to large city where income opportunities concentrated. The large cities of Latin America from the 1940s onwards concentrated at least 30 to 40 percent of the market for basic goods and a substantially higher percentage of the market for luxury goods.

Diversification and the increase in paid work furthered the consumption of industrial goods, and made the rural population more dependent on market fluctuations. "Western" consumption patterns spread to the rural areas of many developing countries, often forcibly, as when colonial powers monetized local economies through taxation and, at times, the forced sale of goods. Over time, these processes meant that there was no substantial cultural barrier to separate the rural inhabitant from the modern city. Under these conditions the growth of income opportunities in the cities and better communications led to the beginnings of exodus. The establishment of bridgehead settlements increased the flow (See the review of the literature in Butterworth and Chance 1981:91-107).

The impact of these changes on the physical environment was mediated by changes in the social environment, resulting in a new environmental system. Consider Diana Wylie's (1989) historical account of how the welfare of two southern African peoples was affected by urbanization and economic growth in South Africa from 1880 to 1980. Her measures of welfare are malnutrition and the incidence of diseases related to malnutrition, such as tuberculosis. She begins with the point that malnutrition should be defined in terms of the environment rather than in terms of individuals: people can lose and gain weight without needing to be classified as undernourished when this loss and gain is part of a broader environmental system designed to adapt to changes in food supply (Ibid.:162). Malnutrition is, rather, the permanent condition that signals the breakdown of this system. The available sets of relations cannot provide the safety nets to defend a population against scarcity.

The peoples she contrasts are rural peoples, the Tswana of Botswana and the southern Nguni of the Transkei, who differ in their ecological situation and agricultural practices. The Transkei was a more varied and richer farming system, sustaining a denser population. This advantage meant, however, that the Nguni were to be more subject to the pressures of economic change than the Tswana. Agriculture was rapidly commercialized and males were likely to seek outside income opportunities when these became available in the mining and other urban centers of South Africa. In both cases, labor migration was to be destructive of the existing environmental system because it removed needed labor from the fields. Its effect was also an indirect one in that it contributed to the increasing monetization and commercialization of local economies. This trend was furthered by the pattern of economic growth in southern Africa, which in the rural areas favored commercial over peasant farming, either directly displacing the peasant farmer or making peasant production uncompetitive on the market.

Wylie shows the progressive disintegration of the old system, using 1880 as a benchmark. In the 1880s, there were differences between the households of commoners and headmen in their access to food and in the amount of labor they could muster, but evidence is cited showing that the logic of the environmental system--based on close local interdependence and the fact that local groups were not tightly controlled from a distant center and had flexibility in changing their subsistence strategies and diet to adapt to season or climate --ensured sufficient sharing to guarantee basic subsistence. By the 1980s, the old environmental system had disappeared in both areas. In Botswana, the local community survives as an economic unit, but is socially and economically fragmented, shows high degrees of spatial mobility, and is dependent on government bureaucracy for welfare. In the Transkei, the situation is more atomized, with a quarter of the population having no land at all and most landholdings being too small to provide subsistence, so that the resident population depends on remittances from labor migration.

A major point made in this study, then, is that the shift from the old environmental system has resulted in the creation of permanent, as opposed to cyclical, disadvantage, even though it is only for a particular category of households--female-headed households and those others without several sources of able-bodied labor. The communal, but shared, insecurity of subsistence farmers is now replaced by a more divisive urban insecurity based on class differentiation. Wylie makes the telling point that the money economy means that the possibility of starvation fades since even the poorest can earn enough to buy some food, but that the children of the poor suffer unprecedented malnutrition (1989:194).

Lourdes Arizpe's study of the villages of Santiago Toxi and San Francisco Dotejiare extends the analysis by showing the basis of the new environmental system that emerges when land and household production are no longer the major source of subsistence (1978, 1982, 1985). The two villages differ from each other in that in Dotejiare land remains an important source of subsistence, whereas in Toxi, greater land scarcity and plot fragmentation, mean that agriculture is increasingly supplementary to off-farm wage labor. In both villages, households use their available labor to ensure the long-term survival of the household unit. But there is a significant difference in the way this is achieved between households in the two villages.

In Dotejiare local agriculture can absorb more household labor than can that of Toxi. The major pattern of migration in Dotejiare is temporary migration in which unmarried children and, less frequently, heads of family move to the city for periods ranging from months to several years (Arizpe 1975:225-6). Parts of households also may move seasonally, leaving someone behind to look after the land, while the others use their kinship and village contacts to find work in the city, the women often working as street vendors--the *Marias* described by Arizpe (1975, 1978)--and the men in construction.

In Toxi, proletarianization is so far advanced that residents have no alternative to seeking wage work, and the village has become in Arizpe's words "a proletarian suburb" (1982:42). The best source of investment for households is not land or craft or service enterprises, but biological reproduction. The main asset of Toxi households is the earning capacity of their members in the labor market, mainly through labor migration to Mexico City. An electrical appliance factory near the village also provides work for Toxi women, creating an incentive for households to invest in their daughters' education to the relatively low levels required by the factory. The earning capacity of household members is tapped sequentially, beginning with the young heads of household, then their adult sons and daughters, resulting in a pattern that Arizpe calls relay migration (Ibid.:Table 7). In contrast to the African case described by Wylie (1989), or that of Dotejiare, there is less need in Toxi for labor in the fields, so that, in effect, women specialize in rearing children for labor migration.

The increasing importance of wage earning to the ways in which a household copes with its environment means, in both the African and the Mexican case studies, a less sustainable pattern of farming, associated with the declining significance of community to welfare outcomes. Inequality, and its sources, become crucial variables in the negative consequences of urbanization for the rural environment. A similar conclusion is reached by Wood and Carvalho when they review the trends in inequality in the 1960s and 1970s in Brazil (1988). They establish that there are sharp differences in life expectation between income groups, between regions, and between races. Moreover, these differences in outcomes remain constant between 1970 and 1980, and there is an increase in some regional differences (Wood and Carvalho 1988:252). The differences in life expectation between income groups is based, they argue, on the particular vulnerability of the poor, both rural and urban, to market forces (Ibid.:115-125). They consider the macrostructural variables that produce inequality and lead to these differential outcomes. The major variable is the capitalist development of an undeveloped country, resulting in a rapid and socially disorganizing process of urbanization, the spatial concentration of wealth in some regions, such as the industrialized Center-South of Brazil. There is a consequent relative underdevelopment of others, such as the North-East of

Brazil, where the rural population exists at near starvation level, supplying migrants to the cities of the region, to those of the Center-South, and to the colonization projects of the Amazon region.

The Disorder of the Cities

Urbanization has resulted in widespread urban poverty and a disordered and polluted city environment. This outcome is not an inevitable result of population pressures, but has political dimensions related to the specific ways in which the urban populations of developing countries cope with their environment. Many of the economic opportunities of cities in developing countries, particularly in small-scale enterprises and self-employment, are based on the spatial heterogeneity of the city that enabled, for example, the food vendor, street trader, or domestic servant to live close to clients. The absence of effective zoning regulation's permitted workshops in residential areas, either in their own building or in part of a dwelling. Even large-scale manufacturing enterprises contributed to what Kowarick, in discussing Sao Paulo, called the "logic of disorder" of the Latin American city (1977).

Located on the outskirts of the city, distant from existing working-class areas, these industries stimulated the irregular settlement of nearby areas. Land speculators often encouraged squatters to occupy land on the expectation that government would subsequently provide infrastructure. In the face of a speculative and poorly developed land market, and the often desperate attempts of both the working and the middle classes to find suitable accommodation, overall urban planning fails (Ward 1990; Gilbert and Ward 1985). One important consequence is that the ecology of most Latin American cities is heterogeneous and the uses of urban space constantly contested by different social groups and economic interests, as Gilbert and Ward (1985) show for Colombian and Mexican cities, Valladares (1989) for Rio de Janeiro, and Rolnik (1989) for Brazilian cities in general. Residential neighborhoods, with housing of different standards, coexist side by side, and there is only a partial emergence of neighborhoods clearly differentiated around social class lines.

To the extent that zoning is absent, the mix of uses in the same local space is likely to increase the exposure of the population to industrial pollution and make waste and contamination a local environmental hazard. Food vending in this environment, whether done formally or informally, is likely to constitute a health hazard. Note, also, as Schteingart (1989b) points out for Mexico City, that this mix of uses is likely to be most pronounced for low-income populations, whose housing is located in and around industrial areas, exposing these populations to the highest concentrations of pollution.

In these ways, the built environment is constructed by the interplay of state, land speculator, construction and real estate companies, and low-income populations searching for cheap housing. Schteingart (1989b) documents the process whereby Mexico City's largest slum, Ciudad Nezahualcoyotl, was put together by a combination of illegal subdivision by landowners, by illicit deals with the *ejiditarios* (small-scale farmers whose land is social property and intended for agricultural use) and, to a lesser extent, by invasion. The various administrations of Mexico City connived in these illegal processes. For them, as for other city administrations in developing countries, informal settlement was both problem and solution at the same time (Mangin 1967). Government priorities were heavily oriented to economic infrastructure as a means of attracting and encouraging industry, and the provision of cheap housing had a much lower priority (see also Gilbert and Ward 1985; Ward 1990).

Schteingart shows that the provision of public housing also involves a heterogeneous and uncoordinated group of actors (1989b). In Mexico City, public housing provision has mainly occurred indirectly through subsidized loans from state agencies. The cost of this housing in Mexico City was such that it had to be rented or sold only to those who earned more than twice the minimum salary--thus excluding the urban poor. The construction of this subsidized housing was undertaken by private consortia on government-owned land that was distributed unevenly throughout the city. Its marketing was also administered by private development companies.

The logic of disorder means that government planning for the urban environment is, at best, ineffectual and, at worst, hypocritical. Though in the 1980s the government of Mexico City moved to decentralize administration and advocated local participation, in reality the only effective participation came from middle-class neighborhoods, mainly intent on protecting their space against encroachment by low-income populations (Aguilar 1988). Keith Pezzoli looks at the development of the Ajusco area to the south of the city (1991). This area is a designated green belt, containing some of the city's last remaining forest cover. Over the years, however, it has been subject to the types of piecemeal occupation that were described earlier: the semilegal sale of *ejidal* land to developers for commercial, industrial, and residential property; invasion by squatters; and appropriation by powerful private and public agencies. The planning authorities attempted to clear some of the irregular settlements, and Pezzoli reports the organizational efforts of residents to counter this threat (1991). In one of these low-income neighborhoods, residents, with the help of outside assessors, developed their own counterproposals for protecting the environment based on a novel system of waste disposal. Their case was made stronger by the city government's practice of using the green belt for ecologically harmful waste disposal practices. However, in the end, the proposals came to nothing due to the expense involved and to the difficulty of sustaining local organization. Since the government came to a compromise with the residents, the movement finally ended, leaving the situation where it had begun.

The Nature of Urban Poverty

In reviewing the data on inequality in Brazil, Wood and Carvalho argue that poverty is not based on passivity, and that poor households actively seek to manage their environment, individually and collectively (1988:126-134). The strategies available to the poor, however, are subsistence strategies, and rather than overcoming inequality they often reproduce it. The poor are paid less for their work than is needed for an adequate family subsistence, and so additional members of the household seek employment, whether these be female heads of household or children. The additional incomes ensure survival, not social mobility, and by taking children out of education or diminishing care for the young, they can perpetuate disadvantage from one generation to another.

Their analysis enables us to extend the analysis of the environmental system that emerges to cope with urban life in Latin America. This system is segmented by class, and the relationships in each segment have a very different capacity to ensure subsistence. One feature, then, of the new environmental system is that class, not community, becomes the major determinant of welfare, with the state providing what safety net exists against deprivation.

In this situation, the household and the household life cycle acquire particular salience in enabling people to cope with the urban environment in most developing countries. In Latin American cities, and in those of other developing areas, the caring and coping capacity of the household--looking after the old or young, pooling resources and helping each other out--is especially important. This creates contradictory pressures, making family cohesion vital while often placing an unequal burden on different family members. Various case studies agree that this burden falls heavily on women, in particular, who are not only the main carers, responsible for domestic chores, but, increasingly, essential contributors to the wage economy of the household. The degree of consensus and of equity in the distribution of household tasks is a crucial variable, as Jelin (1984) argues, in understanding the welfare consequences of household coping strategies between, for example, male-dominated and joint means of allocating household resources. The rates of female labor force participation have increased, particularly in the last decade, and most markedly amongst married women with children. Domestic violence directed against women is one result of the tensions that arise in this situation (Oliveira and García 1991; González de la Rocha 1986).

Two studies in Mexico, one based on samples from the city of Oaxaca and another based on a panel study of poor households in Guadalajara, report the importance of multiple earners to household subsistence (González de la Rocha 1988, 1991; Selby, Murphy and Lorenzen 1990). They show how, in the face of the adverse economic conditions of the Mexican "crisis" of 1982 onwards, households placed more members on the labor market. The increase in the labor supply came from three sources: an increase in female labor force participation; the delayed departure of older children from the family home; and other kin joining the nuclear family. In neither study did households see it as a meaningful option to increase income through the male head of household working longer hours or for more pay. For the Guadalajara sample, Mercedes González de la Rocha (1991) compared the changes between 1982 and 1987 in the income and consumption patterns of households at different stages of their cycle and for households of different sizes. Large households with multiple workers maintained their consumption levels, while smaller households lost ground. The overall result, however, is that by 1987 "working-class households . . . have overall levels of food consumption below the minimum required for light work" (Ibid.:123).

Families benefit from having fewer children, thus not only reducing the mouths to feed, but shortening the period before the ratio of earners to dependents rises sharply. Yet smaller families mean fewer children to care for parents in old age, and since real incomes have stagnated, the drain on resources of young married couples to support parents is likely to be considerable. In the absence of adequate systems of public welfare, the class differentiation in outcomes resulting from the new environmental system is likely to become sharper. Most working-class households are likely to experience poverty at one stage or another of the household cycle. Female-headed households with young dependent children are particularly vulnerable since not only is the mother both earner and carer, but may not have the time to enter into exchange relations with others in the neighborhood or with kin.

The differences in household employment patterns that can be detected in Latin America reflect the changing pattern of industrialization noted earlier. There are major differences in labor markets between "inward-looking" cities of the import-substitution phase of industrialization (such as Guadalajara in Mexico and Belo Horizonte in Brazil) and the cities now based on export-industrialization (such as Manaus in Brazil and the Mexican border cities). In both Guadalajara and Belo Horizonte, women, particularly young unmarried women, are less likely than men to be employed in formal enterprises. The opportunities for women are mainly in local commerce or services, including domestic service. In Manaus and in the Mexican border cities, in contrast, the young, men or women, are more likely to be employed in formal enterprises than their parents, and there is little difference between men and women in the level of autonomous work. The new export-oriented industries of Latin America usually recruit young unmarried females and young males, but at low levels of pay relative to the skilled workers of other industries or even to what can be earned through self-employment.

The Caring Capacity of the Community

What, then, of the broader caring capacity of the community? Can neighborhood relations or those with kin and friends alleviate the difficulties facing low-income households? Studies have shown the importance of extra-household relations to the household economy of the Latin American urban poor. Some studies have emphasized the economic importance of the exchange relationships between neighbors and kin, such as Lomnitz's study of a poor neighborhood in Mexico City (1977). In González de la Rocha's study of poor families in Guatemala, there are detailed accounts of how particularly vulnerable households, such as single females and their children, draw upon kin relations to care for the children, and, at times, for economic assistance (1986). Similar accounts are provided by Chant in her comparison of the ways low-income families cope in two Mexican cities, Puerto Vallarta and Querétaro (1991). Both Chant (1985) and González de la Rocha (1986) emphasize that for many women it is preferable to live alone with their children rather than with an abusive husband, but that they need kin and neighborhood relations to do so.

Most studies have agreed that households are not atomized by the chaos or competition of the cities. The low-income population, whether migrant or native, has been relatively successful in using wider networks to cope with the urban environment.

In an early article, "Urbanization Without Breakdown," Lewis argued for the organizing capacity of rural migrants in the city who used traditional institutions and kinship and village networks to cope with life in the large city (1952). Lomnitz (1977) and Altamirano (1984) show that in some cities, neighborhoods were formed of people tied by kinship, common origin, and, at times, common indigenous language. A related issue is whether the economic and social relations in which households are embedded facilitate community cooperation in managing the urban environment. Some of the first case studies in Latin America showed considerable amounts of neighborhood organization from below (Mangin 1967; Peattie 1968; Leeds 1969). Several variables were important in explaining this degree of activity (Leeds 1969, 1974). First was the illegal nature of the land occupation that gave settlers a common interest in defending their neighborhood and acting together to get legal services. The second was the political conjuncture, specifically whether the regime in power was seeking popular support, as happened with various "populist" governments in Peru and Mexico. The third was the size and complexity of the city's spatial organization, whether, for example, neighborhoods were a mix of tenure and housing types or not.

The first of these factors can be illustrated by my study of community organization in Guatemala City in the mid-1960s (Roberts 1973). One of the communities, San Lorenzo, was a squatter settlement in which the invasion of land had been preceded by meetings of the future squatters. Once invaded, the land was rapidly urbanized by a committee of squatters who allocated space, organized efforts to install drainage and sewage systems, and lobbied to obtain water and other utilities from the authorities. The numbering of houses and the constructing of churches and other public buildings also occurred over a period of five or six years. In addition, there was a gradual process of upgrading housing by replacing cardboard with wood, wood with concrete, adding a second story, and so on. This type of urbanization from below was frequently reported in the studies of the 1960s and 1970s. Leeds, for instance, showed the considerable improvements that *favela* inhabitants had made to their original shacks (1969).

While San Lorenzo illustrates the degree of community organization possible in the squatter settlements of cities of developing countries, a more typical type of urban neighborhood is described by Schmink's study of lower Barreiro in Belo Horizonte (1979). Lower Barreiro was already an established locality when it was formally drawn into Belo Horizonte's orbit in the 1950s to serve as a satellite city for industry and population. The neighborhood has its own municipal services, and its urban development has been a mix of government-subsidized housing, private building, and illegal squatting. By the time of the study, five different sociogeographical zones could be identified within Barreiro that were unequally endowed in services and in quality of housing. Since the five zones are in easy walking distance of each other, Barreiro illustrates dramatically the extent of spatial heterogeneity that occurs in many Latin American cities. This heterogeneity caters both for households in different stages of the domestic cycle and for different income strata of the population. There is consequently considerable mobility both within Barreiro and between Barreiro and other parts of Belo Horizonte.

Schmink differentiates Barreiro households into three groups: stable households, households; moving within Barreiro; and households moving into and out of the neighborhood (1979:269-277). Each group has distinctive characteristics based on a combination of household cycle and occupational statuses. The stable households, for example, tend to have multiple earners and a mix of industrial and autonomous occupations. They also own their own homes, and for them the area allows them to combine effectively the types of income opportunity available to both male and female household members. The second group is households in the second phase of the household cycle, when young children are present. Their internal mobility, argues Schmink (1979:274), is an ongoing search for cheap housing, which since it needs to be near their jobs will still be located in Barreiro. Those moving into and out of Barreiro are different from the other two groups in that they are mainly young households, either single or couples without children, who are formally employed in nearby enterprises. Those leaving tend to be higher-income earners than those entering, since Barreiro's rental accommodation offers a relatively inexpensive starting point for formal workers. When they are successful economically, they can afford to move out to more spacious accommodations elsewhere in the city, even at a distance from their jobs.

More recently, Despres shows how the divergent housing strategies of the social groups present in another Brazilian city, Manaus, contribute to heterogeneity (1991:191-238). There is considerable mobility and little sense of neighborhood cohesion, and households are mainly dependent on their own resources without possessing extended kinship or friendship networks. Despres found clusters of kin-related households in one of the *favelas*, but neighbors who were not kin were strangers (Ibid.:214). In Manaus, small, mainly Protestant religious groupings--and not the parish church--provide one of the main sources of extra-household social integration for residents. In general, Despres reports low levels of interest in community action to improve the environment. Only in one neighborhood was there a high level of formal neighborhood organization, with an active neighborhood association and social center. Residents in this neighborhood were longer established than in the other neighborhoods studied by Despres. It originated in the resettlement of families from the old port area of the city, and though provided with public housing, the settlers had to organize to obtain an adequate infrastructure. Even here the general level of participation in corporate community organization is low.

The nature of the urban neighborhood in Latin America poses particular problems for popular participation in the management of the environment. There are often high levels of commitment to improving the neighborhood and its infrastructure, and there is no shortage of local leadership either in low-income or high-income neighborhoods. The neighborhood movements to secure improvements tend, however, to be intense but short-lived. This was the case in San Lorenzo, Guatemala City. They were short-lived because they were easily co-opted by external authorities ready to grant isolated favors in return for political support. Above all, the heterogeneity and mobility of residents made organization difficult to sustain.

Conclusion

Urbanization is transforming the environment of developing countries more rapidly than it did in the developed world during the nineteenth century. By the year 2000, 45 percent of the population of developing countries will be urban, though regional differences in urbanization remain important, and seven of the world's ten largest cities will be in developing countries (United Nations 1991:Table 2). The environmental consequences of this urbanization are hard to forecast because they depend on policy decisions and on the pace of economic development. In the 1980s, the pace of economic growth in some developing regions, notably Latin America and Africa, slowed considerably, and together with a declining rate of population growth led to a slowing down of urban growth. Despite this slowing down, the urban economies of these regions have not been able to absorb productively the increase in the urban labor supply, resulting in high rates of urban employment and underemployment. Under these conditions, the growth of slums has continued. Also, shortage of capital and dependence on external finance have not allowed developing countries the "luxury" of an environmentally sensitive economic growth. Factories operate without strict control on their waste disposal or pollution. Indeed one incentive for foreign investment in developing countries is less strict environmental standards. Under these conditions, urban growth generally entails a worsening physical environment as factories and automobiles pollute, while populations crowd into cities that do not have an adequate infrastructure to support them.

The unevenness of urban development is an inherent part of a pattern of economic growth based on the integration of developing countries into the world economy through industrialization and through those sectors of their rural economy that can earn export earnings. This pattern of growth has reduced poverty, but has left little room for countries to experiment with alternative spatial patterns of urbanization, other than that of urban concentration. Reduction in tariff protection with the current worldwide move to free trade gives developing countries even less breathing space than they had before. They must find the niches in the world economy, either in manufacturing exports or in agricultural and extractive ones, that enable them to compete and survive. In some countries, these niches have led to changes in the pattern of urbanization, resulting, for example, in rapid urban growth in specially designated export zones, but without making urban growth more even.

Policy is more likely to be effective in mitigating the negative environmental consequences of urbanization if governments accept the trends in urbanization--whether toward concentration or dispersal, depending on the economic conjuncture. Accepting the trends entails working with them by anticipating transport needs or residential expansion, improving infrastructure, and avoiding pricing policies that aggravate the problems of the small-scale farmer. Anti-urban rhetoric is likely to be counterproductive, since it will not change the trends and may result in delaying those policies that can alleviate, to some extent, the negative consequences of rapid urban growth.

Consider the case of Mexico, where government long refused to face up to the issue of Mexico City's growth, basing its policy on a formal rationality of anti-urban concentration and pro-regional development and decentralization (Plan Nacional 1982:1). The substantive rationality that emerged from ad hoc decisions and from permitting the play of market forces was, in contrast, a systematic urban bias in the allocation of public funds and in tariff and pricing policies. Though there has been a major effort to regulate Mexico City and to decentralize administration and services, this effort has had only limited success in face of an entrenched centralism and powerful vested interests (Ward 1990). There are signs that the Mexican population has de-concentrated since the 1970s, along with manufacturing enterprises. This is not, however, the result of the above plans, but of Mexico's macro economic policies, particularly tariff reduction and the encouragement of the maquiladora industrial sector on the northern border. The diseconomies of agglomeration, including pollution, have also played a part in discouraging investment in, and migration to, Mexico City, and in encouraging movement out of the city. Part of that migration has gone to places that are even less well endowed with infrastructure than is Mexico City--the cities of the northern border--causing a further deterioration in housing and utilities.

Anxiety about large, overcrowded, and politically unstable cities has led, at times, to a misplaced concern with overurbanization in developing countries. It is misplaced partly because rural poverty remains a more severe problem than does urban poverty even in areas with high urbanization, such as Latin America (World Bank 1990:29). Given the high rates of population growth, cities, if properly planned, are likely to be more effective in absorbing population productively than are rural areas (Currie 1988:113-126, 145-177). The poor physical environment resulting from urbanization is likely to improve, albeit slowly, if economic growth continues as infrastructure is extended in both urban and rural areas. Tackling the environmental problems associated with urbanization requires us to go beyond a preoccupation with excessive city size. The more serious issue for the future is likely to be the deterioration in the social environment brought by urbanization.

This deterioration consists in a reduction in the caring capacity of the community in both rural and urban areas. The result is to isolate particularly vulnerable groups, such as the aged, female-headed households, and the abandoned young, as population mobility and the disorder of the cities deprive them of wider support relationships. While absolute levels of poverty may have been reduced, the poor continue to bear the brunt of urbanization. They are isolated spatially in urban and rural ghettos, and survive precariously without the security provided by stable jobs or strong community networks. The precarious position of the poor aggravates negative environmental conditions, whether poor housing, contaminated water supply, or industrial pollution, because it inhibits longer-term planning on their part, thus binding them to these conditions. It also contributes to worsening the environment, multiplying informal uses of urban space, and perpetuating unsafe and unhealthy domestic and work activities.

The studies reviewed earlier suggest that there are countertendencies that bode better for the future. The household remains a reasonably effective unit for coping with the strains of urban life, most urban dwellers can call on relatively wide support networks, and people are relatively successful in organizing, though temporarily, to improve their physical environment. But there are severe strains and tensions attending the various survival strategies. These must be addressed by public policy that seeks to build upon grass-roots energies to enhance the caring capacity of the community. One step forward is to recognize that certain groups are made vulnerable by urbanization and to provide help whether by social security transfers or special facilities. Much of this help can be delivered with the participation of local groups. To encourage such participation, local associations of various kinds need to be strengthened through administrative decentralization and being given an effective share in decision making. Ultimately, the care of the environment depends on the actions and concerns of local groups. If these concerns are not harnessed, there is little prospect that public policy alone will prevent the negative impact of urbanization on the environment, particularly since market forces in developing countries draw no profit from environmental sensitivity.

References

Abu-Lughod, Janet (1971). *Cairo: 1001 Years of the City Victorious*. Princeton, N.J.: Princeton University Press.

Abu-Lughod, Janet (1980). *Rabat: Urban Apartheid in Morocco*. Princeton: Princeton University Press.

Aguilar Martínez, G. (1988). "Community participation in Mexico City: a case study." *Bulletin of Latin American Research*, 7,1: 22-46.

Altamirano, T., (1984). *Presencia andina en Lima Metropolitana*. Lima: Fondo Editorial, Pontificia Universidad Católica.

Arizpe, Lourdes (1978). *Migración, etnicismo y cambio económico*. México, D.F.: El Colegio de México.

Arizpe, Lourdes (1982). "Relay migration and the survival of the peasant household." In Helen Safa (ed.), *Towards a Political Economy of Urbanization in Third World Countries*. Delhi: Oxford University Press.

Arizpe, Lourdes (1985). *Campesinado y migración*. México, D.F.: Secretaría de Educación Pública.

Armstrong, William. & Terrence G. McGee (1985). *Theatres of Accumulation*. London & New York: Methuen.

Benería, Lourdes (1991). "Structural adjustment, the labour market and the household: the case of Mexico." In G. Standing and V. Tokman (eds.), *Towards Social Adjustment*. Geneva: International Labour Office, 161-184.

Bradshaw, York W. (1987). "Urbanization and underdevelopment: a global study of modernization, urban bias, and economic dependency." *American Sociological Review*, 52: 224-239.

Bromley, R. and Gerry,C. (eds). (1979). *Casual Work and Poverty in Third World Cities*. Chichester, U.K.: John Wiley.

Butterworth, Douglas and John K. Chance (1981). *Latin American Urbanization*. Cambridge: Cambridge University Press.

Centre for Science and Environment (1989). "The environmental problems associated with India's major cities." *Environment and Urbanization*. I,1:7-15.

Chant, Sylvia (1985). "Family formation and female roles in Queretaro, Mexico." *Bulletin of Latin American Research*. 4:17-32.

Chant, Sylvia (1991). *Woman and Survival in Mexican Cities*. Manchester: Manchester University Press.

Corona, R. (1991). "Confiabilidad de los resultados preliminares del XI Censo General de Población y Vivienda de 1990." *Estudios Demográficos y Urbanos*. El Colegio de México. 16:33-68.

Cortés, Fernando and Rubalcava, Rosa Maria (1991). *Autoexplotación forzada y equidad por empobrecimiento*. México, D.F.: El Colegio de México.

Currie, L. B. (1988). *Urbanización y desarrollo* Bogotá: Cámara Colombiana de la Construcción.

Despres, Leo A. (1991). *Manaus: Social Life and Work in Brazil's Free Trade Zone*. Albany: State University of New York Press.

Deyo, F.C. (1986). "Industrialization and the structuring of asian labor movements: the 'gang of four'." In M. Hanagan and C. Stephenson (eds.), *Confrontation, Class Consciousness and the Labor Process*. New York: Greenwood Press, 167-198.

Figueroa, A. (1984). *Capitalist Development and the Peasant Economy in Peru*. Cambridge: Cambridge University Press.

Garza, G. and Departamento del Distrito Federal (1987). *Atlas de la Ciudad de México*. México, D.F.: Departamento del Distrito Federal/El Colegio de México.

Gershuny, J.I and Ian Miles (1983). *The New Service Economy*. New York: Praeger.

Gilbert, Alan and Ward, Peter (1985). *Housing, the State and the Poor: Policy and Practice in Three Latin American Cities*. Cambridge: Cambridge University Press.

González de la Rocha, Mercedes (1986). *Los recursos de la pobreza: damilias de Bajos ingresos de guadalajara*. Guadalajara, México: CIESAS, El Colegio de Jalisco.

González de la Rocha, Mercedes (1988). "Economic crisis, domestic reorganization and women's work in Guadalajara, Mexico." *Bulletin of Latin American Research*, 7,2:207-223.

González de la Rocha, Mercedes, (1991) . "Family well-being, food consumption and survival strategies during Mexico's economic crisis." In M. Gonzalez and A. Escobar (eds.), *Social Responses to Mexico's Economic Crisis of the 1980*. San Diego, CA.:Center of U.S.-Mexican Studies, 115-127.

Gugler, Josef and William G. Flanagan (1978). *Urbanization and Social Change in West Africa*. Cambridge: Cambridge University Press.

Hardoy, J and D. Satterthwaite (1985). "Third World Cites--the Environment of Poverty." In *World Resources Institute Journal*. 45-56.

Hart, Keith (1987). "Rural-urban migration in West Africa." In J. Eades (ed.), *Migrants, Workers, and the Social Order*. London: Tavistock, 63-81.

Hoskins, W.G. (1970). *The Making of the English Landscape*. Harmondsworth, U.K.: Penguin Books.

Instituto Nacional de Estadística, Geografía e Informática (Inegi) (1990). *Resultados preliminares, XI Censo General de Población y Vivienda, 1990*. Aguascalientes: INEGI.

Jelín, E., (1984). "Familia y unidad doméstica: Mundo público y vida privada." *Estudios Cedes*, Buenos Aires.

Jiménez, Rosano and Sister Aida Velázquez, (1989). "Metropolitan Manila: a framework for its sustained development." *Environment and Urbanization*. 1, 1:51-58.

Johnson, E.A.J. (1970). *The Organization of Space in Developing Countries*. Cambridge, MA: Harvard University Press.

King, Anthony D. (1990). *Urbanism, Colonialism, and the World Economy*. London: Routledge.

Keyfitz, N. (1991). "Population and development within the ecosphere: one view of the literature." *Population Index*. 57,1: 5-22.

Kowarick, L. (1977). "The logic of disorder: Capitalist expansion in the metropolitan area of Greater Sao Paulo." Discussion Paper, Institute of Development Studies at the University of Sussex, United Kingdom.

Lampard, Eric E. (1965). "Historical aspects of urbanization." In P.M Hauser and L. Schnore (eds.), *The Study of Urbanization*. 519-554.

Landes, David (1970). *The Unbound Prometheus*. Cambridge: Cambridge University Press.

Lawton, Richard (1978). *The Census and Social Structure: an Interpretative Guide to Nineteenth Century Censuses for England and Wales*. London: Frank Cass.

Leeds, A. (1969). "The significant variables determining the character of squatter settlements." *América Latina*. 12, 3: 44-86.

Leeds, A. (1974). "Housing-settlement types, arrangements for living, proletarianization, and the social structure of the city." In W. Cornelius and F. Trueblood (eds.), *Latin American Urban Research 4* Beverly Hills and London: Sage.

Lewis, Oscar, (1952). "Urbanization without Breakdown: a Case Study." *Scientific Monthly*, 75,1 (July).

Lomnitz, Larissa (1977) Networks and Marginality: Life in a Mexican Shantytown. New York: Academic Press.

Long, N. and B. Roberts (eds.) (1978). *Peasant Cooperation and Capitalist Expansion in Central Peru*. Austin and London: University of Texas Press.

Lowry, Ira (1991). "World Urbanization in Perspective." In K. Davis and M. Bernstam (eds.), *Resources, Environment, and Population*, pp148-178. New York: Oxford University Press.

Mangin, William   (1967). "Latin American squatter settlements: a problem and a solution." *Latin American Research Review*. 2,3:75.

Mitchell, B. R. (1976).   *European Historical Statistics, 1750-1970*. New York: Columbia University Press.

Mitchell, B. R. (1983) *International Historical Statistics: the Americas and Australasia*. Detroit: Gale Research Co.

Morse, R. (1962).   "Latin American cities: aspects of function and structure." *Comparative Studies in Society and History*. 4.

Nemeth, Roger J. & David A. Smith (1985).   "The political economy of contrasting urban hierarchies in South Korea and the Phillipines." In M. Timberlake (ed.), *Urbanization in the World-Economy*. Orlando: Academic Press, 183-206.

Oberai, A.S. (1991). "Urban population growth, employment and poverty in developing countries." In United Nations, *Consequences of Rapid Population Growth in Developing Countries*. New York: Taylor and Francis, 191-218.

Oliveira, Orlandina y Brígida García "Jefes de hogar y violencia doméstica." Paper presented at the Congreso de Población y Salud Mental, México, 18 al 23 de agosto, 1991.

Oliveira, Orlandina de and Bryan R. Roberts (1989). "Los antecedentes de la crisis urbana: urbanización y transformación ocupacional en América Latina: 1940-1980." In M. Lombardi y D. Veiga (comps.), *Crisis Urbana en el Cono Sur*. Montevideo: CIESU.

Peattie, Lisa R. (1968). *The View from the Barrio*. Ann Arbor: The University of Michigan Press.

Pezzoli, Keith (1991). "Environmental conflicts in the urban milieu: the case of Mexico City."In M. Goodman and M. Redclift, *Environment and Development in Latin America: the Politics of Sustainability*. Manchester and New York: Manchester University Press, 205-229.

Portes, Alejandro and John Walton (1981). *Labor, Class and the International System*. New York: Academic Press.

Raza, M. et al. (1981). "India: urbanization and national development," In M. Honjo (ed.), *Urbanization and Regional Development*. Nagoya, Japan: Maruzen Asia, 71-96.

Roberts, Bryan (1973). *Organizing Strangers: Poor Families in Guatemala City*. Austin: University of Texas Press.

Roberts, Bryan (1978). *Cities of Peasants*. London: Edward Arnold.

Roberts, Bryan (1989). "Urbanization, Migration and Development." *Sociological Forum*. 4,4: 665-691.

Rodgers, Gerry (ed.) (1989). *Urban Poverty and the Labour Market: Access to Jobs and Incomes in Asian and Latin American Cities*. Geneva: International Labour Office.

Rolnik, Raquel (1989). "El Brasil Urbano de los Años 80: un retrato." in M. Lombardi and D. Veiga (comps), *Las Ciudades en Conflicto*. Montevideo: Ediciones de la Banda Oriental/Cieu, 174-194.

Schmink, Marianne, (1979). "Community in Ascendance: Urban industrial growth and household income strategies in Belo Horizonte, Brazil." Ph. D. Dissertation, University of Texas at Austin.

Schteingart, Marta (1989a). "The environmental problems associated with urban development in Mexico City." *Environment and Urbanization*. 1, 1:40-50.

Schteingart, Martha (1989b). *Los productores del espacio habitable: estado, empresa y sociedad en la Ciudad de México*. México, D.F. El Colegio de México.

Selby, Henry, Murphy, Arthur D. and Stephen A. Lorenzen (1990). *The Mexican Urban Family: Organizing for Self-Defense*. Austin: University of Texas Press.

Skinner, George W.(ed.) (1977). *The City in Late Imperial China*. Stanford: Stanford University Press.

Standing, Guy and Tokman, Victor (eds.) (1991). *Towards Social Adjustment*. Geneva: International Labour Office.

United Nations (1987). *The Prospects of World Urbanization*. New York: United Nations.

United Nations (1991). *World Urbanization Prospects 1990*. New York: United Nations

United Nations (1992). *Demographic Yearbook 1990*. New York: United Nations.

U.S. Bureau of the Census (1975). *Historical Statistics of the US. Colonial Times to 1970*, Part One. Washington, D.C.: Bureau of the Census.

Valladares, Licia (1989). "Rio de Janeiro: la visión de los estudiosos de lo urbano." in M. Lombardi and D. Veiga (comps), *Las ciudades en conflicto*. Montevideo: Ediciones de la Banda Oriental/Cieu, 195-222.

Wallerstein, Immanuel M. (1976). *The Modern World-System: Capitalist Agriculture and the Origins of the European World Economy in the Sixteenth Century*. New York: Academic Press.

Ward, Peter (1990). *Mexico City: the Production and Reproduction of an Urban Environment*. London: Belhaven Press.

Wood, Charles and J. A. M. de Carvalho (1988). *The Demography of Inequality in Brazil*. Cambridge: Cambridge University Press.

World Bank (1990). *World Development Report 1990* New York: Oxford University Press.

Wylie, Diana (1989). "The changing face of hunger in Southern African History, 1880-1980." *Past and Present*, 122: 159-199.

PART THREE

Population and Environment: Conclusions

11

Conclusions: Rethinking the Population-Environment Debate

This concluding chapter summarizes some of the essential perspectives on the population-environment debate that arise from the studies reported here and suggests research directions that flow from them.

Terms of the Debate

The most important lesson of the work reported here is that the terms and models of the debate on population and the environment have been oversimplified and decontextualized; they need to be recast to better reflect the complexities of the real world. Even the terms "population" and "environment" themselves must be broken down into their constituent elements in order to attempt the construction of a more complex matrix.

The more complex definitions of population and environment, and the study of their interaction, that are needed depend on two basic observations. The first is that population, given its fundamental biological dimension, is embedded in social relations. Demographic transitions, for example, are part of a more generalized set of socioeconomic transitions and must be understood in that context. The second assumption is that there will be no simple, one-to-one correspondence between the population as defined above and the physical space people occupy. Thus, through such processes as migration, urbanization, and trade, population growth may be less important for the study of environmental degradation than such processes as extraction, consumption and waste generation. It is important, however, not to include all economic and social problems, such as general problems of urban poverty or rural under-development, into a definition of "environmental degradation."

Population must not be reified as if the simple numbers of human bodies were all that mattered. From the standpoint of population/environment analysis, people are significant in terms of what these humans do in a matrix of social and environmental interactions. This means that population must be understood as a process of biological and social reproduction whose components are principally interactions. Research must focus on these interactions as they relate to the three determinants of population--fertility, mortality and migration.

In analyzing the characteristics of populations, the most important aspects to understand in this particular debate include: the inertial growth of populations due to continued growth in cohorts of young women; the aging of populations prevalent now in industrialized countries but which may also characterize developing countries in the relatively near future; the health and medical factors which influence offspring survival; women's education and labor participation; and cultural perceptions of the value of children and their role in ensuring old age security for their parents.

Studies focusing on the interaction between migration and environment would seem especially timely and relevant in this regard. Migration, for example, may work to the advantage of both sending and receiving societies when labor migrates to places where capital investments are creating employment. It may also, however, work to the special disadvantage of the poor through such processes as dispossession or urban crowding in conditions of homelessness and unemployment. The gender dimension of migration, the nature and scope of international migration and the mobility of people in such frontier areas as Amazonia must be given research priority.

The view of the environment as a set of objects out there in nature to be handled--or mishandled--by social actors is also an oversimplification. Defining environment as anything around us, however, or as a synonym for context, as in "social environment", becomes too inclusive to be of use. To narrow down the analytical field, it is helpful to begin by defining the object of study as "environmental transformations." Some environmental transformations can be positive for human welfare. Land management techniques that deter land erosion and provide higher agricultural yields may be, at least in the short term, beneficial to both humans and the environment. In empirical situations, it is often difficult to determine exactly at what point environmental change becomes environmental degradation. Heuristically, environmental degradation, or negative environmental outcomes, must be analyzed in terms of the effects on sustainability of such changes. The challenge becomes to define the cutting points of sustainability in different natural settings and for different groups of social actors.

Central Questions

Rethinking the population-environment debate means more than redefining the terms. It also means recasting some of the basic questions. Instead of asking whether population is the principal cause of environmental degradation, a more insightful question would be: How are the factors that negatively affect the quality of life of people and the sustainability of the natural environment dynamically interrelated? Framed in this way, research can be focussed on how populations use natural resources and deposit certain kinds of pollutants or waste in the landscape.

The starting point for research would be to determine the use of local resources by local people. This should include local perceptions of environmental degradation, through what may be called the "user's perspective." Indeed, there is beginning to be a substantial body of work under the labels of community-based development and participatory development that is contributing to this starting point.

The next step would be to proceed to an understanding of how other, non-local, people use those resources given national and international exchanges. This means studying how the access and use of natural resources by local users is related to, and affected by, the direct or indirect uses of those resources made by others at other levels of society. A variety of models could then be built to incorporate mediating mechanisms--institutions, codes, practices--linking the patterns of resource use by specific resource users (who may be individuals, families, communities, corporations or even states in the case of the creation of natural reserves or public sector corporations) to global trends. These models could incorporate different levels of social organization: micro (local), meso, macro and global. Thresholds of sustainability of different resources, and the ecosystems in which they are embedded, would be linked to these multi-level consumption patterns and to their cumulative and systemic effects.

The mediating mechanisms include, but are not necessarily limited to, the following: a) rural and urban land tenure; b) rural and urban resource distribution; c) agricultural intensification and kinds of inputs used; d) legal-institutional frameworks; e) use of technologies; f) labor relations, gender and other divisions of labor; g) *de jure* and *de facto* political relations; h) market mechanisms in exchange systems; and i) cultural knowledge, values and practices, including social perceptions of environmental transformations.

These are not discrete entities, boxes in a chart to be filled in, but components of sets of social relations that may be cooperative or conflicting and that are resolved through negotiation. Thus "labor relations" describes a relation between employers and employees, and "markets" describes relations among producers, traders and consumers. Gender is important to all of the mechanisms--structuring many of the relations just described in significant ways. Few existing studies analyze these mediating mechanisms in depth, especially using the longitudinal methods so crucial for understanding long-term environmental transformations.

Linking the processes generated by these mediating mechanisms to global phenomena certainly cannot be done in a mechanical way. The same macroeconomic forces may have different outcomes in different local settings. For example, the generalization that decay in rural livelihoods results from macroeconomic processes (including the fall in agricultural prices) may hold true as an overall pattern. But there may be different outcomes in different places and times due to local variations in the social and physical environment. Thus, it is critical to make a theoretical distinction between generalization of phenomena which are widespread geographically and globalization of phenomena which affect all the inhabitants of the planet. The first may be primarily a quantitative process that requires surveys and statistical methods. The second, in contrast, may be a more qualitative process that can only be understood by recasting the concepts, methods and frameworks with which the social sciences-- and in many cases the natural sciences as well--have dealt with such processes before. Examples of such processes include the irreversible loss of biodiversity or changes in the biochemical cycles or climate systems due to human action. The effects of some of these global processes may in fact bypass mediating mechanisms and have direct effects on local settings. For example, change in micro climates will occur no matter what land tenure, agricultural, political or social systems a region may have.

Competition for Natural Resources

A useful model for beginning to analyze these complex relations is a set of concentric circles with the relations of population and environment at the center, nestled in social relations as the first concentric circle, with macroeconomic forces the second circle and global factors the third. This emphasizes the embeddedness of the population-environment interactions. The social relations, however, cannot be modelled as if they are static but should rather be dealt with as the shifting strategies of social actors within boundaries (which may themselves shift) established by the mediating mechanisms.

Although the studies that should be encouraged would focus on the local situations, they must also bring in macroeconomic forces. Policy decisions made at a macro level, related to trade, financial and political relationships between nations, may have important effects on the local situation. These macroeconomic forces have a dynamic of their own and are affected by a variety of sectoral interests. This makes it difficult to establish common patterns of policy influence on population and environment since these may vary between regions, countries and local settings according to economic influence and cultural values. Despite these complications, understanding policy decisions relating to environmental phenomena at the local, national and international levels should be a research priority.

Since there are ways of using resources which are more or less damaging to people or to the natural environment, the gathering of data in a particular locality or on a particular pattern of resource use can effectively be organized by having as a central question: how can conflicting use of resources be negotiated to the benefit of all the social groups involved while ensuring the sustainability of the natural environment? In such negotiations, special attention must be paid to the differential impacts that changing access, use or behavior towards certain natural resources have on different social actors. For example, restrictions on the cutting of trees in the rainforest may mean a loss of profits for ranchers, who have other income-generating activities, while it may mean hunger for and threaten the survival of poor shifting cultivators. When trade-offs are considered in making policy decisions, these groups will require different levels of support. Local communities, indigenous groups or poor women may often be the best managers of local resources if they fully understand and agree that this management is a priority. Providing the necessary support for such actions may, in the end, be the most effective way of introducing sustainable use of resources.

To achieve a fuller understanding, however, empirical research must ask specific questions, such as: what is the economic centrality of the resource in question to the groups involved? Once data are available to answer this question, the systemic links to mediating mechanisms and to policy decisions made at different levels can be explored. Data on the various mediating mechanisms can then be tailored to this question. This permits links to global processes driving investment (capital restructuring at the global level, debt relations) and to environmental outcomes at the site of resource management; it orders the mediating mechanisms for the purpose of investigating the problems at hand and suggests the aspects of each which may be most relevant. Policy recommendations can then emerge which take into account the margins for change that different economic relations, including income situations, will allow in a given locality. Insuring that the analyses take into account inequality enhances the opportunities to promote fair, and also successful, policies.

The impact of population growth on such processes cannot be predicted *a priori* but needs to emerge empirically in each research situation. The relationships are often reciprocal and circular. Population growth, for example, may change *de facto* patterns of land distribution and livelihoods. These changes may, in turn, affect migration patterns. But while rural outmigration may be prompted by such economic processes, the size of migratory cohorts depends on rates of population growth, which, in turn, influence the distribution of lands. The participation of men and women in the labor market, and in trade and consumption, are also affected. To further complicate the scenario, demographic patterns of labor supply in different regions also have an impact on the kinds of jobs and labor relations that are created.

Conflict and Cooperation

Consumption of natural resources may increase with population growth, with development expectations or with artificially-created needs, leading to greater conflict over scarce resources. Such conflict may, in the coming decades, deepen, overlap or reshape local or regional social differentiation. In some cases, the poor or local resource users may pay more than their share of costs in adapting to new environmental regulations which result from decisions made at the macro level. Thus, given the urgency of survival needs of poor families, especially in rural areas, short term depletion of resources may result, and in some cases may be exacerbated by population growth. This suggests that the best way to ensure population stabilization and long-term sustainability of the global system is to promote broad redistributive policies and reforms.

An understanding of conflict resolution over resource use must take into account two processes. The first pertains to information and attitudes. A critical obstacle to the participation of local groups in environmental protection is the lack of reliable, accurate information. Fundamental attitudes on the use and management of natural systems will have to change. A long-term perspective on development must be built, accepted and implemented through effective policy and participatory action.

The second process is people's awareness and willingness to act on new strategies. This is affected by social perceptions of environmental change and by the relative bargaining position of different social actors in effecting these transformations. Perceptions and mechanisms for negotiation must be analyzed both at the local and regional levels.

Recovering and documenting valuable local knowledge related to agronomy, botany, traditional medicine, house construction and other topics should be a research priority, especially since this indigenous knowledge often pertains to ecosystems which have not been researched by established science. Very often, such knowledge is held by women of rural communities, especially information related to the pharmacopeia, traditional therapeutic practices, identification of plants and animals and their uses, domestic technologies, child care and horticulture.

Non-governmental organizations are increasingly playing an important role in raising awareness of and finding negotiated solutions to environmental problems. In some countries they are moving into the spaces left open by the retraction of government through privatization policies. In general, ecologists are increasingly influencing public opinion and pressuring for policy changes which will affect the access, distribution, and management of natural resources.

Gender and Resource Management

The gender dimension of environmental transformation is only beginning to be adequately investigated. Women are taking on the new tasks of gathering and processing of forest products, for example, as markets for these products are proposed as a curb on deforestation. Such new tasks, though, may add to women's overall burdens in situations in parts of Africa and Latin America where women are increasingly being pulled into the labor market. In many countries, adjustment policies have also increased women's work burdens through cuts in social services previously provided by the state. These relationships require more study, especially to examine how changes in the gender division of labor affect the use and management of natural resources.

In many communities, such as in India and in the poor suburban belts of Latin American cities, women are already providing leadership in conservation and resource management. This opens up a new field of participation and empowerment for women. Despite this, women are typically omitted from research on natural resource management and consumption. Resources are often discussed in terms of social production and not domestic production. Studies of domestic life, however, have shown that crucial changes take place when women have access to resources such as potable water. Much more research is needed on environment and resources for domestic production, such as fuelwood.

Another salient issue is the differential perception that groups of men and women may have regarding resource use which may, in turn, have an impact on consumption. As women increase their participation in the labor market they become direct consumers of marketable goods. Advertising may increasingly target women and children, which can change attitudes towards resources and waste production.

Research must also focus on the role of women in redefining spatial patterns of rural settlements. This is especially evident in rainforest regions with high immigration flows. In such frontier settings, women are more vulnerable to certain kinds of environmental stress such as lack of fuelwood, drinking water, sewage and waste disposal systems, although in the rainforests they appear to be less subject to work hazards than men. Some women seem to prefer urbanized settings since their domestic work burdens are reduced by reducing the time spent gathering fuel, water and food. In health terms, women seem to fare better in these newly urbanized settings then men although the burden of caring for sick or injured family members often falls primarily on women.

The gender dimension cannot be modeled simply as lying outside or parallel to other processes or characteristics of environmental change. It must be integrated into every question at each level of inquiry.

Conclusions and Thoughts on Future Research

The realization of sustainable societies requires an improved understanding of the complex forces which cause global environmental problems and hinder social and economic development. The impacts of population pressure, over-consumption, land degradation, deforestation, climate change, loss of biological diversity, industrialization, waste accumulation, and water and energy use on human society and the environment create major problems for which the expertise of the scientific community will be crucial in helping to clarify remedial policies and options for action.

It is still widely believed that much of the earth's environmental degradation --such as the widespread destruction of the rainforests--is due mainly to the South's material poverty and its population growth. In this sense, the developing countries have borne much of the burden and the blame for environmental degradation. Similarly the focus on population issues has been centered on the need to reduce population growth levels in the Third World. But this prevailing approach disguises serious causes of environmental damage; among these causes are many which directly affect women's health and livelihoods. In the meantime, problems of great urgency in the Third World (such as land distribution, education and the appropriate use of the environment) are not addressed as they should be, and the connections between these problems and a development model which emphasizes economic growth are ignored.

Development must be measured by the capacity of a society to provide sustainable livelihoods for all. Sustainable livelihoods become a primary basis of human well-being, population self-regulation and sustainable development. This concept of sustainable livelihoods links three major themes--environment, reproductive rights and population, and alternative economic frameworks--all of which directly affect the condition and position of women from the South.

To achieve this sustainable pattern of resource use and population growth, a much deeper understanding of the interactions of population and per capita resource consumption as mediated by technology, culture and values must be developed. It is important to differentiate between economic growth--an increase in the amount of production and consumption--and economic development--the enhancement of human well-being without a necessary increase in consumption. Instead of focusing primarily on continued economic growth, research needs to be focused on the linkages between population, consumption and distribution patterns. Particularly, the impact of population growth on environmental change must be understood to act along with other mediating factors, such as poverty, consumption levels, access to resource use, gender equity, and technology. The crux of the problem of assuring a sustainable world is understanding the full range of possible interactions between and among humans and their natural environment and choosing from this spectrum those forms of interaction that sustain life. Sustainability is only then seen not only as a global aggregate process but also as one that can assure sustainable livelihoods for the vast majority of local peoples.

About the Editors and Contributors

Bina Agarwal is an economist with the Institute of Economic Growth at Delhi University. Her most recent book is entitled *Who Sows? Who Reaps? Gender and Land Rights in South Asia*. Her research encompasses issues of poverty, gender, rural development and environmental degradation with a focus on Asia.

Lourdes Arizpe is director of the Institute of Anthropological Research, Universidad Nacional Autónoma de México and a vice-president of the International Social Science Council. She has authored numerous books and articles on women, development, and migration issues. Her most recent book is *La mujer en el desarrollo de México y de América Latina*.

Richard E. Bilsborrow is an economist/demographer at the Carolina Population Center at the University of North Carolina-Chapel Hill. His recent work has focused on interrelations among demographic processes, rural development, and environment including deforestation. An edited volume entitled *Population and Deforestation in the Humid Tropics* is forthcoming.

Stephen G. Bunker is in the Department of Sociology at the University of Wisconsin-Madison. His recent work has focused on Latin America, as in the volume *Underdeveloping the Amazon*, while earlier work on Africa included the volume *Peasants Against the State*.

Martha Geores is a geographer at the University of Maryland-College Park. Her publications include a 1992 volume with Richard Bilsborrow entitled *Rural Population Dynamics and Agricultural Development: Issues and Consequences Observed in Latin America*. Her research investigates the relation of population dynamics to protected forests in the United States and Latin America.

Peter D. Little is a senior research associate with the Institute for Development Anthropology and associate professor with the Department of Anthropology at SUNY Binghamton. His most recent book is entitled *Elusive*

Granary: Herder, Farmer, and State in Northern Kenya. His research interests include the interaction of East African pastoralism with wider economic and environmental systems.

Wolfgang Lutz is with the International Institute for Applied Systems Analysis in Laxenburg, Austria. Author of *Distributional Aspects of Human Fertility: A Global Comparative Study*, Dr. Lutz has published widely his quantitative analyses of population trends and their potential impact upon the environment.

David C. Major, an economist and natural resources planner, is program director for the Program on Global Environmental Change at the Social Science Research Council. His most recent book is *Large-Scale Regional Water Resources Planning* (with Harry E. Schwarz).

Alberto Palloni is in the Department of Sociology at the University of Wisconsin-Madison. His research focuses on issues of population in Latin America. His forthcoming book is entitled *Population and Society in Latin America, 1900-1990.*

Bryan R. Roberts is a sociologist who directs the Population Research Center at the University of Texas-Austin. Author of *Cities of Peasants*, Professor Roberts researches the dynamic interaction between urbanization, migration, and development.

Marianne Schmink is an anthropologist with the Center for Latin American Studies at the University of Florida-Gainesville. Professor Schmink is the author of the 1992 book *Contested Frontiers in Amazonia.* Her work focuses on the political ecology of the Amazon and she has published in both scholarly journals and edited volumes on this topic.

Gita Sen is an economist with the Indian Institute of Management in Bangalore, India. Co-author (with Caren Grown) of *Development, Crises and Alternative Visions: Third World Women's Perspectives*, Dr. Sen focuses primarily on issues of gender and third world development.

M. Priscilla Stone is program director for the Africa Program at the Social Science Research Council. She is an anthropologist whose research focuses on agricultural intensification and gender dynamics in West African farming systems.

Margarita Velázquez is a researcher with the Center for Interdisciplinary Research at the Universidad Nacional Autónoma de México. She is currently a fellow with the LEAD program at El Colégio de México.

Index

Agriculture
 fertilizer, 185-187, 193-197
 intensification, 171-202
 irrigation,185-187, 193-196,
 225, 239, 268
Amazonia, deforestation in, 6,
 257-266
 colonial history, 262
Biodiversity, 23, 42, 56, 95, 130,
 174-175, 342
Carrying capacity, 18, 20, 277
Case studies, methods for, 8,
 125-161
Chipko movement, in India, 111-
 115, 269, 270
Common property, 97-98, 216
 and colonialism, 97, 270
Community resource management,
 96, 98-99, 343
Consumption levels, 4-5, 56, 277
Deforestation, 6, 125-161, 253-271
Demographic transitions, 17
Desertification, *see* land degradation
Development Alternatives with
 Women for a New Era
 (DAWN), vii
Ecology, 58, 222, 227-228, 323
Economic factors and population
 and environment, 3, 4, 28, 343
Economic growth, 75, 81-83, 346
Education, 75-76, 346
Environmental degradation, 339-
 340
 and poverty, 21, 226
Environmentalism, 67-84

Erhlich/Simon debate, 3, 21-24,
 277-278
Ethnicity, in Lacandón rainforest
 case study, 29-31
Extractive projects, demographic
 and ecological effects, 286-297
 see also land degradation
Family planning, 15, 29, 71,
 74-79, 81, 84, 310
Fertility, 17, 340
 in Africa, 45-46
 levels, 47
 transitions, 43
 see also family planning
Fuelwood, 104-107, 229, 236, 256
Gender, 67-84, 87-118, 340-341,
 345
Health and population, 76-77, 83-
 84, *see also* family planning
Households, 226-230, 238-239, 253-
 271
India,
 deforestation in, 266-270
 women and environment in,
 87-118
Industrialized nations, *see* North-
 South divide
Institutions, as mediating forces, 8,
 213, 341
Integrated Project in Arid Lands in
 Kenya (IPAL), 230-234
Intensification, agricultural,
 171-202
International Social Science Council
 (ISSC), vii

351

352